Microbiology Research Advances

nova
science publishers

www.novapublishers.com

Microbiology Research Advances

Microbial Technologies and Their Applications
Santosh Kumar Karn, PhD
and Anne Bhambri (Editors)
2023. ISBN: 979-8-88697-913-8 (Hardcover)
2023. ISBN: 979-8-88697-937-4 (eBook)

New Research on Mycorrhizal Fungus
Qiang-Sheng Wu (Editor)
Ying-Ning Zou (Editor)
Yue-Jun He (Editor)
Nong Zhou (Editor)
2023. ISBN: 979-8-88697-637-3 (Hardcover)
2023. ISBN: 979-8-88697-662-5 (eBook)

The Rhizosphere: Structure, Ecology and Significance
Somboon Tanasupawat, PhD (Editor)
Yong-Jie Xu, PhD (Editor)
2023. ISBN: 979-8-88697-674-8 (Hardcover)
2023. ISBN: 979-8-88697-810-0 (eBook)

Endophytes: Types, Potential Uses and Mechanism(s) of Action
Pragya Tiwari, PhD (Editor)
2022. ISBN: 979-8-88697-045-6 (Hardcover)
2022. ISBN: 979-8-88697-205-4 (eBook)

Endotoxins and their Importance
Arif Pandit, PhD (Editor)
R. S. Sethi, PhD (Editor)
2022. ISBN: 978-1-68507-839-3 (Softcover)
2022. ISBN: 978-1-68507-913-0 (eBook)

More information about this series can be found at
https://novapublishers.com/product-category/series/microbiology-research-advances/

Santosh Kumar Karn
and Anne Bhambri
Editors

Microbial Technologies and Their Applications

nova
science publishers
www.novapublishers.com

DOI: https://doi.org/10.52305/JUTX4763

NOTICE TO THE READER

The Publisher has taken reasonable care in the preparation of this book but makes no expressed or implied warranty of any kind and assumes no responsibility for any errors or omissions. No liability is assumed for incidental or consequential damages in connection with or arising out of information contained in this book. The Publisher shall not be liable for any special, consequential, or exemplary damages resulting, in whole or in part, from the readers' use of, or reliance upon, this material. Any parts of this book based on government reports are so indicated and copyright is claimed for those parts to the extent applicable to compilations of such works.

Independent verification should be sought for any data, advice or recommendations contained in this book. In addition, no responsibility is assumed by the Publisher for any injury and/or damage to persons or property arising from any methods, products, instructions, ideas or otherwise contained in this publication.

This publication is designed to provide accurate and authoritative information with regards to the subject matter covered herein. It is sold with the clear understanding that the Publisher is not engaged in rendering legal or any other professional services. If legal or any other expert assistance is required, the services of a competent person should be sought. FROM A DECLARATION OF PARTICIPANTS JOINTLY ADOPTED BY A COMMITTEE OF THE AMERICAN BAR ASSOCIATION AND A COMMITTEE OF PUBLISHERS.

Library of Congress Cataloging-in-Publication Data

Names: Karn, Santosh Kumar, editor. | Bhambri, Anne, editor.
Title: Microbial technologies and their applications / editors, Santosh Kumar Karn, PhD, Associate Professor & Head of Department of Biochemistry & Biotechnology, Sardar Bhagwan Singh University, Balawala, Dehradun, Uttarakhand, India, Anne Bhambri, Research Scholar, Department of Biochemistry & Biotechnology, Sardar Bhagwan Singh University, Balawala, Dehradun, Uttarakhand, India; Department of Biotechnology, Shri Guru Ram Rai University, Patel Nagar, Dehradun, Uttarakhand, India.
Description: New York : Nova Science Publishers, [2023] | Series: Microbiology research advances | Includes bibliographical references and index. |
Identifiers: LCCN 2023023428 (print) | LCCN 2023023429 (ebook) | ISBN 9798886979138 (hardcover) | ISBN 9798886979374 (adobe pdf)
Subjects: LCSH: Microbial biotechnology.
Classification: LCC TP248.27.M53 M538 2023 (print) | LCC TP248.27.M53 (ebook) | DDC 660.6--dc23/eng/20230622
LC record available at https://lccn.loc.gov/2023023428
LC ebook record available at https://lccn.loc.gov/2023023429

Published by Nova Science Publishers, Inc. † New York

Contents

Preface

Microbial biotechnology is an essential area which promotes advanced research by using microbes for value-added products. This book deals with recent microbial technologies used in the field of environmental biotechnology such as climate change, bioremediation, biodiesel, oil pollution, extremophiles etc. Environmental biotechnology is a very vast and rapidly growing field that deals with various aspects. The scope of this book is to provide recent information regarding microbial technologies and their applications.

Chapter 1 - Depleting fossil fuel resources, global warming concerns, and rising pollution levels necessitate the development of more environmentally friendly alternative renewable fuels. Bioethanol, as a liquid biofuel, has received greater consideration as a fuel additive and fuel quality enhancer. Currently, first-generation bioethanol is derived from food crops. Because these are major food crops, the question of food vs. fuel is always raised. As a result, the utilization of second generic non-edible lignocellulosic biomass for bioethanol production has recently gained attention. Cellulosic waste biomasses, in particular, gain advantages in bioethanol production. Cellulose is recorded as the world's most prevalent natural polymer, and cellulose-based agro-waste, industrial solid waste such as paper waste and paper pulp waste is accelerated by human activities and contributed to pollution. Utilizing waste to create value-added products ensures economic sustainability while also promoting environment management stewardship. Cellulose hydrolysis is the process that breakdown of cellulose into its monomers. Hydrolysis employs physical, chemical and biological methods. Sugar fermentation is mainly mediated by microbes such as fungi and bacteria. To integrate the bioprocesses of hydrolysis and fermentation, different configurations are used. Separate Hydrolysis and Fermentation (SHF), Simultaneous Saccharification and Fermentation (SSF) and Consolidated Bioprocessing (CPB) are the most common configurations. Though bioethanol is regarded as a potential fuel additive, it is partially miscible with diesel and forms an emulsion when mixed. However, bioethanol production and utilization as a fuel additive require further advancement before they can be used without requiring vehicle modifications. Bioethanol must be sustainable to become a viable alternative for the future energy crisis. This chapter will discuss bioethanol production mainly from second-generation biomass, cellulosic solid waste such as paper waste, and agricultural waste such as corncobs and so on, as well as different pre-treatment methods, cellulose hydrolysis methods, sugar fermenting techniques, and challenges.

Chapter 2 - Extremophiles are organisms that inhabit and thrive in environments that show permanent values of environmental parameters close to the lower and upper limits for life. Although the general perception is that extreme environments are rare, a more significant fraction of the earth is covered with extreme environments. Therefore, extreme environments provide habitats for a considerable fraction of the biomass and biodiversity on the earth. Extremophiles are very useful in various applied aspects such as biotechnology, industrial application, bioremediation, and fundamental research areas such as astrobiological studies, origin, evolution, and future of life on earth. In this chapter the biodiversity of extremophiles, adaptations of extremophiles, applied aspects of extremophiles, extremozymes and the future with extremophiles are discussed in detail.

Chapter 3 - The majority of enzymes are used for catalyzing a biological reaction with standard conditions. Extremophilic enzymes showed as a relatively undefined and understudied class of enzymes, characterized by complex adaptation mechanisms. Extremophilic enzymes can catalyze reactions under unusual conditions, such as high and low temperatures, high salinity, exposure to radiation, high ionic strength, extreme pH values, and exposure to toxins. Thus, extremophile enzymes are a novel unexplored, and undiscovered class of enzymes due to their specific survival capacity in extreme conditions. Extremophile enzymes are produced from the extremophile microbes that belong to the archaea and bacteria groups. Thus, an extremophile bacterium can serve as a novel source of industrial enzymes. Extremophile enzymes have great potential for industrial applications as well as some therapeutic applications like anti-cancer molecules and anti-cell signaling molecules. Currently, studies are focusing on extremophilic microbes and enzymes due to their unique and unexplored characteristics. As a result, this topic has become a highly relevant and promising area of research with industrial and therapeutic applications. In this chapter, the authors are discussing the initial growing information about extremophile microbes and extremophile enzymes and also summarize up-to-date knowledge on the genomic level with various application domains.

Chapter 4 - Sulphur is one of the most interesting as well as challenging elements to work with chemically; it is plentiful geochemically and is crucial biochemically. Sulphur and its compounds have been the subject of more than 100 years of scientific inquiry, resulting in a large volume of literature that is difficult to search for credible information. However, there are several credible studies in the literature that critically and comprehensively assess the chemical literature. Increased SO2 concentrations in the atmosphere have been linked to fossil fuel combustion, moist deposition, and acid rain in the previous century. Despite recent reductions in SO2 emissions, urban and industrial areas continue to be vulnerable to excessive sulphur contamination, necessitating further research into the deleterious effects of excess sulphur on environmental parameters and components. With increased human activities, more sulphur is introduced into the environmental system and put the ecosystem and human health in jeopardy. The sources, effects and remedial techniques of sulphur pollution are reviewed and discussed in this chapter. Moreover, a typical study on sewer deterioration has also been discussed in this context.

Chapter 5 - Pollution has increased due to urbanization, industrialization, and population pressure. Around 1.7 to 8.8 million metric tons of oil are released into the world's water annually, wreaking havoc on many ecosystems. Diesel, crude oil, petroleum, lubricating oil, bitumen, bunker oil, and other types of oil pollute the environment. Oil spills, urban runoff, wastewater treatment plants, and other sources account for more than 90% of oil pollution. Oil pollution harms the environment, economy, and society. Contaminants must be dealt with in three different ways: Identify the issue, determine the nature and amount of the pollutant, and select the most appropriate remediation method. As a result of the need to clean up polluted sites, new technologies have emerged that focus on the detoxification and destruction of contaminants rather than the traditional disposal method. Oil remediation can be accomplished using a variety of chemicals, biological, and mechanical methods. Solidification, soil washing, dehalogenation, thermal technologies, flotation, and chemical reduction are among the physicochemical remediation methods available. There are technologies for bioaccumulation, bioaugmentation, biotransformation, bioaccumulation, biosorption, bioleaching, attenuation, and other biological remediation. Micro/nanotechnologies have sparked tremendous interest in oil remediation as a novel technology due to the synergistic effect of unique motion characteristics and inherited physicochemical properties of nanomaterials. Another novel challenge is integrating electrochemical methods with biological routes, microbial fuel cells, enzyme technologies, and bio-electrochemical systems. Integrated remediation techniques have recently received more attention than single-technology remediation. On the other hand, bioremediation is a green, appealing, and promising option for oil remediation at a contaminated site, as it is more cost-effective and selectively degrades pollutants without harming the site or its native flora and fauna. The development of remediation processes must integrate various disciplines such as microbial ecology, biochemistry, microbial physiology, and biochemical and bioprocess engineering in order to be commercially successful.

Chapter 6 - Hydrocarbons are organic chemicals based on carbon (C) and hydrogen (H) compounds. Pollution of hydrocarbons (primarily petroleum hydrocarbons) in the environment is becoming a major issue worldwide. This is most likely due to the world's heavy reliance on petroleum as a primary energy source, population growth, rapid industrialization and a total disregard for environmental health. Polycyclic Aromatic Hydrocarbons (PAHs) and related compounds such as acridine, benzopyrene, fluoranthene, naphthalene, quinolone and phenanthridine have been found in high concentrations in the environment all over the world. The unintentional or intentional release of hydrocarbons into the environment is a major source of pollution in the water, soil, and air. The natural equilibrium that exists between living species and their surroundings is usually disrupted by these hydrocarbon pollutants. Carcinogens and neurotoxic organic pollutants have been identified in hydrocarbon components. Hydrocarbons in the environment enter the human food web via plants due to their proclivity for bioaccumulation and transfer from one food chain to another, posing a threat to the ecosystem. Hydrocarbons have been found in various food chains, with disastrous results for both wildlife and humans. Man-made hydrocarbons are resistant to degradation and have a long half-life in the environment. Because hydrolysis and photolysis are ineffective for all hydrocarbon types, different degradation techniques of physical and chemical treatment methods are used. However, most of these methods have limitations when it comes to entirely remediating hydrocarbon contaminated sites. Some of these methods result in daughter compounds that are more harmful to the environment than the parental compounds. Bioremediation is the most environmentally friendly way for removing hydrocarbon contamination as it relies on natural microorganisms' ability to break down hydrocarbons into harmless substances. Contaminants from the environment are transformed or degraded using microbial enzymatic activities. Because it is a natural procedure that does not often produce harmful by-products, it is a cost-effective remediation strategy compared to other treatments. Due to the complete mineralization of toxins from the environment, it also provides a long-term solution. Biological remediation has several advantages over other treatment methods, including the ability to destroy pollutants rather than transfer them to another medium, low worker exposure to contaminants, long-term public health protection and the possibility of a shorter remediation time.

Chapter 7 - The upsurge of living standards and life expectancy has given rise to a multitude of industries that alter the soil, water, and air by adding hazardous materials in a direct or indirect manner. Most of these chemicals are resistant to natural degradation processes, allowing them to persist in the environment for decades, disrupting natural biogeochemical cycles and threatening human health. Therefore, waste management containing various hazardous chemicals must remain a top priority. Physicochemical treatment methods such as activated carbon, membrane filtration, photochemical and electrochemical destruction, irradiation, and ion exchange are expensive, produce secondary pollution, and have limited applicability. As a result, the development of cost-effective, efficient, and environmentally friendly remediation technologies is critical. Phytoremediation, the use of plants and/or their associated microflora for the treatment of contaminants, is a growing concern to remove pollutants or reduce their bioavailability in the environment. It is an aesthetically pleasing, low-cost, easily monitored technique that preserves topsoil, reduces erosion, and pollution leaching. Plants and endophytes have the ability to uptake pollutants from contaminated soil or liquid media at low concentrations and then degrade, transform, or immobilize the pollutants through various mechanisms such as phytodegradation, rhyzodegradation, phytostabilization, phytoextraction, and phytovolatilization. Even though phytoremediation has a long history in contaminated sites, the current concern and approach have been to find novel and developed enzyme-based phytoremediation pathways to overcome limitations such as the need for a long incubation time plant necrosis due to toxicity, and so on. Understanding the basic physiological and biochemical processes in plants that underpin different phytoremediation processes is a promising way to improve plants' practical applicability in phytoremediation. As a result, this chapter will focus on phytoremediation of pollutants, specifically its applications, remediation mechanisms, plant associated endophytes, synergistic strategies, enzymes, benefits, limitations, and future directions.

Chapter 8 - Plant growth promoting rhizobacteria (PGPR) are soil bacteria that have the ability to influence plant growth both directly and indirectly. These organisms may be able to reduce or eliminate the use of chemical fertilisers and toxic chemical inputs. Exploring PGPR in agriculture is thus one of the more promising techniques for increasing agricultural production while minimising environmental damage. Simultaneously, nano-technological applications are having a significant impact on agriculture. Plant growth promoting rhizobacteria (PGPR) play an important role in the sustainable agriculture industry. The increasing demand for crop production while reducing the use of synthetic chemical fertilisers and pesticides is a major challenge today. The use of PGPR has been shown to be an environmentally friendly method of increasing crop yields by facilitating plant growth via either a direct or indirect mechanism. PGPR mechanisms include regulating hormonal and nutritional balance, inducing resistance against plant pathogens, and solubilizing nutrients for easy uptake by plants. Furthermore, PGPR exhibits both synergistic and antagonistic interactions with microorganisms in the rhizosphere and beyond in bulk soil, which indirectly increases plant growth rate. Many bacteria species have been described in the literature as being effective at improving plant growth when acting as PGPR. However, there is a gap between the PGPR's mode of action (mechanism) for plant growth and its role as a biofertilizer—hence the importance of nano-encapsulation technology in improving PGPR efficacy. As a result, this review fills the gap and summarises the mechanism of PGPR as a biofertilizer for agricultural sustainability.

Chapter 9 - The increasing world population is one of the significant threats to food security. Currently, the cultivable land for agriculture is gradually decreasing, and the soil is becoming infertile. The vigorous use of chemical fertilizers and pesticides is increasing continuously for higher crop productivity to meet the growing populations global food demand. Nowadays, chemical farming is causing many environmental problems that directly and indirectly affect human health. Therefore, organic agriculture (bio-fertilizer, biopesticide, Integrated Crop Management (ICM), Integrated pest management (IPM), crop rotation, green manuring, Farmyard manure (FYM), mixed cropping sequence, etc.) is the only alternative practice for saving our environment and producing quality food for the human being. This book chapter focuses on biofertilizers and their application for crop improvement. Bio-fertilizer contains microorganism providing adequate nutrients to the host plant for proper growth and development. Living microorganisms have a specific function used to prepare biofertilizers to augment plant growth and reproduction. It has great potential to sustain long-term soil fertility and crop production. Nowadays, the application of liquid biofertilizer shows a very effective response in various crops, which has overcome the previous restrictions of biofertilizer applications. However, we need efficient and indigenous biofertilizers to enhance crop productivity, nutritional quality and soil health. These biofertilizers can replace chemical fertilizers. Biofertilizers are eco-friendly, cost-effective and socially acceptable for sustainable agricultural productivity.

Chapter 10 - Phytoplankton is the foundation of aquatic food webs and is critical to the functioning and services of ecosystems all over the world. As a result of global climate change affected by increased wind speed, concentrations of greenhouse gas, precipitation, incoming solar radiation, and air temperature are expected to change (primarily CO2). These critical climate variables have a direct influence on lake water quality by changing the flow and temperature balance. Periodic blooms of phytoplankton and a shift in the neutral trophic balance can emerge from high nutrient enrichment concentrations mixed with predictable climate variability. As a result, dissolved oxygen concentrations at low concentrations might fluctuate significantly. Algal production could also reach excessive levels. Climate change also affects phytoplankton size structure and species composition, favoring traits best altered to fluctuating climate conditions. Understanding the mechanisms that govern phytoplankton adaptation to warming will improve the realism of models that describe the ecological and biogeochemical effects of climate change. Higher temperatures initially hampered population growth because respiration is more temperature-sensitive than photosynthesis, resulting in less carbon available for development. Tolerance to high temperatures evolved after 100 generations as a result of greater down-regulation of respiration in comparison to photosynthesis. By down-regulating respiration and allocating fixed carbon more efficiently to growth, phytoplankton overcame the metabolic constraint imposed by respiration's higher temperature sensitivity. The rapid evolution of phytoplankton carbon-use efficiency suggests a potentially broad mechanism for thermal adaptation. It implies that as temperatures rise, evolutionary responses in phytoplankton will alter biogeochemical cycles, thereby changing the structure and function of food webs. Phytoplankton has a significant evolutionary capacity in response to climate change.

Chapter 11 - Pollutants in aqueous solutions have negative environmental, economic, and societal consequences. In rural and urban regions, pollution caused by nitrite, nitrate, ammonia, and phosphorus is a significant concern.

Nitrate is one of the most prevalent contaminants found in groundwater. Increased use of artificial fertilizers, waste disposal, and changes in land use are the primary drivers of the gradual increase in nitrate levels in groundwater supplies. Methemoglobinemia is caused by the presence of nitrate in drinking water. Furthermore, nitrate and nitrite levels in water and food have increased cancer risk. In commercial fertilizers, Ammonia is a critical component. In an aquatic environment, Ammonia and ammonia-related chemicals are abundantly available. Ammonia's biological effects in humans are dose-dependent and influenced by ambient concentrations. Phosphate ion discharge also leads to eutrophication in water, with significant environmental consequences. Algal blooms are triggered when there is too much phosphorus in the water. Algal blooms eventually release poisons that can kill wildlife. Algae, fungus, and bacteria are among the microorganisms that may convert nitrate ions into organic material. The detoxification process that releases Ammonia involves several microbial enzymes, including nitrate and nitrite reductases. Microorganisms including algae, bacteria, and fungi have been employed successfully in bioremediation. Bacteria and fungi use a several ways to extract nitrogen and phosphorus. These processes include adsorption, biosorption, bioaccumulation, mineralization, ammonia adsorption, nitrogen fixation, ammonia volatilization, nitrification, and immobilization. These bacteria transform inorganic pollutants into organic matter through an assimilatory reduction mechanism. As a result, microbe-mediated biological treatment could be recommended as a viable option for treating Nitrite, Nitrate, Ammonia, and Phosphate-contaminated waters with minimum environmental impact.

Chapter 12 - Anthropological activities associated with urbanization result in significant wastewater generation. Xenobiotic pollutants in wastewater treatment have resulted in severe environmental issues. Improper discharge of wastewater into the natural environment has several adverse effects on aquatic and terrestrial ecosystems. Significantly, wastewater contributes to the spread of infectious and non-infectious diseases, which ultimately negatively impact human health. Thus, practical and cost-effective wastewater treatment methods are required for long-term wastewater disposal. Advanced physicochemical methods are being used to treat wastewater in the modern world. However, biological wastewater treatment, as opposed to traditional physicochemical treatment methods, is mainly due to its cost-effective and eco-friendly approach. Biological treatment is an essential step in nearly all wastewater treatment processes. Bioprocessing of waste is taken by syntrophic microbial communities followed by metabolism and co-metabolism of specific waste substrates. In biological wastewater treatment, the partial conversion of complex waste substrates into non-toxic simpler compounds by interconnected fermentative, acetogenic, and methanogenic microbial groups is critical. Aerobic and anaerobic microbial application techniques are used for wastewater treatment as traditional methods. In contrast, membrane bioreactors, microbial fuel cells, activated slug systems, sequencing batch reactors, and aerobic and anaerobic lagoon systems are considered novel treatment technologies. The use of extracellular cell-free hydrolytic enzymes to degrade organic waste materials is a novel new technology that has some advantages over the use of microbial biomass. Hence, biological treatment has been in the limelight in recent wastewater treatment processes.

Chapter 13 - Bacterial cellulose is an aggregated event to form a ribbon like structure from β-1,4-glucan chain and form hydrogel film at air surface of culture media. Cellulose is a food source for herbivores, and it requires microbial degradation for the digestion in the alimentary tract. Bacterial cellulose has very high economic value due to its various usages in industries like paper, plastics, synthetic polymers, adhesive, explosive, moisture proof coating and thickening agents for foods. Cellulose is most abundant in plants due to its presence in plant cell walls but its separation and isolation from plants is very cost effective. During isolation of cellulose some other organic monomers and polymers are also filtered out and it is very difficult to separate them. The isolation of cellulose does not meet the requirements of industries. The dependency on plats for cellulose is also harmful for the environment. To meet with the industrial requirement of cellulose we need another sustainable source of production of cellulose. The exploration and exploitation of bacterial cellulose can reduce the dependency on plants. The bacterial species that belong to the Gluconacetobacter genus produce cellulose. The bacterial cellulose is a nanomaterial that possesses exclusive properties like high polymerization degree, crystallinity, surface area, tensile strength and flexibilities, water holding capacity. Bacterial cellulose is pure form and free from lignin and hemicelluloses compared to plants. The bacterial cellulose has wide application due to their non-cytotoxic behavior and attracting interest in therapeutics industries. Nowadays the bacterial cellulose captures more strength in industries and to sustain the industrial demand it also requires bacterial strains that produce highly pure and more amount of cellulose. The Komagataeibacter xylinusis an acetic acid bacterium able to produce bacterial cellulose from various carbon sources. The K2G30 strain has the ability to produce much higher amounts of bacterial cellulose in mannitol and glucose-based media. The genomic analysis of K2G30 confers presence of one bcsAB gene and three bcs operons that are part of bacterial cellulose production. The bacterial cellulose can be produced from various carbon sources and various types of organic waste. The optimization of the upstream process on the basis of genomic analysis can produce a higher concentration of bacterial cellulose. The downstream process also determines the end quality and quantity of the product. By the help of genome editing and recombinant DNA technology can modify the genome of various bacterial cellulose production bacteria and develop new strains that produce higher concentration of bacterial cellulose.

Acknowledgments

This book is the result of a study that started approximately two years ago. Several authors have contributed to its development and refinement in one way or another. We would like to take the opportunity to thank them. All of them went through several valuable rounds of thorough critiques on drafts of every chapter. The Editors are very grateful to God for being with us and giving us the wisdom and ability to do this assignment. Editors are thankful to Nova Science Publisher, USA for providing us this opportunity.

Editors like to express gratitude to the Management, Sardar Bhagwan Singh University, Balawala, Dehradun, Uttarakhand, India, for providing facility, space and resources to conduct this work successfully. Next, Editors like to thank the parents and families always trusting in us and teaching us to believe in our abilities and strengths and never give up until the goal is achieved, further who extended their moral support. Prof. Karn and Bhambri like to acknowledge the publisher, editors, and support staff at Nova Science Publishers, they were very helpful in the various stages of developing and producing the book.

Dr. Santosh Kumar Karn
Anne Bhambri
Editors

Chapter 1

Bioethanol: Emphasis on the Conversion of Second-Generation Cellulose to Energy

Y. S. Chandrasiri[1]
W. M. L. I. Weerasinghe[1]
D. A. T. Madusanka[1]
and P. M. Manage[1,2,*]

[1]Centre for Water Quality and Algae Research, Department of Zoology, University of Sri Jayewardenepura, Gangodawila, Nugegoda, Sri Lanka
[2]Faculty of Graduate Studies, University of Sri Jayewardenepura, Gangodawila, Nugegoda, Sri Lanka

Abstract

Depleting fossil fuel resources, global warming concerns, and rising pollution levels necessitate the development of more environmentally friendly alternative renewable fuels. Bioethanol, as a liquid biofuel, has received greater consideration as a fuel additive and fuel quality enhancer. Currently, first-generation bioethanol is derived from food crops. Because these are major food crops, the question of food vs. fuel is always raised. As a result, the utilization of second generic non-edible lignocellulosic biomass for bioethanol production has recently gained attention. Cellulosic waste biomasses, in particular, gain advantages in bioethanol production. Cellulose is recorded as the world's most prevalent natural polymer, and cellulose-based agro-waste, industrial solid waste such as paper waste and paper pulp waste is accelerated by human activities and contributed to pollution. Utilizing waste to create value-added products ensures economic sustainability while also

[*] Corresponding Author's Email: pathmalal@sjp.ac.lk.

In: Microbial Technologies and Their Applications
Editors: Santosh Kumar Karn and Anne Bhambri
ISBN: 979-8-88697-913-8

promoting environment management stewardship. Cellulose hydrolysis is the process that breakdown of cellulose into its monomers. Hydrolysis employs physical, chemical and biological methods. Sugar fermentation is mainly mediated by microbes such as fungi and bacteria. To integrate the bioprocesses of hydrolysis and fermentation, different configurations are used. Separate Hydrolysis and Fermentation (SHF), Simultaneous Saccharification and Fermentation (SSF) and Consolidated Bioprocessing (CPB) are the most common configurations. Though bioethanol is regarded as a potential fuel additive, it is partially miscible with diesel and forms an emulsion when mixed. However, bioethanol production and utilization as a fuel additive require further advancement before they can be used without requiring vehicle modifications. Bioethanol must be sustainable to become a viable alternative for the future energy crisis. This chapter will discuss bioethanol production mainly from second-generation biomass, cellulosic solid waste such as paper waste, and agricultural waste such as corncobs and so on, as well as different pre-treatment methods, cellulose hydrolysis methods, sugar fermenting techniques, and challenges.

Keywords: cellulose, bioethanol, cellulosic waste, cellulose hydrolysis, sugar fermentation

Introduction

Bioethanol as a Global Trend

The rapid unprecedented advances in human civilization during the last decades have exponentially increased the global energy demand. The amount of energy obtained from each and every energy source and used by humans worldwide in nearly every sector is referred to as global energy production and global energy consumption respectively. [1]. World energy consumption is mainly influenced by per capita energy consumption, the technical level of energy conversion, energy saving, and world population [2]. The exponential global population rate, economic growth and increased per capita energy consumption drive up the world's energy consumption. At the same time, technological advancements boost energy efficiency and reduce global energy consumption [3].

Main energy sources in the world include fossil fuel, nuclear energy, and biofuels while geothermal energy, wind energy, tide and wave energy, ocean thermal energy conversion, solar etc. are considered minor energy sources [4].

Thus, the demand for global energy is mainly addressed by using fossil fuels that are accounted for more than 80% of the total energy requirement. United Nations stated that the global population reached 8 billion people in 2022 and the global population will be nearly 9.157 billion people by 2040, which in turn causes a challenge in supplying energy [1]. Therefore, it has been predicted that the current fossil fuel reserve will be depleted within the next 40-50 years.

The main contributor to the increased levels of atmospheric CO_2 is fossil fuel burning [5]. Therefore, the inadequate fossil fuels, economic concerns and rising global temperature due to fossil fuel burning have accelerated the interest in novel alternative renewable energies [6]. Usually, renewable energy includes energies produced using natural and biosphere origins without requiring irreversible utilization of mineral energy feed stock [7].

Biomass has the potential in meeting future global energy needs sustainably. In addition, biomass is the largest and cheapest source that easily gets converted into other energy forms, such as heat, electricity and fuels [8]. The term "biomass" represents any material of biological origin, including various feedstock ranging from energy crops, agricultural residues, forestry residues, organic solid wastes, and algae. The utilization of biomass for bioenergy production provides a sustainable and carbon-neutral energy supply, as well as reduces dependence on non-renewable energy resources. For the sustainable transition of non-renewable resources into renewable resources, novel technologies and initiatives are required [9]. On average, biomass contributes to primary energy consumption by 15%. However, in developing countries biomass provides up to 48%, and in industrialized countries, an average of 2% - 3% of primary energy consumption is provided by biomass [1].

Bioethanol is a main biomass-based fuel type. Considering all the facts like dwindling fossil fuel reserves, increasing oil prices and climate change caused by global warming resulting from fossil fuel burning, the world is now more concerned about bioethanol as an alternative fuel [10].

Biofuel Production

The current decline of the reserves of fossil fuels and the impacts due to the consumption of fossil fuels on the environment has led the research interest in renewable fuel in a sustainable manner. High demand for fuel with the current growing population, rapid industrialization, and urbanization have also accelerated this interest [11, 12].

Biofuels are a type of bioenergy which could be utilized as the most promising alternative to fossil fuels, because of their ability to derive from the abundant natural renewable sources [13, 14]. Since biofuels are renewable and eco-friendly, the replacement of biofuel instead of fossil fuel has been uplifted in the present [15]. Biofuels are mainly categorized as primary biofuel and secondary biofuel. Organic matters which are used directly as fuel without any processing are known as primary biofuel (e.g., fuel woods used for cooking), whereas, secondary biofuels are produced through the processing of organic biomasses for industrial applications or as a transportation fuel etc. (e.g., bioethanol, biodiesel, etc.) [16].

Biofuels are produced from solid biomass through different chemical and biological processes. Bioethanol, biodiesel, syngas, and biogas are some of the common biofuels. Biofuels have become more popular and are used mainly for the replacement of conventional transport fuels [17].

Bioethanol

Bioethanol (CH_3CH_2OH) is a clear, colourless, and inflammable polar solvent and also known as ethyl alcohol, which is obtained through the fermentation of biomasses [18]. Hence, bioethanol reduces fossil fuel consumption and environmental pollution, it has become the most widely used liquid biofuel worldwide [19]. Bioethanol is directly used as pure ethanol or as "gasohol". Gasohol can be produced by blending bioethanol with gasoline or with diesel, which is called e-diesel or ethanol-diesel blends [20]. Bioethanol is investigated as a fuel extender, gasoline or diesel improver, and as an octane enhancer to minimise the discharge of exhaust gasses in bioethanol-diesel blends [21]. Bioethanol provides a higher octane value, broad limits of flammability, high speeds of flame, and accelerated heat of vaporization, when compared with gasoline and is having the qualities of low level of toxicity, ease of biodegradability, production of lesser air-borne pollutants, which makes it more environmentally friendly than petroleum fuels [20, 22].

Potential Biomass for Bioethanol Production

Depending upon the biomass source, bioethanol is broadly differentiated as first generation, second generation, third generation and fourth generation [23]. Bioethanol which is acquired from edible food crops like sugar and starch is known as the first generation bioethanol, while bioethanol acquired from non-edible lignocellulosic feedstock is called second-generation bioethanol [24]. Bioethanol obtained from algae and genetically modified microalgae are categorized as third-generation bioethanol and fourth-generation bioethanol respectively [25].

First-Generation Bioethanol

In the production of first-generation bioethanol raw materials used are categorized as;

- Sugar-based food crops such as sweet sorghum, sugarcane, and sugar beet [26];
- Fruits like grapes, dates, watermelon and apples etc. [27];
- Crops which contain starch-like grains such as corn, rice, wheat, barley, sweet sorghum, etc. and root plants like potato, cassava etc. [28].

Due to high sugar yield and low conversion costs, sugar-based food crop biomasses are considered more advantageous. Nevertheless, their seasonal availability is the main limiting factor [29].

Crops which contain sugar, fruits and sugar refinery wastes contain substantial amounts of saccharose, which can be extracted and ultimately fermented to bioethanol. Crops which contain only starch can be fermented into ethanol by conventional fermentation. Since starch is having a long-chain glucose polymer and needs to be hydrolysed into glucose by enzymatic hydrolysis and then subjected to fermentation [30].

The addition of α-amylase results in the hydrolysis of starch at first to dodge the gelatinization and heating at elevated temperatures (141^0C - 180^0C). Then, glucose could be derived from the liquefied starch through a hydrolysis step with the use of the enzyme glucoamylase. The sugars are fermented by using microorganisms [31]. The microorganisms used for ethanol production are mainly categorized into three groups; yeast (*Saccharomyces* sp.), bacteria and mold. *Saccharomyces cerevisiae* is the prime species employed in fermentation and *Zymomonas mobilis* is the second most studied fermenting microbe [30].

To produce first-generation bioethanol wheat, corn, and sugarcane are widely used. Corn is the leading feedstock for first-generation bioethanol and it is the dominant biomass type used for bioethanol production in the US. Sugarcane is considered the second dominant feedstock and Brazil is the largest ethanol producer using sugarcane in the world [9]. Ethanol production in Sri Lanka is mainly from first-generation bioethanol production using sugarcane molasses. Approximately, per year only 12 million litres of ethanol is being produced, hence the lack of sugar molasses [32].

Although first-generation biofuels have a higher potential for large-scale production, the utilization of edible crop foods as energy sources causes the "Food vs fuel" debate. Due to competition for agricultural lands, an increase in food prices and regional limitations causes problems in sustainable biofuel production. Therefore, scientists looking forward to a variety of conventional biomasses other than crop foods for bioethanol production, especially lignocellulose are being considered.

Second-Generation Bioethanol

Second-generation bioethanol is derived from non-food/lignocellulosic biomass that has more sustainability due to lignocellulosic materials being considered economical, and renewable and are present in ample amounts required for the making of biodiesel, bioethanol, and other biochemicals in a cost-effective manner [33]. The yield of bioethanol and conversion rate are depending on the biomass type, the presence of cellulose and hemicellulose in generous amounts and low levels of lignin content [34]. The technology behind this generation is allowing for separate lignin, hemicellulose and cellulose in a plant because alcohol is conceivable through cellulose fermentation.

Lignocellulosic materials are categorized by resource type as (1) waste paper, (2) Municipal Solid Waste (MSW), (3) forest residues, and (4) resources of crop residues [24, 33].

Lignocellulosic biomasses include lignocellulosic by-products or wastes of agriculture products or its associated industries, comprised of wheat, rice husks, rice straw, jute sticks, cotton stalks, corn cob, bagasse, maize cobs and coconut shells or fermentable organic content of MSW, sewage treatment sludge, or waste from forest materials, such as peel, leaves, sawdust, wood chips, and bark [5].

The key advantage of using lignocellulosic biomass, in the bioconversion of ethanol is that it avoids competition with food production, as all lignocellulosic biomass are non-food materials. Furthermore, they are more abundant sources and all most all of them are waste products [35]. The three main polymers that comprise lignocellulosic biomass include; cellulose, hemicellulose, and lignin [36]. Poor porosity, elevated crystallinity, and content of lignin in lignocellulosic biomass make it difficult in using them as feedstock for bioethanol production. To overcome these difficulties and utilize the lignocellulosic biomass for bioethanol production, pre-treatments are required [37]. Then the pre-treated biomass is hydrolyzed into monosaccharaides and then, the monosaccharides could be fermented into ethanol by different microorganisms [38].

When lignocellulose is converted into bioethanol, should consider the following matters to make sure the bioethanol production is cost-effective [39].

- Whether holocellulose components convert into monomeric sugars completely;
- Enhanced fermentation in presence of toxins derived from pre-treatment of biomass;
- Lignin valorization;
- Low waste-generating rate;
- High energy efficiency.

Third-Generation Bioethanol

The third-generation bioethanol production is attentive to the utilization of algae as the feedstock [40]. Algae are photosynthetic organisms that may be unicellular (microalgae) or multicellular (macroalgae) [41]. Bioethanol production from algae includes major steps like harvesting or collecting algae, handling feedstock, carbohydrate extraction, hydrolysis, fermentation, and product recovery [42].

The employment of microalgae as a feedstock for bioethanol production is more advantageous, as higher starch contents can be obtained for optimized culture conditions and algae are capable of all-year-round production [40]. The major components of the cell walls of microalgae include; cellulose, hemicelluloses, pectin, protein, lignin, and other carbohydrate types and through the enzymatic hydrolysis or action of an acid, these components are feasible to convert into monomers during the production of bioethanol [43].

Macroalgaehave been identified as a promising feedstock, due to having the ability to convert the feedstock into bioethanol easily, as they have low or free lignin composition. Also, for macroalgae cultivation, there is no need for fresh water and no interference with the utilization of agricultural lands [42]. In brown macroalgae (Phaeophyceae), carbohydrates; such as alginates, cellulose, laminaran, fucoidan, and mannitol are present. The cell wall of the red macroalgae (Rhodophyceae) consists of polysaccharides like cellulose, agar, mannan, carrageenan and xylene, while green macroalgae (Chlorophyceae) has cellulose, mannose, and xylene, which can be converted to simple carbohydrate monomers to produce bioethanol [41, 44].

Hence, the high content of lipids, rapid growth rate and CO_2-fixing capacity, algae serve as probable biomass for biofuel production. As another advantage, the presence of only basic needs such as CO_2, nutrients, water, and the sunlight is adequate for the growth and survival of algae. Whereas some algae cultures also can be grown in wastewater, and due to that it reduces the cost, as no need for a freshwater medium for their growth. There is no need for a land area for algae cultivation since algae have the ability to grow in aquatic environments, as well as in unutilized dry lands. In addition, when compared to traditional fuel, renewable biofuels have a very favorable energy yield [45, 46].

Fourth-Generation Bioethanol

Fourth-generation bioethanol is an emerging field, which uses genetic engineering of microalgae. In this technology, the existing, natural biological systems are redesigned to enhance the CO_2 intake for photosynthesis while making an artificial carbon sink to obtain high yields of bioethanol [25]. In the fourth generation, mainly microalgae are genetically modified because they are considered efficient converters of solar energy. Furthermore, microalgae are abundant and have efficient access to water, CO_2, and nutrients, and have a massive surface area to mass ratio due to their simple cellular morphology [47].

Some of the microalgal species such as *Botryococcus braunni, Schizochytrium* sp. *Chlorella* sp., and *Scenedesmus* sp. had refitted to accelerate the rate of growth and increase the resilience to multiply in environments with nutrients [48]. Production of fourth-generation bioethanol is deliberated, as a carbon-negative process hence the production of carbon from this technology is less than the carbon captured [49].

The Cellulose to Ethanol: Conversion Principle

In the act of bioethanol production, lignocellulosic biomass is considered a principal feedstock and has gained great attention, through the high lignin content in biomass, interfering with the hydrolysing process [50]. Several groups of lignocellulosic biomass like aquatic plants (water hyacinth), energy crops (perennial grasses and other dedicated energy crops) forest materials (pruning, sawdust, softwood, hardwood, and bark thinning residues), agricultural waste residues (bagasse, stoves and cereal straws), and organic fraction of MSW can be easily collected. Some of these are available seasonally and some biomass can be collected throughout the year [29].

In lignocellulosic biomass, 40% - 60% of the total dry weight is composed of cellulose, while 20% - 40% is hemicellulose, 10% - 25% is having lignin and small amounts of extractives and ashes [5, 36]. The major obstacle to the use of lignocellulosic biomass for ethanol production is its natural recalcitrant to microbial and enzymatic degradation. Factors that cause recalcitrance in biomass include; structural factors like pore size, cellulose specific surface area, degree of polymerization, the crystallinity of cellulose, and the volume and chemical factors like content and composition values in lignin, hemicelluloses, and acetyl groups. Due to the strong interconnection of these factors, lignocellulosic biomasses have difficulties in the hydrolysis process [50, 51].

In plant cell walls, cellulose consists of the major structural polymer and also the amplest biopolymer in nature is known to be cellulose. Cellulose is a linear, crystalline homopolymer containing β D-glucose repeating subunits linked by β-1, 4 glycosidic bonds and packed into micro fibril bundles, having an average molecular weight of about 1 00 000 Da [29, 52].

In lignocellulosic biomass, cellulose is present in crystalline forms besides the amorphous forms. A high percentage of cellulose present in biomass is crystalline, while a little amount presents as amorphous cellulose [50, 53].

The crystalline structure of the cellulose chain provides its considerably high tensile strength with minimum flexibility and insolubility in water. Therefore, it is less resistant to biological attacks and also insoluble in most solvents, which is the major limitation of cellulose hydrolysis [53]. Therefore, fermentable D-glucose can only be produced by either action of acid or enzyme by breaking β-1, 4 glycoside bonds. A dimer of two glucose units known as cellobiose is the basic building block of cellulose [54].

The amorphous form of cellulose can be easily degraded by enzymes, while crystallinity reduces the enzymatic degradability of cellulose [55]. Due to having hydrogen bonds, covalent bonds, and forces of Van der Waals in cellulose, it makes a rigid structure. A tightly packed backbone and high crystallinity result from the rigid structure of cellulose, which results in insolubility to water and resistance against de-polymerization [56].

In nature, hemicellulose is present as the second most ample polymer containing approximately 20% - 35% in plant material. It has a lower molecular weight than cellulose with a molecular weight of 30,000 Da on average [29]. They consist of shorter, linear, and chains of highly branched, consisting of different monosaccharide subunits to produce glucomannans, xyloglucan, mannans, xylans, and others. They are the most thermo-chemically sensitive polymer. Hemicelluloses are linked to lignin via covalent bonds [57].

It is a hetero-polymer containing monosaccharides that are categorized into pentoses (d-xylose and l-arabinose) and hexoses (d-mannose, d-galactose, and d-glucose) [58]. Also, there are uronic acids which are sugar acids, which typically include d-glucuronic, 4-O-methyl glucuronic and D-galacturonic acids [59]. When compared with cellulose, hemicellulose contains branches with short lateral chains having various sugars and easily hydrolysed polymers. Even when they form co-crystals with cellulose chains, they are not forming aggregates. The presence of hemicellulose helps to form a connection between fibres of lignin and cellulose, giving more rigidity networks [60].

Lignin is a highly branched phenolic compound. Phenyl propane is the basic structure of lignin monomers that are distinguishable only through having methoxyl group substitution on the aromatic rings. The main polymers which construct the lignin polymer include; coniferyl alcohol, phydroxyphenyl alcohol, and synapyl alcohol [61].

It is a hydrophobic heterogeneous phenolic polymer and mainly functions as a "cement" to the cellulose fibres in plants [62]. Therefore, it provides impermeability, structural support, protection from microbial attacks and conditions of oxidative stress [5, 63]. Also, it has the ability to irreversible adsorption of cellulase enzyme which affects enzymatic hydrolysis [64].

Lignin has the ability to give rigidity to the cell wall of the plant and to resist attacks from microbes [5]. Hydrophobicity and structural rigidity of lignocellulosic biomass is given by lignin. Lignin binds the other two major components of hemicelluloses to cellulose in the cell wall. Lignin provides a physical barrier to inhibit the digestion of cellulose by enzymes [50]. Also, harmful effects such as adsorption of hydrolytic enzymes non-specifically to "sticky" lignin, non-productive binding and interference with carbohydrate-lignin complexes and toxic effects of lignin derivatives on the microorganisms are caused by lignin [57]. However, before utilizing the lignocellulosic biomass, pre-treatment must be carried out to lower the biomass recalcitrance and increase the enzymatic hydrolysis of the biomass [38].

Pre-Treatment Effects and Different Pre-Treatment Techniques

Cellulose, hemicellulose, and lignin which are associated very closely compose the structure of the lignocellulosic biomass. Hence, there is a need for intense pre-treatment procedures to segregate cellulose from lignin and hemicellulose. The pre-treatment process increases the solubility or degradation of cellulose, through the decreasing crystallinity of cellulose and increasing the porosity of the lignocelluloses, so that the remaining cellulose can be hydrolysed more efficiently by enzymes [65]. Figure 1 shows the diagramatic representation of the pre-treatment process.

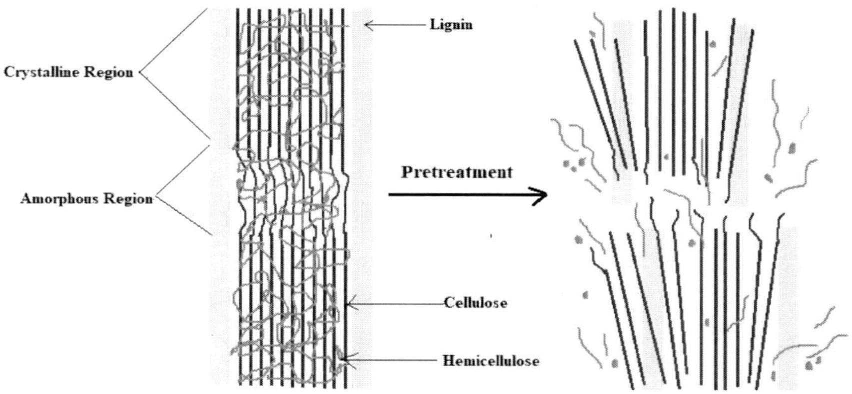

Figure 1. Schematic representation of the role of pre-treatment.

According to Seidl and Goulart., (2016) [66], the same procedures of pre-treatment cannot be applied to various types of lignocellulosic biomass, due to having variations in the percentage and cellulose, hemicellulose and lignin composition and the variety of bonds that they form. Therefore, different types of pre-treatment methods are being promoted to enhance the hydrolysis of cellulose and thus can obtain increased yields of fermentable sugars [67].

Pre-Treatment Methods

As shown in Figure 2, pre-treatment methods that have been used nowadays are broadly classified as biological, physical, chemical and physio-chemical pre-treatment [68]. The following characteristics define whether the selected pre-treatment method is a systematic pre-treatment process or not [69].

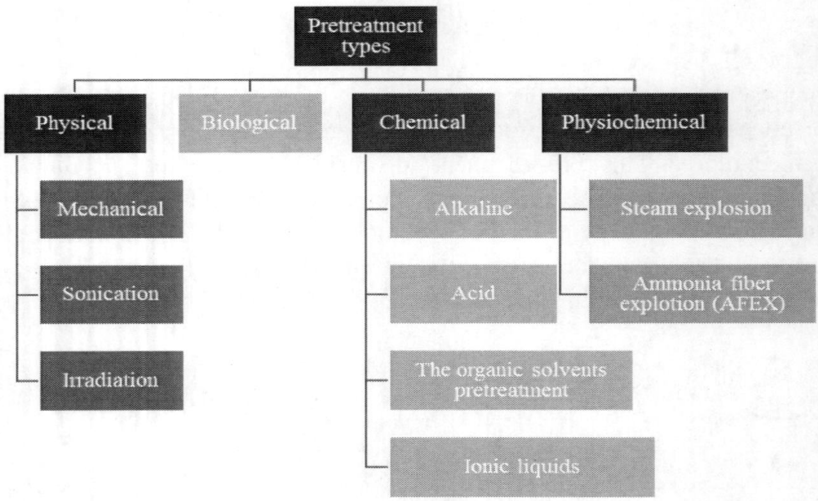

Figure 2. Different pre-treatment methods.

- A higher rate of recovery for all carbohydrates;
- Higher digestibility rate of the cellulose in the process of enzymatic hydrolysis;
- Production of no or deficient amounts of products released from sugar and lignin degradation;
- Pre-treated biomass should be possible to ferment without detoxification and should require a low energy demand;
- Low capital and operational cost.

Physical Pre-Treatment

As the first step in biomass pre-treatment, a physical or mechanical process is carried out in order to minimize the size of the material by breaking the physical structure. The physical pre-treatments result in enhancing the hydrolysis efficiency since it increases the approachable specific surface area of lignocellulosic biomass to enzymes by reducing the particle size or disturbing the structural regularity of lignocellulosic materials [68]. The main physical pre-treatment methods include mechanical comminution, sonication and irradiation [70].

Mechanical Comminution

Mechanical comminution pre-treatments carry grinding, chipping, or milling methods of biomass, which leads to a physical reduction of the size of particular biomass and increases the surface area that can be accessible. The purpose of this is to reduce the biomass crystallinity thus enhancing the hydrolysis [71]. Because of that, this method should be carried out first before exposing biomass to any other pre-treatment methods [72].

Sonication

The application of ultrasounds on biomass loses, swells, and ruptures the fibres in lignocellulosic biomass and disintegrates the lignocellulosic structures. The pre-treatment efficiency is affected by the choice of the oscillation frequency, processing temperature, biomass type and time. Usually, 20 – 40 kHz of sonication frequency is applied to break the internal hydrogen bonds of lignocellulose [73, 70].

Irradiation

Irradiation of the biomass pre-treatment is done mostly using gamma rays, electron beams or microwaves [68]. Usually, when microwaves are used, the radiation energy applied is about 300 - 700W/m^3. Application of microwaves on lignocellulosic biomass results in the evaporation of intermolecular water and the breaking of hydrogen bonds thus, decreasing the lignocellulose molecule stability. Also, the crystal size of cellulose decreases considerably resulting in enhanced efficiency of hydrolysis of the biomass [70]. Irradiation of lignocellulosic biomass by gamma rays or electron beams cleaves the β-1, 4 -glycosidic bonds and provides a bigger surface area and a reduced crystallinity. However, this is a somewhat costly method [74].

Biological Pre-Treatment Methods

Biological pre-treatment is achieved through the employment of either enzymes or microorganisms in the lignocellulose materials. Mostly used microorganisms are soft rot, white rot and brown rot fungi that degrade the lignin, which acts as a barrier to cellulose hydrolysis through the enzymatic pathway and causes the binding of cellulases unproductively [69]. Brown rot fungi are capable of degrading cellulose, while soft rot and white rot fungi are capable of degrading both cellulose and lignin. However, white rot fungi are known as the most favorable agent useful in biological pre-treatment [71]. *Punctualaria* sp. *Irpex lacteus, Pleurotus ostreatus, Eucalyptus grandis* and *Ceriporiopsis subvermispora* are known to be effective in the biological pre-treatment of various lignocellulosic materials [75].

Biological pre-treatment is considered an environmentally friendly and energy-saving method since it is possible to perform it at low temperatures and without the use of chemicals. It is a slow and time-consuming process, and it is mostly followed by some other pre-treatment methods [74].

Chemical Pre-Treatment Methods

After pre-treating lignocellulosic biomass physically and biologically they are exposed to chemical treatments. Popular chemical pre-treatment methods include alkaline or acid pre-treatment methods; furthermore, oxidation agents, ionic liquids (ILs) and organic solvents are also used as methods for chemical pre-treatment [70].

Alkaline Pre-Treatment

Alkaline pre-treatment is employed to remove lignin and hemicellulose, which increases the enzymatic digestibility of lignocellulosic materials. Under alkaline conditions, alkyl aryl linkages in lignin and hemicellulose are readily cleaved and cause swelling of cellulose, which results in reduced crystallinity, disruption of the structure of lignin, expansion of the internal surface area, and separate structural linkages between carbohydrates and lignin to the cellulose result in a noticeable enhance in the yields of saccharification [76]. Numerous alkaline reagents like KOH, $Ca(OH)_2$, NaOH, and aqueous ammonia are utilized for alkaline pre-treatments [33]. This process required lower temperature and pressure than other pre-treatment techniques.

Acid Pre-Treatment

In acid hydrolysis acids like sulphuric, carbonic, hydrochloric, hydrofluoric, phosphoric, and nitric, acids are used. Acid pre-treatment results in the partial or complete hydrolysis of hemicelluloses, the crystallinity of cellulose, and delignification of lignocellulosic biomass [56]. A widely practiced method is the use of dilute-acid than concentrated acids for pre-treatment of lignocellulosic materials [77].

The Organic Solvents Pre-Treatment

Organic solvents like methanol, ethanol, acetone, glycols, ethylene glycol, and tetrahydrofurfuryl alcohol or phenols are utilized to pre-treat lignocellulosic biomass. To induce the efficiency of the pre-treatment procedure, a small amount of HCl or H_2SO_4 is added and performed at temperatures above room temperature. This method causes the solubilisation of lignin and hemicellulose which gives treated cellulose that is perfect for enzymatic hydrolysis [69, 67].

Ionic Liquids

Ionic liquids typically have a larger organic cation and a smaller anion. Due to anion and cations' synergistic effect, the ionic liquid is capable to initiate the cellulose cleavage process. During the formation of hydrogen bonds between anions and cellulose, instead of forming hydrogen bonds with cellulose, cations are affected towards the process of dissolution. Ionic liquids exhibit characteristics such as recyclability, high ionic conductivity, thermal stability, low volatility, and high solvation power [78].

Physiochemical Pre-Treatment

Physiochemical pre-treatment methods are processes composed of in-between or as a combination of physical pre-treatment and chemical pre-treatment methods. Steam explosion, Ammonia fiber explosion (AFEX) and Hydrothermolysis or liquid hot-water pre-treatment are the methods that belong to this category [74].

Steam Explosion

During the method of steam explosion, through a high pressure of saturated steam lignocellulosic material is treated, and then suddenly the reduction of pressure has been done. This results in an explosive decompression of the biomass. Usually, the initial temperature of the process is between 160^0C - 260^0C (corresponding pressure, 0.69-4.83 MPa) which is maintained for a few seconds to several minutes. After that material is exposed to atmospheric pressure. Degradation of hemicellulose and transformation of lignin because of elevated temperature and thus enhancing the hydrolysis of cellulose are the key results of this process [77]. It required a lower energy requirement than mechanical milling, however, sugars cannot completely recover due to the disruption of the lignin-carbohydrate matrix incompletely. In the existence of a catalyst (alkali or acid) reaction rate could be improved.

Ammonia Fiber Explosion (AFEX)

The AFEX method is a thermochemical treatment, which leads lignocelluloses into highly digestible products. In the AFEX technique, under elevated pressures and moderate temperatures liquid anhydrous ammonia is applied to biomass and then depressurized rapidly. This expands the ammonia gas rapidly and makes biomass fibres with swellings and physical disruptions. AFEX method disrupts the linkage between lignin and carbohydrate, and ammonolysis of the cross-linked bonds of glucuronic, while partially recrystallizing the cellulose structure. All of these results in improving the surface area that can be accessed by the enzymatic attack during hydrolysis [67]. This technique is a dry-to-dry process and a significant difference caused to the composition of the biomass is not visible. This method of pre-treatment is being advantages since the washing step is not needed with its associated production of waste and expenses [79, 80].

Cellulose Hydrolysis

Principle

Hydrolysis or saccharification is known as a crucial step in the bioethanol production process from cellulosic biomass, as cellulose and hemicellulose are being hydrolyzed into freely available monomer molecules. These monomers can be easily fermented through the cleavage of glycosidic bonds in lignocellulosic substrates [81]. This facilitates the liquefaction of the chemical by reacting with water, as well as it is fruitful for some of the organic compounds which are resistant to degradation and fermentation [82] (Figure 3).

Hydrolysis of starch, cellulose or similar polysaccharides results in hexose sugars (glucose and fructose);

$$(C_6H_{10}O_5)\,n + nH_2O \longrightarrow nC_6H_{12}O_6$$

While hemicellulose hydrolyses into pentose sugars (xylose, mannose, arabinose, etc.);

$$(C_5H_8O_4)\,n + nH_2O \longrightarrow nC_5H_{10}O_5$$

Figure 3. Hydrolysis of starch, cellulose or similar polysaccharides and hemicellulose.

Various Cellulose Hydrolysis Methods (Chemical/Physical/Biological)

Hydrolysis is performed either by biological (enzymatic hydrolysis) or chemical (acid hydrolysis) hydrolysis [36].

Chemical Hydrolysis/Acid Hydrolysis

In chemical hydrolysis, concentrated or diluted acids are used. Acids like formic, sulphuric, sulphurous, hydrofluoric, hydrochloric, phosphoric, and nitric are commonly used [83]. During the acid hydrolysis, without the use of any preliminary pre-treatments, acids are capable of penetrating lignin and cellulose and hemicelluloses are hydrolysed into the fermentable sugar molecules [84].

Acid hydrolysis can be performed in two different ways that are commonly used; at an elevated temperature with the presence of dilute acid or at a low temperature with the presence of concentrated acid [20]. Acids like Sulphuric (H_2SO_4) and Hydrochloric (HCl) acids are the most widely used acids to hydrolyse cellulose and hemicelluloses [36].

Dilute Acid Hydrolysis

Dilute acid hydrolysis is the oldest technology used to produce ethanol from cellulosic biomass. In this process, a sulphuric acid solution of about 1% concentration is used at a high temperature (about 488K) inside a continuous flow reactor [85]. The benefits of the dilute acid hydrolysis process include the fast reaction rate, which facilitates continuous processing but it gives a low sugar yield [86]. The formation of inhibitory compounds like furfural, acetic acid, phenols and HMF which affect the following fermentation step can be induced by high temperatures [87].

Concentrated Acid Hydrolysis

Concentrated acid hydrolysis has the capability to convert cellulose into glucose and convert hemicellulose into 5-carbon sugar rapidly and completely [88]. 70% sulphuric acid is used at 40^0C to 50^0C for about 2 - 4 hours in a reactor and the sugar yield is high, due to high acid concentrations (30% - 70%). High acid concentration, leads to equipment corrosion and entails economic and environmental problems [89].

Biological Hydrolysis/Enzymatic Hydrolysis

The hydrolysis process is highly specific which converts the complex lignocellulosic biomasses into simple monomers [36]. It is performed at mild temperatures and pH conditions (i.e., pH - 4.8 and temperature – 45^0C-50^0C), which are optimum conditions for the enzyme activity of cellulase [36]. This process resulted in about 80% – 95% of high yields of sugars and environmental impacts are greatly reduced [90].

Usually, cellulase enzymes and hemicellulase enzymes are employed to hydrolyse the lignocellulosic biomasses [91]. Cellulases usually comprise three complementary enzyme groups; endo-cellulase, exo-cellulase or cellobiohydrolase, and β-glucosidases. With the combined activities of these 3 enzymes, complex cellulose polymer can be hydrolysed into simple glucose monomers. Several ancillary enzymes attack hemicellulose like β-xylosidase, acetylesterase, glucuronidase, xylanase, glucomannanase and galactomannanase, other than three major groups of cellulase enzymes [92]. Enzymatic hydrolysis does not cause corrosion problems, but it is a slow process that takes several days [83].

- Endoglucanase – Breaks the non-covalent interactions available in cellulose;
- Exoglucanase – Chain ends are being hydrolysed in order to break the polymer into smaller sugars;
- β-glucosidase – Finally hydrolyse the disaccharides and tetra-saccharides into glucose [93].

Involvement of Microorganisms in Cellulose Hydrolysis

Sources for the extraction of cellulose enzyme systems are the best suitable for the microbial system [94, 95]. The enzymes obtained from microorganisms are applied as an individual enzymes or a mixture of enzymes as purified enzymes or crude enzyme extracts in the hydrolysis process [96]. A vast variety of bacteria and fungi have the tendency of producing cellulase enzymes to hydrolyse the lignocellulosic materials.

The gut of animals that feed on cellulosic biomasses contains microorganisms, which are producing cellulase enzymes. e.g., microorganisms in the insects' guts like termites' guts (Isopteran), and bookworms (Lepidoptera) [97]. Also, the digestive system of cows, horses, and goats like ruminants contains microorganisms which are responsible for the degradation of cellulosic wastes. A vast variety of microorganisms have been characterized and identified with the capability of cellulose hydrolysing, including many fungal and bacterial strains.

Fungal species responsible for cellulosic enzyme activity are *Chaetomium*, *Fusarium*, *Myrothecium*, *Aspergillus*, *Trichoderma*, *Penicillium* etc [94]. *Clostridium*, *Actinomycetes*, *Butyrivibrio fibrisolven*, *Bacteroides succinogenes*, *Methanobrevibacterrum inantium*, and *Bacillus* sp. are the recorded bacteria species responsible for cellulase enzyme production [98, 99, 100, 101]. When compared with fungi, bacteria have a high capability to be used in the production of cellulase enzyme as their growth rate is relatively high growth, as well as they can adapt to any environmental conditions quickly and easily and are able in making enzyme complements which are highly stable [102].

There are several studies that yeasts also can be used in hydrolysis as some of them have the ability to make extracellular α-amylase and glucoamylase. These yeast species include *Filobasisium capsuligenum, Trichosporon pullulans*, *Lipomyces starkeyi, Saccharomycopsis bispora, Schwanniomyces castelli, Lipomyces kononenkoae, Saccharomycopsis capsularis, Schwanniomyces alluvius, Saccharomycopsis fibuligera,* and *Candida tsukubaensis* CBS 6389. The lignocellulose-degrading fungus such as *Trichoderma reesei*is are also useful for cellulolytic enzyme production that can be utilized in hydrolysis [31].

Sugar Fermentation

Principle

The fermentable sugars released from hydrolyzing of lignocellulosic biomass are then used as the substrate, which is subjected to fermentation [96]. Fermentation is a biological pathway, which produces ethanol from sugars through the diversity of fermentation agents; either by fungi or bacteria through their metabolic activity. The reactions of converting hexoses and pentoses into ethanol are anaerobic and can be expressed as follows [36] (Figure 4).

Fermentation of hexoses;

$$C_6H_{12}O_6 \longrightarrow 2C_2H_5OH + 2CO_2$$

Fermentation of pentoses;

$$3C_5H_{10}O_5 \longrightarrow 5C_2H_2OH + 5CO_2$$

Hexoses including glucose, galactose, and mannose; pentoses including xylose and arabinose are fermented to ethanol.

Figure 4. Conversion of hexoses and pentoses in to ethanol.

Sugar Fermentation and Acetic Acid Formation

A major hindrance in second-generation bioethanol production is the formation of acetic acid. This is mainly due to Acetic Acid Bacteria (AAB), which oxidise ethanol under aerobic conditions [103]. Due to microbial contaminants, the type and number of microbes during the fermentation changes and this is known as microbial succession. In the initial stages of fermentation fermenting microbes like yeast are dominated and can be finally replaced by AAB which results in acetic acid formation. This results in reduced ethanol percentages and also inhibits ethanol production as acetic acid is toxic to yeast growth and metabolism [104].

Acetic acid formation;

$$C_2H_5OH + O_2 + H_2O \longrightarrow CH_3COOH$$

Currently, 19 genera of AABs are classified, which include; Ameyamaea, *Asaia*, *Acidomonas*, *Acetobacter*, *Endobacter*, *Commensalibacter*, *Bombella*, *Granulibacter*, *Gluconacetobacter*, *Kozakia*, *Gluconobacter*, *Komagataeibacter*, *Neokomagataea*, *Nguyenibacter*, *Swingsia*, *Tanticharoenia*, *Saccharibacter*, *Neoasaia*, and *Swaminathania* [103].

It has been observed that the use of culture media with complex ingredients reduces AAB growth while promoting ethanol-fermenting yeast growth thus the complex media provide a sufficient amount of nutrients and other non-nutritional components to enhance yeast growth and survival [104].

The Different Types of Microorganisms Involved in Sugar Fermentation

Microbes perform a noteworthy role in the fermentation process. Therefore, selecting the most suitable strain is very crucial. The ability to consume pentoses and hexoses, the ability to tolerate high ethanol concentrations, and inhibitory compounds, and also the necessity of nutrients in minimum amounts are the key features of an exquisite microbe [105]. A wide range of microorganisms comprising bacteria, fungi and yeast have been involved during the fermentation.

The yeast *Saccharomyces cerevisiae* is a vastly studied and widely utilized microorganism for fermentation. *Escherichia coli, Kluyveromyces marxianus,* and *Candida brassicae* are also known to be used in fermentation [96]. *Zymomonas mobilis* is the second most widely used fermenting microorganism and is considered an ideal organism for bioethanol production on a large scale due to its high ethanol production capability, excellent ethanol tolerance, as well as the capability of surviving in a broad range of pH [106].

As stated by the study done by Elshaghabee et al. (2016) [107], selected intestinal microorganisms and lactic acid bacteria *Bacteroides thetaiotao-micron*, *Anaerostipes caccae*, *Bifidobacterium longum*, *Escherichia coli*, *Enterococcus fecalis*, *Lactobacillus fermentum*, *Lactobacillus plantarum*, *Lactobacillus acidophilus*, *Weissella confuse* were identified as ethanol-producing microorganisms by fermenting hexoses, pentoses, disaccharides, and inulin. *Achromobacter* sp. is also known as a fermenting bacterium [100, 101].

Anaerobic thermophilic bacteria, like some filamentous fungi, such as *Neurospora* sp., *Monilia* sp., *Aspergillus* sp., *Zygosaccharomyces rouxii*, *Trichoderma viride*, *Clostridium thermocellum* and *Paecilomyces* sp. were also identified as fermenting bacteria [31].

Bio-Processes Mediated by Microorganisms for Bioethanol Production

In order to integration of hydrolysis and fermentation bioprocesses, a variety of configurations are used. The most common configurations are Separate Enzymatic Hydrolysis and Fermentation (SHF), Simultaneous

Saccharification and Fermentation (SSF) and Consolidated Bioprocessing (CPB) [100] (Figure 5).

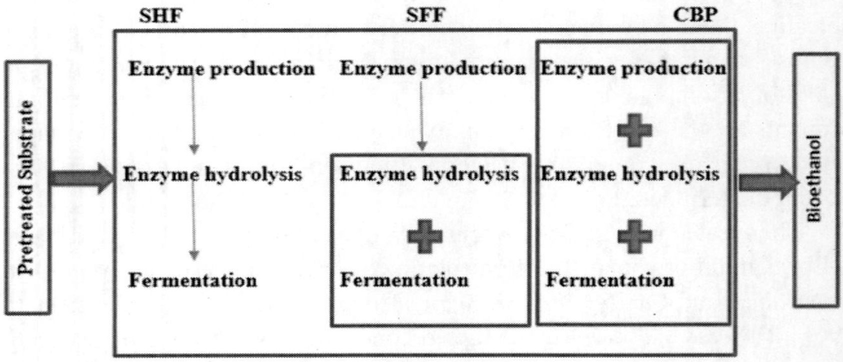

Figure 5. Overview of various fermentation strategies.

Simultaneous Saccharification and Fermentation (SSF)

Reducing sugars resulting in hydrolysis are fermented simultaneously into ethanol during the SSF process. Since hydrolysed sugars released from cellulosic biomass through enzymatic activity are consumed immediately by the fermentative microorganism for the production of ethanol, product inhibition is greatly reduced in hydrolysis. In SSF a single reactor is used to combine the hydrolysis of cellulosic biomass with the simultaneous fermentation of released sugars. This method is easy to operate and needs low equipment [36]. Usually, in SSF, hydrolysis is the rate-limiting process.

Separate Enzymatic Hydrolysis and Fermentation (SHF)

SHF method has two steps in the process where hydrolysis and fermentation processes are performed in separate units to take full advantage of each of the processes [71]. At first through enzymes, the pre-treated cellulosic biomass is hydrolysed into glucose and xylose, and then resulted in sugars are fermented into ethanol. The functioning of both enzymatic hydrolysis and fermentation at their relevant optimum conditions is the key benefit of the SHF process.

Nevertheless, the ethanol yield is affected by enzyme activity inhibition due to sugar accumulation [5].

Consolidated Bio-Processing (CBP)

In CPB all the reactions which are needful for the bioconversion of biomass into ethanol are integrated, as enzyme production, enzymatic hydrolysis and fermentation all are performed in one step. This process does not need operational costs for the production of enzymes and to release of sugars, all the substrates are utilized [5]. This approach is more appropriate when compared with SSF and SHF processes when considering its energy requirement, complexity, efficiency of yield, and cost of production for second-generation bioethanol production [48].

Cellulosic Solid Waste Management and Bioethanol Production

Since the last century, municipal solid waste becoming a critical problem in both developing and developed countries [108]. Even though the countries implement policies and legislation to reduce the waste production, the population growth along with urbanization increases the waste generation. As estimated, the solid waste is generated by about 1.3 billion tons throughout the world and it has anticipated that this will reach 2.2 billion tons by 2025 [109]. Due to increased solid waste production low and middle-income countries are severely affected as waste generation, waste disposal and waste management are not properly regulated.

The best practice for managing solid waste is the zero waste concept and to meet this many schemes were introduced including the 3R's concept; Reduce, Reuse and Recycle [110]. However, disposal of waste at landfills remains the most popular option worldwide, which leads to a shrinking of landfill capacity. Construction of new landfill sites leads to several issues such as; the averse of communities to build new landfill sites nearby, high construction costs, the tightness of environmental regulations for the closure of landfills, and the operation and construction of existing ones [111]. Nowadays, as eco-friendly solutions incineration, composting and biofuel production is taking place.

Among solid waste generated across the world cellulosic solid waste is a major component. Generally, this cellulosic biomass was either converted into organic fertilizers (compost), landfilled or burned [112]. The increasing generation of waste, limited landfilling spaces and environmental concerns that occur due to the burning of waste have made people find new methods to manage cellulosic waste products [113]. When cellulosic waste is burned, it releases pollutants including nitrogen dioxide, nitrous oxide, carbon monoxide, and particles which can produce ozone and nitric acid, which causes severe health and ecological problems [114]. Also, when large proportions of these wastes are dumped into landfills may lead to the emission of carbon dioxide and methane, surface and groundwater contamination, odour formation, and soil contamination. This emission of methane from landfills contributes to global warming, as methane is a greenhouse gas. Moreover, leachate with heavy metals, high amounts of NH_3-N, organic substances and some other hazardous substances may release from landfills which will cause problems in irrigation and soil [115].

Hence, cellulosic waste causes severe environmental impacts, people are concerned about more environmentally friendly waste management strategies. In that case, using this cellulosic waste for energy production is a major concern at the present.

Global Trend on Waste-Derived Bioethanol Production

Currently, in the world, cellulosic waste including agricultural waste, forest materials and organic fractions of MSW is used for bioethanol production [116]. According to recent estimations, about 2 billion tons of MSW are generated and are anticipated to grow up to 3.4 billion tons by 2050. Also, 33% of the generated MSW are handled in unhygienic ways that will result in many problematic issues [115, 117]. The use of an organic fraction of the MSW for the production of bioethanol now is trending [118].

Ethanol production from agro wastes is now popular across the world. Agro waste, such as sugarcane bagasse, sugarcane bark, corn cob, corn husk, and corn stalk are used [119]. Corncob is used to produce bioethanol because they are abundant and widely distributed around the world. The United States Department of Agriculture (USDA) reported that in 2018-2019 global corn production was 1098.95 million tons [120]. As a result, corncob residues are being simultaneously generated annually in large amounts and conventionally discarded as solid wastes, which causes a considerable waste of resources and terrible impacts on the environment [121].

Paper waste, which is a part of the degradable fraction of MSW is a promising lignocellulosic feedstock for bioethanol production [111]. That is because wastepaper is highly abundant, cost-effective, have a relatively high abundance of carbohydrates, and can be digested easily without using any aggressive physical or chemical pre-treatments. Most importantly utilization of waste paper for bioethanol production provides a very effective and critical way for paper waste management [95]. According to the literature, in Sri Lanka, a considerable percentage of paper waste consider solid waste without any proper usage (75% -80%) and hence recycling is still at a preliminary level [122].

The paper waste fraction of MSW consists of a large number of recycled magazines, books, newspapers and paper sludge, the solid waste produced in the pulp and paper industry. This paper sludge differs from wastes resulting from numerous pulp and paper industries, including spent sulphite liquors, Kraft pulp and paper sludge. All this paper waste is used as feedstock for the production of bioethanol [36, 95]. Higher yields of bioethanol need to be obtained from waste paper in order to implement it as a feasible environmentally sustainable waste management strategy,

Among MSW, a notable amount is of food waste hence, one-third of food which is grown globally is lost or wasted. Food manufacturers and food processors discard an immeasurable amount of food waste annually as direct food products or as by-products. These food wastes contain many valuable components and especially they are rich in sugars that can be converted into bioethanol [123]. While the production and storage stages of fruits and vegetables, fruit and vegetable waste are produced. Due to overproduction, insufficient skills, climate change, natural catastrophes and shortages in proper infrastructure, the production of fruit and vegetable waste can be exalted [114]. Fruit wastes are an ideal feedstock for the production of bioethanol since they are rich in sugar [124].

However, at present, a vast amount of commercial bioethanol is produced using food crops as the substrate for first-generation bioethanol. In Brazil sugarcane, in the United States, corn, sugar beet and wheat in European Union are used as feedstock for bioethanol production [125].

Challenges and Solutions

Bioethanol is attractive as an alternative fuel and utilized as an automotive fuel. Bioethanol is a potential solution for future energy crises [126] because it can be either used directly in engines that are specially designed or used as lower blends in CI engines without any modifications in the engines as blends. Bioethanol is mainly used as blends, blended with gasoline and diesel, which makes it important as a fuel extender. This helps to increase the lifetime of the remaining fossil fuel resources [36]. Consequently, ethanol-diesel blends (or e-diesel) have gained much interest recently because of many benefits including;

- Reduce petroleum dependency and enhance the use of renewable fuels.
- The cost of ethanol production has become more favourable compared to petroleum fuel costs.
- Reduce emission of diesel particulate matter (PM),
- Reduce emission of CO from engines.

According to Waterland et al. (2003) [127], about 20% to 40% reductions in particulate matter (PM) emissions and about 20% to 30% reductions in CO emissions have been recorded. Therefore, utilizing bioethanol as an alternative fuel has both economic and environmental benefits.

Bioethanol-Diesel Emulsion

Partial or full replacement of fossil fuels with bioethanol has gained worldwide attention because excessive consumption of fossil fuels releases high proportions of particulate matter (PM), oxides of nitrogen (NOx), carbon monoxide (CO), and other hazardous ingredients, which are considered as air pollutants. Moreover, it has been estimated that around 63% of the emission of greenhouse gases, due to carbon dioxide (CO_2) emission occurs from the direct combustion of fuels [106]. So, to reduce the greenhouse effect, and air pollution, and as a fuel extender ethanol is used blended with different fuels [128]. Blending ethanol with diesel fuel and utilizing it as fuel was a fascinating research interest since the 1980s and was identified that technically ethanol-diesel blends can be accepted for the diesel engines that are currently existing [129]. Ethanol-diesel blends are prepared by mixing ethanol up to 15 volumes per cent with diesel. The frequently used blends contain 5% ethanol mixed with 95% diesel (E5), 10% ethanol mixed with 90% diesel (E10) and 85% ethanol and 15% diesel (E15) [21].

Diesel fuel is a derivative of petroleum and is a blend of hydrocarbons, mainly aliphatic hydrocarbons comprising 8-28 carbon atoms with a lower H/C mass ratio. Adding ethanol to diesel produces physio-chemical changes in diesel [130].

As cited in Parthasarathi et al. (2014) [131], numerous studies have stated that ethanol is miscible with diesel, while some other various studies have shown that ethanol is partially immiscible with diesel. The temperature, content of water, hydrocarbon composition of the diesel and additives are the main factors that decide the solubility of ethanol in diesel [57]. The water content is a crucial factor responsible for the miscibility of ethanol in diesel [132]. When the water content is higher than 1% (vol/vol) phase separation can be occurred [133]. Stable solutions of ethanol-diesel blends can be obtained by mixing anhydrous ethanol with diesel at warm ambient temperatures. But the blends separate into two phases at temperatures below about 10^0C [130].

As mentioned in Lapuerta et al. (2007) [134], with the increasing temperature of ethanol-diesel blend the stability and the solubility of blends increase. The hydrocarbon composition is another critical factor that determines the diesel-ethanol blend stability. It has been found that the aromatic content of the diesel reduces the solubility of ethanol and reduction of the aromatic content in the diesel can increase ethanol miscibility in diesel [133].

Normally commercially available diesel is made up of fossil diesel and various types of additives. These additives are added to commercial diesel in different amounts for different purposes such as the reduction of pernicious emissions, the improvement of stability of the fluid over a wide range of conditions, the improving of the viscosity index, and reducing the change in viscosity rate with temperature, for the improvement of ignition by the reduction of its delay time and flash point etc. [135].

Lapuerta et al. (2017) [136], have described three stages in phase separation of a mixture of ethanol and diesel; initial turbidity, interface appearance and displacement of the interface until equilibrium. At first, samples turned quickly into a turbid solution with no phase separation. Then the more transparent less dense phase of bioethanol was in the superior part of the mixture, while the lower phase remained turbid. In the third stage, the width of the bioethanol phase was increased until an equilibrium position, the interface moved towards the inferior part and finally, the diesel phase became transparent.

When ethanol is partially miscible with diesel, ethanol-diesel emulsions are made. Emulsions are made when two or more fluids which are immiscible are blended and do not have a stable internal structure whereas the droplets are statistically dispersing in a matrix of the liquid. The emulsions are classified into three categories; Macro emulsions - cloudy emulsions with particles which are greater than 400nm (0.4μm), Microemulsions - thermodynamically stable, semi-opaque, a bluish-white solution with particles ranging between 100 and 400nm (0.1 - 0.4μm), and Nanoemulsions - kinetically stable solution with particle size less than 100nm (0.1μm) and are transparent [137]. Using co-solvents and emulsifiers can enhance the miscibility of ethanol in diesel. The co-solvents are organic solvents that easily mix with diesel, as well as with ethanol and make the blend homogeneous [21]. A co-solvent should have polar groups to establish intermolecular interactions with polar molecules, like ethanol and water, and nonpolar groups to establish intermolecular interactions with nonpolar molecules, like diesel hydrocarbons to improve the blend stability [129]. Common co-solvents used are fatty acid esters, such as methylic, ethylic, cyclic or aromatic alcohols as pure combustibles and add about 1% to 2% (vol/vol) anhydrous ethanol into ethanol-diesel blends up to 15% [138].

Emulsifiers enhance the stability of ethanol-diesel blends, when the water content is higher than 1% (vol/vol) and also at lower temperatures with 15% or even 20% ethanol concentrations [139]. Inhibiting phase separation in ethanol-diesel blends is the main function of emulsifiers. However, they also improve lubricity, detergency, and low-temperature properties in ethanol-diesel blends [140].

When considering ethanol-diesel blends as an alternative fuel, several critical factors should be concerned prior to checking the potential commercial usage of the ethanol-diesel blends. Those factors are properties in the blends including; stability, viscosity, density, flashpoint, lubricity, sulfur content, safety and compatibility of materials, the engine performance as the effect of the fuel, emissions and durability. This is because the blending of ethanol with diesel fuel can dominate those major properties [141]. Therefore, The American Society for Testing and Materials (ASTM) has developed and introduced a standard for denatured ethanol for blending with gasoline/diesel; ASTM D4806 [139]. In addition to that, considering the materials' compatibility and corrosiveness is very important. Properties such as flashpoint and flammability are very important when considering safety issues. Finally, fuel biodegradability is evaluated as it is vital to reduce any environmental pollution [141].

Lowering the Flashpoint Incorporating Bioethanol into Diesel

Flashpoint is called the lowest temperature where the fuel vapours will ignite under a low flame. This is important for safety as the vapour produced due to the evaporation of fuel creates flammable conditions especially during refuelling the partially filled fuel tanks or if there is damage/leakage that happens in tanks or any other component of the fuel system [129]. Therefore, lower flashpoints increase flammability in the engines. Normally diesel fuel's flashpoint is about $52^0C - 65^0C$ [142] and ethanol has a flashpoint of about 37.8^0C [129].

Lower flashpoint values may cause accidental ignitions very easily. Therefore, ethanol-diesel blends should be stored and handled with more caution [138, 143]. Torres-Jimenez et al., (2011) [143], describe that flashpoint decreases greatly due to more volatile ethanol. Because ethanol evaporates before diesel during the heating process, and the start of ignition is reported at this particular temperature for all ethanol-diesel blends. Even the presence of a small fraction of methanol also can reduce the flash point greatly [144].

Additional Fuel Properties

The mass of a substance per unit volume is known as density. The density of the fuel is important for consistency and good fuel economy [145]. Usually, in ethanol-diesel blends, density is decreased with the increase of ethanol content [101, 140]. According to Torres-Jimenez et al. (2011) [143], reduced density can increase pollutant emissions, as well as deteriorate the engine performance, while higher density can produce more power and more smoke.

The capability of lubricating metal surfaces with fuel is known as Lubricity. This is important to maintain good conditions in the engine components, such as fuel pumps that are lubricated by the fuel [140]. As denoted by Torres-Jimenez et al. (2011) [143] and Chandrasiri et al. (2022) [101], the lubricity of the prepared blends has increased with the increasing ethanol percentage of the blends. This is because the losses due to the evaporation of ethanol from the lubricating layer increase lubricities [136].

Sulfur content is important to minimize engine wear. Sulfur can produce acidic by-products during combustion which are harmful to both environment and the engine. Reduced sulfur content is a good quality fuel [146]. The sulfur content in ethanol-diesel blends decreases with increasing ethanol percentage. Singh and Bharj, (2019) [146], have demonstrated the reduction in sulfur content in ethanol-diesel blends and have stated that it is because the addition of sulfur-free ethanol reduces the total sulfur content in the blends.

Viscosity is the measurement of resistance to the flow of a liquid. This is important for injector flow, consistency, and atomization of the fuel. There are always maximum and minimum viscosity limits. The flow of fuel during cold starting is ensured by the upper viscosity limit, and the lower limit is important at high temperatures to avoid the chance of a consequential power loss [145]. The viscosity of the ethanol-diesel blends is decreased with the rising ethanol percentage and many studies have observed the same variation [136, 143].

The calorific value is the amount of energy released when 1kg of fuel burns. As emphasized by Nandakumar and Malayalamurthi, (2015) [140], calorific values decrease when ethanol is added. Singh and Bharj, (2019) [146], have mentioned this is because the addition of ethanol which has a low calorific value, reduces the overall calorific values of the blends.

Other Applications

Bioethanol is a platform chemical that can be make utilized in the making of numerous molecules such as propylene, acetaldehyde, ethylene, diethyl ether, and ethyl acetate in pharmaceuticals, cosmetics and beverages [125]. Also, ethanol and mixtures of ethanol are used as promising sources for hydrogen production for fuel cell applications [147, 148].

Conclusion

Second-generation bioethanol production using cellulosic biomass has become an interesting research interest since the world is about to face a crisis in supplying the energy in near future. As cellulosic biomass is abundant and is considered a solid waste product, the production of second-generation bioethanol is an outstanding way out of the energy crisis as well as can be considered a way of managing several types of solid wastes. The bottleneck in converting cellulose into bioethanol is the rigid structure of the cellulose and the use of several pre-treatment methods can induce the feasibility of hydrolysing, and saccharification and enhance the bioethanol yield. Microbial activities play a vital role during pre-treatment, hydrolysis and saccharification processes in second-generation bioethanol production. As a fuel blending ethanol with commercial diesel fuel is a widely accepted way to lessen environmental pollution, and the greenhouse effect and extend the lifetime of the current fossil fuel sources. Density, viscosity, sulfur content and calorific value are the important physicochemical characteristics in ethanol-diesel blends. The addition of ethanol into diesel makes several changes in the ethanol-diesel blends and researchers have identified the use of lower ethanol-diesel blends as a hopeful solution for the energy crisis.

References

[1] Lizunkov, V., 2018. Population of the world and regions as the principal energy consumer. *International Journal of Energy Economics and Policy,* 2018, 8(3), pp.250-257.

[2] Kober, T., Schiffer, H.W., Densing, M. and Panos, E., 2020. *Energy Strategy Reviews.* Elsevier, 31, pp.1-19.

[3] Newell, R.G. and Raimi, D., 2020. Global energy outlook comparison methods: 2020 update. *Resources for the Future: Washington, DC, USA*, pp.1-33.

[4] International Energy Agency, 2019, October. *Key World Energy Statistics* 2019. Paris, France: IEA.

[5] Rastogi, M. and Shrivastava, S., 2017. Recent advances in second generation bioethanol production: an insight to pretreatment, saccharification and fermentation processes. *Renewable and Sustainable Energy Reviews,* 80, pp.330-340.

[6] Ahmad, T. and Zhang, D., 2020. A critical review of comparative global historical energy consumption and future demand: The story told so far. *Energy Reports, 6,* pp.1973-1991.

[7] Arutyunov, V.S. and Lisichkin, G.V., 2017. Energy resources of the 21st century: problems and forecasts. Can renewable energy sources replace fossil fuels? *Russian Chemical Reviews,* 86(8), pp.777.

[8] Bioenergy, I.E.A., 2009. Bioenergy–a sustainable and reliable energy source. *International Energy Agency Bioenergy,* Paris, France.

[9] Naqvi, M. and Yan, J., 2015. First-generation biofuels. *Handbook of clean energy systems,* pp.1-18.

[10] Khandaker, M.M., Qiamuddin, K., Majrashi, A. and Dalorima, T., 2018. Bioethanol production from fruit and vegetable waste by using saccharomyces cerevisiae. *Bioethanol Technologies*, pp.37-53.

[11] Bhan, C., Verma, L. and Singh, J., 2020. Alternative fuels for sustainable development. In *Environmental Concerns and Sustainable Development*.Springer, Singapore, pp.317-331.

[12] Vasić, K., Knez, Ž. and Leitgeb, M., 2021. Bioethanol production by enzymatic hydrolysis from different lignocellulosic sources. *Molecules*, *26*(3), pp.753.

[13] Alalwan, H.A., Alminshid, A.H. and Aljaafari, H.A., 2019. Promising evolution of biofuel generations. Subject review. *Renewable Energy Focus*, *28*, pp.128-140.

[14] Zentou, H., Abidin, Z.Z., Yunus, R., Awang Biak, D.R. and Korelskiy, D., 2019. Overview of alternative ethanol removal techniques for enhancing bioethanol recovery from fermentation broth. *Processes,* 7(7), pp.458.

[15] Moustakas, K., Loizidou, M., Rehan, M. and Nizami, A.S., 2020. A review of recent developments in renewable and sustainable energy systems: Key challenges and future perspective. *Renewable and Sustainable Energy Reviews*, *119*, pp.109418.

[16] Leong, W.H., Lim, J.W., Lam, M.K., Uemura, Y. and Ho, Y.C., 2018. Third generation biofuels: a nutritional perspective in enhancing microbial lipid production. *Renewable and sustainable energy reviews*, *91*, pp.950-961.

[17] Madusanka, D.A.T. and Manage, P.M., 2018. Potential utilization of Microcystis Sp. for biodiesel production; Green Solution for Future Energy Crisis. Asian Jr. of Microbiol. *Biotech. Env. Sc,* vol.20. no.2, pp.506-512.

[18] Burnete, N.V., Filip, N. and Barabas, I., 2015. Diesel-ethanol blends and their use in diesel engines. *J. Automot. Eng,* 89, pp.88-106.

[19] El-Sheekh, M.M., Bedaiwy, M.Y., El-Nagar, A.A., ElKelawy, M. and Bastawissi, H.A.E., 2022. Ethanol biofuel production and characteristics optimization from wheat straw hydrolysate: Performance and emission study of DI-diesel engine fueled with diesel/biodiesel/ethanol blends. *Renewable Energy*.

[20] Azhar, S.H.M., Abdulla, R., Jambo, S.A., Marbawi, H., Gansau, J.A., Faik, A.A.M. and Rodrigues, K.F., 2017. Yeasts in sustainable bioethanol production: A review. *Biochemistry and Biophysics Reports,* 10, pp.52-61.

[21] Odziemkowska, M., Matuszewska, A. and Czarnocka, J., 2016. Diesel oil with bioethanol as a fuel for compression-ignition engines. *Applied Energy,* 184, pp.1264-1282.

[22] Sandalcı, T., Karagöz, Y., Orak, E. and Yüksek, L., 2014. An experimental investigation of ethanol-diesel blends on performance and exhaust emissions of diesel engines. *Advances in Mechanical Engineering,* 6, pp.409739.

[23] Singh, A., Prajapati, P., Vyas, S., Gaur, V.K., Sindhu, R., Binod, P., Kumar, V., Singhania, R.R., Awasthi, M.K., Zhang, Z. and Varjani, S., 2022. A Comprehensive Review of Feedstocks as Sustainable Substrates for Next-Generation Biofuels. *BioEnergy Research*, pp.1-18.

[24] Yaser, D., Kashif, S., Sarkar, B. and Pallavi, R.B.M., 2019. 14-Biofuels: Their characteristics and analysis. In *Woodhead Publishing Series in Composites Science and Engineering, Biomass, Biopolymer-Based Materials, and Bioenergy (Deepak Verma, Elena Fortunati, Siddharth Jain, XZ, ed.)*, *277*, pp.325.

[25] Mat Aron, N.S., Khoo, K.S., Chew, K.W., Show, P.L., Chen, W.H. and Nguyen, T.H.P., 2020. Sustainability of the four generations of biofuels–A review. *International Journal of Energy Research,* 44(12), pp.9266-9282.

[26] Ortiz-Reyes, E. and Anex, R.P., 2022. Economic and environmental performance of non-cellulosic fermentable carbohydrates production for biofuels and chemicals. *Journal of Cleaner Production, 353*, pp.132526.

[27] Imteaz, M.A., Hossain, A.S. and Ahsan, A., 2022, March. Data-driven modelling of bioethanol fuel production from rambutan fruit waste. In *Proceedings of the Institution of Civil Engineers-Waste and Resource Management.* Thomas Telford Ltd, pp.1-7.

[28] Saini, A., Sharma, D. and Sharma, S., 2022. Bioethanol: Substrates, Current Status, and Challenges. In *Status and Future Challenges for Non-conventional Energy Sources Volume* 2Springer, Singapore, pp.231-269.

[29] Zabed, H., Sahu, J.N., Suely, A., Boyce, A.N. and Faruq, G., 2017. Bioethanol production from renewable sources: Current perspectives and technological progress. *Renewable and Sustainable Energy Reviews,* 71, pp.475-501.

[30] Naik, S.N., Goud, V.V., Rout, P.K. and Dalai, A.K., 2010. Production of first- and second-generation biofuels: a comprehensive review. *Renewable and sustainable energy reviews,* 14(2), pp.578-597.

[31] Lin, Y. and Tanaka, S., 2006. Ethanol fermentation from biomass resources: current state and prospects. *Applied microbiology and biotechnology,* 69(6), pp.627-642.

[32] Gunawardena, S., 2009. *Liquid biofuel for transportation in Sri Lanka.,* pp.23.

[33] Balat, M., 2011. Production of bioethanol from lignocellulosic materials via the biochemical pathway: a review. *Energy conversion and management,* 52(2), pp.858-875.

[34] Tye, Y.Y., Lee, K.T., Abdullah, W.N.W. and Leh, C.P., 2016. The world availability of non-wood lignocellulosic biomass for the production of cellulosic ethanol and potential pretreatments for the enhancement of enzymatic saccharification. *Renewable and Sustainable Energy Reviews, 60*, pp.155-172.

[35] Guerrero, A.B., Ballesteros, I. and Ballesteros, M., 2018. The potential of agricultural banana waste for bioethanol production. *Fuel,* 213, pp.176-185.

[36] Branco, R.H., Serafim, L.S. and Xavier, A.M., 2019. Second generation bioethanol production: on the use of pulp and paper industry wastes as feedstock. *Fermentation,* 5(1), pp.4.

[37] Onuki, S., Koziel, J.A., van Leeuwen, J.H., Jenks, W.S., Grewell, D. and Cai, L., 2008. Ethanol production, purification, and analysis techniques: a review. In 2008 Providence, Rhode Island, June 29–July 2, 2008. *American Society of Agricultural and Biological Engineers,* pp.1-12.

[38] Zhao, C., Qiao, X., Shao, Q., Hassan, M. and Ma, Z., 2020. Evolution of the lignin chemical structure during the bioethanol production process and its inhibition to enzymatic hydrolysis. *Energy & Fuels,* 34(5), pp.5938-5947.

[39] Kang, Q., Appels, L., Tan, T. and Dewil, R., 2014. Bioethanol from lignocellulosic biomass: current findings determine research priorities. *The Scientific World Journal,* 2014, pp.1-14.

[40] Teixeira, A.C.R., Sodré, J.R., Guarieiro, L.L.N., Vieira, E.D., de Medeiros, F.F. and Alves, C.T., 2016. A Review on Second and Third Generation Bioethanol Production (No. 2016-36-0515). *SAE Technical Paper,* pp.1-7.

[41] Jambo, S.A., Abdulla, R., Azhar, S.H.M., Marbawi, H., Gansau, J.A. and Ravindra, P., 2016. A review on third generation bioethanol feedstock. *Renewable and sustainable energy reviews, 65*, pp.756-769.

[42] Tan, I.S., Lam, M.K., Foo, H.C.Y., Lim, S. and Lee, K.T., 2020. Advances of macroalgae biomass for the third generation of bioethanol production. *Chinese Journal of Chemical Engineering,* 28(2), pp.502-517.

[43] Velazquez-Lucio, J., Rodríguez-Jasso, R. M., Colla, L. M., Sáenz-Galindo, A., Cervantes-Cisneros, D. E., Aguilar, C. N., Fernandes, B. D., Ruiz, H. A. (2018). 'Microalgal biomass pretreatment for bioethanol production: a review', *Biofuel Research Journal,* 5(1), pp.780-791.

[44] Alia, K.B., Rasul, I., Azeem, F., Hussain, S., Siddique, M.H., Muzammil, S., Riaz, M., Bari, A., Liaqat, S. and Nadeem, H., 2019, March. Microbial production of ethanol. In *Microbial Fuel Cells: Materials and Applications. Materials Research Forum LLC., Pennsylvania*, pp.307-334.

[45] Javed, F., Aslam, M., Rashid, N., Shamair, Z., Khan, A.L., Yasin, M., Fazal, T., Hafeez, A., Rehman, F., Rehman, M.S.U. and Khan, Z., 2019. Microalgae-based biofuels, resource recovery and wastewater treatment: a pathway towards sustainable biorefinery. *Fuel, 255*, pp.115826.

[46] Kaniapan, S., Hassan, S., Ya, H., Patma Nesan, K. and Azeem, M., 2021. The utilisation of palm oil and oil palm residues and the related challenges as a sustainable alternative in biofuel, bioenergy, and transportation sector: A review. *Sustainability*, *13*(6), pp.3110.

[47] Moravvej, Z., Makarem, M.A. and Rahimpour, M.R., 2019. The fourth generation of biofuel. In *Second and Third Generation of Feedstocks.* Elsevier,pp.557-597

[48] Prasad, R.K., Chatterjee, S., Mazumder, P.B., Gupta, S.K., Sharma, S., Vairale, M.G., Datta, S., Dwivedi, S.K. and Gupta, D.K., 2019. Bioethanol production from waste lignocelluloses: A review on microbial degradation potential. *Chemosphere,* 231, pp.588-606.

[49] Halder, P., Azad, K., Shah, S. and Sarker, E., 2019. Prospects and technological advancement of cellulosic bioethanol ecofuel production. In *Advances in eco-fuels for a sustainable environment.* Woodhead Publishing, pp.211-236.

[50] Zhang, M., Wang, F., Su, R., Qi, W. and He, Z., 2010. Ethanol production from high dry matter corncob using fed-batch simultaneous saccharification and fermentation after combined pretreatment. *Bioresource Technology,* 101(13), pp.4959-4964.

[51] Zoghlami, A. and Paës, G., 2019. Lignocellulosic biomass: understanding recalcitrance and predicting hydrolysis. *Frontiers in chemistry,* 7, pp.874.

[52] de Vries, R.P. and Visser, J.A.A.P., 2001. Aspergillus enzymes involved in degradation of plant cell wall polysaccharides. *Microbiology and molecular biology reviews*, *65*(4), pp.497-522.

[53] Gray, K.A. and Zhao, L., 2006. Bioethanol. *Current opinion in Chemical Biology,* vol.10. no.2, pp.142-147.

[54] O'sullivan, A.C., 1997. Cellulose: the structure slowly unravels. *Cellulose*, vol.4. no.3, pp.173-207.

[55] Bajpai, P., 2016 (a). Structure of lignocellulosic biomass. In *Pretreatment of lignocellulosic biomass for biofuel production,* Springer, Singapore, pp.7-12.

[56] Conde-Mejia, C., Jimenez-Gutierrez, A. and El-Halwagi, M., 2012. A comparison of pretreatment methods for bioethanol production from lignocellulosic materials. *Process Safety and Environmental Protection,* 90(3), pp.189-202.

[57] Bajpai, P., 2016. Pretreatment of lignocellulosic biomass. In *Pretreatment of Lignocellulosic Biomass for Biofuel Production.* Springer, Singapore, pp.17-70.

[58] Cardona, C.A. and Sánchez, Ó.J., 2007. Fuel ethanol production: process design trends and integration opportunities. *Bioresource technology*, vol.98. no.12, pp.2415-2457.

[59] Peng, F., Peng, P., Xu, F. and Sun, R.C., 2012. Fractional purification and bioconversion of hemicelluloses. *Biotechnology advances*, vol.30. no.4, pp.879-903.

[60] Hendriks, A.T.W.M. and Zeeman, G., 2009. Pretreatments to enhance the digestibility of lignocellulosic biomass. *Bioresource technology*, vol.100. no.1, pp.10-18.

[61] Duval, A. and Lawoko, M., 2014. A review on lignin-based polymeric, micro-and nano-structured materials. *Reactive and Functional Polymers*, vol.85, pp.78-96.

[62] Binod, P., Gnansounou, E., Sindhu, R. and Pandey, A., 2019. Enzymes for second generation biofuels: recent developments and future perspectives. *Bioresource Technology Reports*, *5*, pp.317-325.

[63] Pérez, J., Munoz-Dorado, J., De la Rubia, T.D.L.R. and Martinez, J., 2002. Biodegradation and biological treatments of cellulose, hemicellulose and lignin: an overview. *International microbiology*, vol.5. no.2, pp.53-63.

[64] Ravindran, R. and Jaiswal, A.K., 2016. A comprehensive review on pre-treatment strategy for lignocellulosic food industry waste: challenges and opportunities. *Bioresource technology*, vol.199, pp.92-102.

[65] Dias, M.O., da Cunha, M.P., Maciel Filho, R., Bonomi, A., Jesus, C.D. and Rossell, C.E., 2011. Simulation of integrated first- and second-generation bioethanol production from sugarcane: comparison between different biomass pretreatment methods. *Journal of industrial microbiology & biotechnology,* 38(8), pp.955-966.

[66] Seidl, P.R. and Goulart, A.K., 2016. Pretreatment processes for lignocellulosic biomass conversion to biofuels and bioproducts. *Current Opinion in Green and Sustainable Chemistry,* 2, pp.48-53.

[67] Xu, Z. and Huang, F., 2014. Pretreatment methods for bioethanol production. *Applied biochemistry and biotechnology,* 174(1), pp.43-62.

[68] Sun, S., Sun, S., Cao, X. and Sun, R., 2016. The role of pretreatment in improving the enzymatic hydrolysis of lignocellulosic materials. *Bioresource technology,* 199, pp.49-58.

[69] Galbe, M. and Zacchi, G., 2012. Pretreatment: the key to efficient utilization of lignocellulosic materials. *Biomass and bioenergy,* 46, pp.70-78.

[70] Kucharska, K., Rybarczyk, P., Hołowacz, I., Łukajtis, R., Glinka, M. and Kamiński, M., 2018. Pretreatment of lignocellulosic materials as substrates for fermentation processes. *Molecules,* 23(11), pp.2937.

[71] Aditiya, H.B., Mahlia, T.M.I., Chong, W.T., Nur, H. and Sebayang, A.H., 2016. Second generation bioethanol production: A critical review. *Renewable and sustainable energy reviews,* 66, pp.631-653.

[72] Keskin, T., Abubackar, H.N., Arslan, K. and Azbar, N., 2019. Biohydrogen production from solid wastes. In *Biohydrogen.* Elsevier, pp.321-346.

[73] Bussemaker, M.J.; Xu, F.; Zhang, D., 2013. Manipulation of ultrasonic effects on lignocellulose by varying the frequency, particle size, loading and stirring. Bio resource. *Technology.* 148, pp.15-23.

[74] Galbe, M. and Zacchi, G., 2007. Pretreatment of lignocellulosic materials for efficient bioethanol production. In *Biofuels.* Springer, Berlin, Heidelberg, pp.41-65.

[75] Sindhu, R., Binod, P. and Pandey, A., 2016. Biological pretreatment of lignocellulosic biomass–An overview. *Bioresource technology,* 199, pp.76-82.

[76] Park, Y.C. and Kim, J.S., 2012. Comparison of various alkaline pretreatment methods of lignocellulosic biomass. *Energy,* 47(1), pp.31-35.

[77] Kumar, P., Barrett, D.M., Delwiche, M.J. and Stroeve, P., 2009. Methods for pretreatment of lignocellulosic biomass for efficient hydrolysis and biofuel production. *Industrial & engineering chemistry research,* 48(8), pp.3713-3729.

[78] Mohd, N., Draman, S.F.S., Salleh, M.S.N. and Yusof, N.B., 2017, February. Dissolution of cellulose in ionic liquid: A review. In *AIP conference proceedings.* AIP Publishing LLC. Vol. 1809, No. 1, pp.20-35.

[79] Blümmel, M., Teymouri, F., Moore, J., Nielson, C., Videto, J., Kodukula, P., Pothu, S., Devulapalli, R. and Varijakshapanicker, P., 2018. Ammonia Fiber Expansion (AFEX) as spin off technology from 2nd generation biofuel for upgrading cereal straws and stovers for livestock feed. *Animal Feed Science and Technology,* 236, pp.178-186.

[80] Chundawat, S.P., Pal, R.K., Zhao, C., Campbell, T., Teymouri, F., Videto, J., Nielson, C., Wieferich, B., Sousa, L., Dale, B.E. and Balan, V., 2020. Ammonia fiber expansion (AFEX) pretreatment of lignocellulosic biomass. JoVE (*Journal of Visualized Experiments*), (158), pp.57-88.

[81] Limayem, A. and Ricke, S.C., 2012. Lignocellulosic biomass for bioethanol production: current perspectives, potential issues and future prospects. *Progress in energy and combustion science,* 38(4), pp.449-467.

[82] Yang, B. and Wyman, C.E., 2008. Pretreatment: the key to unlocking low-cost cellulosic ethanol. *Biofuels, Bioproducts and Biorefining: Innovation for a sustainable economy,* vol.2. no.1, pp.26-40.

[83] Binod, P., Janu, K.U., Sindhu, R. and Pandey, A., 2011. Hydrolysis of lignocellulosic biomass for bioethanol production. In *Biofuels.* Academic press, pp. 229-250.

[84] Verardi, A., De Bari, I., Ricca, E. and Calabrò, V., 2012. Hydrolysis of lignocellulosic biomass: Current status of processes and technologies and future perspectives. *Bioethanol,* pp.95-122.

[85] Graf, A. and Koehler, T., 2000. Oregon cellulose-ethanol study. *Oregon Office of Energy*. Salem OR USA, pp.1-10.

[86] Badger, P.C., 2002. Ethanol from cellulose: a general review. *Trends in new crops and new uses,* vol.1, pp.17-21.

[87] Haldar, D., Sen, D. and Gayen, K., 2016. A review on the production of fermentable sugars from lignocellulosic biomass through conventional and enzymatic route—a comparison. *International journal of green energy,* vol.13. no.12, pp.1232-1253.

[88] Demirbas, A., 2007. Producing and using bioethanol as an automotive fuel. *Energy Sources,* Part B, vol.2. no.4, pp.391-401.

[89] Kennes, D.; Abubackar, H.N.; Diaz, M.; Veiga, M.C.; Kennes, C. 2016. Bioethanol production from biomass: Carbohydrate vs. syngas fermentation. *J. Chem. Technol. Biotechnol.,* vol.91. pp.304-317.

[90] Kamzon, M.A., Abderafi, S. and Bounahmidi, T., 2016. Promising bioethanol processes for developing a biorefinery in the Moroccan sugar industry. *International Journal of Hydrogen Energy,* vol.41. no.45, pp.20880-20896.

[91] Harris, P.V., Xu, F., Kreel, N.E., Kang, C. and Fukuyama, S., 2014. New enzyme insights drive advances in commercial ethanol production. *Current opinion in chemical biology,* vol.19, pp.162-170.

[92] Wang, M., Li, Z., Fang, X., Wang, L. and Qu, Y., 2012. Cellulolytic enzyme production and enzymatic hydrolysis for second-generation bioethanol production. *Biotechnology in China III: Biofuels and Bioenergy,* pp.1-24.

[93] Karmakar, M. and Ray, R.R., 2011. Current trends in research and application of microbial cellulases. *Res J Microbiol,* vol.6. no.1, pp.41-53.

[94] Gupta, P., Samant, K. and Sahu, A., 2012. Isolation of cellulose-degrading bacteria and determination of their cellulolytic potential. *International journal of microbiology,* 2012, pp.1-6.

[95] Nair, R.B., Lennartsson, P.R. and Taherzadeh, M.J., 2017. Bioethanol production from agricultural and municipal wastes. In *Current developments in biotechnology and bioengineering*. Elsevier, pp.157-190.

[96] Maitan-Alfenas, G.P., Visser, E.M. and Guimarães, V.M., 2015. Enzymatic hydrolysis of lignocellulosic biomass: converting food waste in valuable products. *Current Opinion in Food Science,* 1, pp.44-49.

[97] Gupta, R., Yadav, G., Kumar, G., Yadav, A., Saini, J.K. and Kuhad, R.C., 2019. Second generation bioethanol production: the state of art. *Sustainable approaches for biofuels production technologies, pp*.121-147.

[98] Schwarz, W., 2001. The cellulosome and cellulose degradation by anaerobic bacteria. *Applied microbiology and biotechnology,* vol.56. no.5-6, pp. 634-649.

[99] Rajwade, J.M., Paknikar, K.M. and Kumbhar, J.V., 2015. Applications of bacterial cellulose and its composites in biomedicine. *Applied microbiology and biotechnology,* vol.99. no.6, pp.2491-2511.

[100] Weerasinghe, W. M. L. I., Madusanka, D. A. T., and Manage, P. M., 2021. Isolation and Identification of Cellulase Producing and Sugar Fermenting Bacteria for Second-Generation Bioethanol Production. *International Journal of Renewable Energy Development, Volume* 10(4), pp.699-711.

[101] Chandrasiri, Y.S., Weerasinghe, W.M., Madusanka, D.A. and Manage, P.M., 2022. Waste-Based Second-Generation Bioethanol: A Solution for Future Energy Crisis. *International Journal of Renewable Energy Development,* 11(1), pp.275-285.

[102] Patagundi, B.I., Shivasharan, C.T. and Kaliwal, B.B., 2014. Isolation and characterization of cellulase producing bacteria from soil. *International Journal of Current Microbiology and Applied Sciences,* vol.3. no.5, pp.59-69.

[103] Gomes, R.J., de Fatima Borges, M., de Freitas Rosa, M., Castro-Gómez, R.J.H. and Spinosa, W.A., 2018. Acetic acid bacteria in the food industry: systematics, characteristics and applications. *Food technology and biotechnology,* 56(2), pp.140.

[104] Apriyanto, M. and Umanailo, M., 2019. Decrease polyphenols, ethanol, lactic acid, and acetic acid during fermentation with addition of cocoa beans innoculum. *International Journal of Scientific & Technology Research,* 8(10), pp.461-465.

[105] Van Zyl, W.H., Lynd, L.R., den Haan, R. and McBride, J.E., 2007. Consolidated bioprocessing for bioethanol production using Saccharomyces cerevisiae. In *Biofuels.* Springer, Berlin, Heidelberg, pp. 205-235.

[106] Liu, H., Hu, B. and Jin, C., 2016. Effects of different alcohols additives on solubility of hydrous ethanol/diesel fuel blends. *Fuel,* 184, pp.440-448.

[107] Elshaghabee, F.M., Bockelmann, W., Meske, D., de Vrese, M., Walte, H.G., Schrezenmeir, J. and Heller, K.J., 2016. Ethanol production by selected intestinal microorganisms and lactic acid bacteria growing under different nutritional conditions. *Frontiers in microbiology,* 7, pp.47.

[108] Serge Kubanza, N. and Simatele, M.D., 2020. Sustainable solid waste management in developing countries: a study of institutional strengthening for solid waste management in Johannesburg, South Africa. *Journal of Environmental Planning and Management,* 63(2), pp.175-188.

[109] Ghosh, P., Sengupta, S., Singh, L. and Sahay, A., 2020. Life cycle assessment of waste-to-bioenergy processes: a review. *Bioreactors,* pp.105-122.

[110] Vignesh, K.S., Rajadesingu, S. and Arunachalam, K.D., 2021. Challenges, issues, and problems with zero-waste tools. In *Concepts of Advanced Zero Waste Tools.* Elsevier., pp.69-90.

[111] Byadgi, S.A. and Kalburgi, P.B., 2016. Production of bioethanol from waste newspaper. *Procedia Environmental Sciences,* 35, pp.555-562.

[112] Lee, J.T., Ee, A.W.L. and Tong, Y.W., 2018. Environmental impact comparison of four options to treat the cellulosic fraction of municipal solid waste (CF-MSW) in green megacities. *Waste Management,* 78, pp.677-685.

[113] Zhang, Z., Gonzalez, A.M., Davies, E.G.R., Liu, Y., 2012. Agricultural wastes. *Water Environ. Res.,* 84, pp.1396-1416.

[114] Cho, E.J., Trinh, L.T.P., Song, Y., Lee, Y.G. and Bae, H.J., 2020. Bioconversion of biomass waste into high value chemicals. *Bioresource Technology,* 298, pp.122-386.

[115] Nanda, S. and Berruti, F., 2021. Municipal solid waste management and landfilling technologies: a review. *Environmental Chemistry Letters,* 19(2), pp.1443-1466.

[116] Abo, B.O., Gao, M., Wang, Y., Wu, C., Ma, H. and Wang, Q., 2019. Lignocellulosic biomass for bioethanol: an overview on pretreatment, hydrolysis and fermentation processes. *Reviews on environmental health,* 34(1), pp.57-68.

[117] Abdallah, M., Talib, M.A., Feroz, S., Nasir, Q., Abdalla, H. and Mahfood, B., 2020. Artificial intelligence applications in solid waste management: A systematic research review. *Waste Management,* 109, pp.231-246.

[118] Vyas, S., Prajapati, P., Shah, A.V. and Varjani, S., 2022. Municipal solid waste management: Dynamics, risk assessment, ecological influence, advancements, constraints and perspectives. *Science of The Total Environment,* pp.152802.

[119] Braide, W., Kanu, I.A., Oranusi, U.S. and Adeleye, S.A., 2016. Production of bioethanol from agricultural waste. *Journal of fundamental and Applied Sciences,* 8(2), pp.372-386.

[120] Yang, L., Ru, Y., Xu, S., Liu, T. and Tan, L., 2021. Features correlated to improved enzymatic digestibility of corn stover subjected to alkaline hydrogen peroxide pretreatment. *Bioresource Technology,* 325, pp.124688.

[121] Chen, M., Xia, L. and Xue, P., 2007. Enzymatic hydrolysis of corncob and ethanol production from cellulosic hydrolysate. *International Biodeterioration & Biodegradation,* 59(2), pp.85-89.

[122] Wipulasena, A. (14 October, 2018). Country generates 7,500MT waste daily: Strategic solutions key to fixing garbage crisis. *Sunday Observer.* Retrieved from https://www.sundayobserver.lk/2018/10/14/news-features/country-generates-7500mt-waste-daily-strategic-solutions-key-fixing-garbage.

[123] Ul-Islam, M., Ullah, M.W., Khan, S. and Park, J.K., 2020. Production of bacterial cellulose from alternative cheap and waste resources: a step for cost reduction with positive environmental aspects. *Korean Journal of Chemical Engineering,* 37(6), pp.925-937.

[124] Iag, L.K., Tjokro, A.J. and Pahari, D., 2021. The Efficiency of Using Biowaste as a Greener Alternative to Conventional Fuel. In *IRC-SET 2020.* Springer, Singapore, pp. 247-255.

[125] Bajpai, P. 2020. Developments in bioethanol. Springer *Nature,* pp.1-215.

[126] Manieniyan, V., Thambidurai, M. and Selvakumar, R., 2009. Study on energy crisis and the future of fossil fuels. *Proc. SHEE,* 10, pp.2234-3689.

[127] Waterland, L.R., Venkatesh, S. and Unnasch, S., 2003. *Safety and performance assessment of ethanol/diesel blends* (e-diesel) (No. NREL/SR-540-34817). National Renewable Energy Lab., Golden, CO. (US), pp.1-48.

[128] Rakopoulos, D.C., Rakopoulos, C.D., Kakaras, E.C. and Giakoumis, E.G., 2008. Effects of ethanol–diesel fuel blends on the performance and exhaust emissions of heavy duty DI diesel engine. *Energy Conversion and Management,* 49(11), pp.3155-3162.

[129] Hansen, A.C., Zhang, Q. and Lyne, P.W., 2005. Ethanol–diesel fuel blends—a review. *Bioresource technology,* 96(3), pp.277-285.

[130] de Menezes, E.W., da Silva, R., Cataluna, R. and Ortega, R.J., 2006. Effect of ethers and ether/ethanol additives on the physicochemical properties of diesel fuel and on engine tests. *Fuel,* 85(5-6), pp.815-822.

[131] Parthasarathi, R., Gowri, S. and Saravanan, C.G., 2014. Effects of ethanol-diesel emulsions on the performance, combustion and emission characteristics of DI diesel engine. *American Journal of Applied Sciences,* 11(4), pp.592.

[132] Gerdes, R., and Suppes, G. J., 2001. Miscibility of Ethanol in Diesel Fuels, *Ind. Eng. Chem. Res. 40*, pp.949–956.

[133] Hajba, L., Eller, Z., Nagy, E. and Hancsok, J., 2011. Properties of diesel-alcohol blends. *Hungarian Journal of Industry and Chemistry,* 39(3), pp.349-352.

[134] Lapuerta, M., Armas, O. and Garcia-Contreras, R., 2007. Stability of diesel–bioethanol blends for use in diesel engines. *Fuel,* 86(10-11), pp.1361-1367.

[135] Ribeiro, N.M., Pinto, A.C., Quintella, C.M., da Rocha, G.O., Teixeira, L.S., Guarieiro, L.L., do Carmo Rangel, M., Veloso, M.C., Rezende, M.J., Serpa da Cruz, R. and de Oliveira, A.M., 2007. The role of additives for diesel and diesel blended (ethanol or biodiesel) fuels: a review. *Energy & fuels,* 21(4), pp.2433-2445.

[136] Lapuerta, M., Hernández, J.J., Fernández-Rodríguez, D. and Cova-Bonillo, A., 2017. Autoignition of blends of n-butanol and ethanol with diesel or biodiesel fuels in a constant-volume combustion chamber. *Energy,* 118, pp.613-621.

[137] Debnath, B.K., Saha, U.K. and Sahoo, N., 2015. A comprehensive review on the application of emulsions as an alternative fuel for diesel engines. *Renewable and Sustainable Energy Reviews,* 42, pp.196-211.

[138] Reyes, Y., Aranda, D.A.G., Santander, L.A.M., Cavado, A. and Belchior, C.R.P., 2009. Action principles of cosolvent additives in ethanol− diesel blends: stability studies. *Energy & Fuels,* 23(5), pp.2731-2735.

[139] McCormick, R.L. and Parish, R., 2001. *Milestone report: technical barriers to the use of ethanol in diesel fuel* (No. NREL/MP-540-32674), pp.1-22.

[140] Nandakumar, S. and Malayalamurthi, R., 2015. Properties of Diesel− Ethanol Blended Fuels with Surfactant. *Int. J. Eng. Res. Technol,* 4(11), pp.72-75.

[141] Yahuza, I and Dandakouta, H., 2015. A performance review of ethanol-diesel blended fuel samples in compression-ignition engine. *Journal of Chemical Engineering & Process Technology,* 6(5), pp.1-6.

[142] Kumar, N. and Chauhan, S.R., 2013. Performance and emission characteristics of biodiesel from different origins: A review. *Renewable and Sustainable Energy Reviews,* 21, pp.633-658.

[143] Torres-Jimenez, E., Jerman, M.S., Gregorc, A., Lisec, I., Dorado, M.P. and Kegl, B., 2011. Physical and chemical properties of ethanol–diesel fuel blends. *Fuel,* 90(2), pp.795-802.

[144] Romano, S.D., Sorichetti, P.A. and Buesa Pueyo, I., 2009. Methanol content in biodiesel estimated by flash point and electrical properties. In *Bioethanol: production, benefits and economics.* NOVA Science Publishers Inc New York, pp.136-147.

[145] Niculescu, R., Iosub, I., Clenci, A., Zaharia, C. and Iorga-Simăn, V., 2017, October. Development of a test method for distillation of diesel-biodiesel-alcohols mixtures at reduced pressure. In *IOP Conference Series: Materials Science and Engineering.* IOP Publishing. Vol. 252, No. 1, pp.1-10.

[146] Singh, G.N. and Bharj, R.S., 2019. Study of physical-chemical properties for 2nd generation ethanol-blended diesel fuel in India. *Sustainable Chemistry and Pharmacy,* 12, pp.100-131.

[147] Badmaev, S.D. and Snytnikov, P.V., 2008. Hydrogen production from dimethyl ether and bioethanol for fuel cell applications. *International journal of hydrogen energy,* 33(12), pp.3026-3030.

[148] Snytnikov, P.V., Badmaev, S.D., Volkova, G.G., Potemkin, D.I., Zyryanova, M.M., Belyaev, V.D. and Sobyanin, V.A., 2012. Catalysts for hydrogen production in a multifuel processor by methanol, dimethyl ether and bioethanol steam reforming for fuel cell applications. *International journal of hydrogen energy,* 37(21), pp.16388-16396.

Chapter 2

Extremophiles and Enzymes

Kosala Sirisena[1,2,*]
Gayani Yasodara Liyanage[2,4]
Dilini Sadeepa[2,3]
and Pathmalal Manage[2,3]
[1]Department of Environmental Technology, Faculty of Technology, University of Colombo, Sri Lanka
[2]Centre for Water Quality and Algae Research, Department of Zoology, University of Sri Jayewardenepura, Gangodawila, Nugegoda, Sri Lanka
[3]Faculty of Graduate Studies, University of Sri Jayewardenepura, Gangodawila, Nugegoda, Sri Lanka
[4]Department of Aquatic Bioresources, Faculty of Urban and Aquatic Bioresources, University of Sri Jayewardenepura, Gangodawila, Nugegoda, Sri Lanka

Abstract

Extremophiles are organisms that inhabit and thrive in environments that show permanent values of environmental parameters close to the lower and upper limits for life. Although the general perception is that extreme environments are rare, a more significant fraction of the earth is covered with extreme environments. Therefore, extreme environments provide habitats for a considerable fraction of the biomass and biodiversity on the earth. Extremophiles are very useful in various applied aspects such as biotechnology, industrial application, bioremediation, and fundamental research areas such as astrobiological studies, origin, evolution, and future of life on earth. In this chapter the biodiversity of extremophiles, adaptations of extremophiles, applied aspects of extremophiles, extremozymes and the future with extremophiles are discussed in detail.

* Corresponding Author's Email: kosala@et.cmb.ac.lk.

In: Microbial Technologies and Their Applications
Editors: Santosh Kumar Karn and Anne Bhambri
ISBN: 979-8-88697-913-8

Keywords: extremophiles, biotechnology, extremozymes, astrobiology, bioremediation

Introduction to Extremophiles: History and Discovery

Extremophiles can be simply described as organisms that can survive and thrive in extreme environments. This broad definition as the term "extreme environment" can be very subjective. According to our current understanding, the earth is the only planet that has life and that possesses conditions favorable for the survival. All the organisms that inhabit earth are hugely influenced by various environmental factors around them. These key environmental factors are: temperature, salinity, pH, atmospheric pressure, oxygen stress, humidity and radiation. Different organisms have different tolerance limits for each parameter. For example, the optimum temperature at which the desert animals thrive could be intolerable for human. In contrast, the optimum temperature for human metabolic activities could be inhospitable for some microorganisms that inhabit Antarctic Sea ice. For all these environmental parameters, there are upper and lower limits beyond which no organisms could survive. Environments in which physicochemical parameters possess permanent values close to those lower and upper limits for life can be defined as extreme environments [1]. Therefore, extremophiles can be comprehensively defined as organisms that inhabit and thrive in environments that show permanent values of environmental parameters close to the lower and upper limits for life.

Extremophiles can be broadly categorized into three groups based on their responses to harsh environmental conditions: (a) certain organisms show hostile behaviors when they expose to extreme environmental conditions for example, some bacteria form endospores when exposed to conditions that are not favorable for their growth and survival. These endospores show resistance to various harsh conditions such as UV radiation, desiccation, extreme ends of temperature and pH and chemical disinfectants. These endospores can remain dormant for extended periods of time and will be reactivated to vegetative stage when the environmental parameters become favorable for their growth; (b) certain organisms possess the ability to develop complex molecular and cellular mechanisms to combat harsh environmental conditions if they have to face such conditions; (c) certain organisms desperately require having extreme environmental conditions for their survival and growth. Such organisms cannot complete their life cycle without having harsh conditions. For example, hyperthermophilic microorganisms couldn't grow below 65°C. Among all three groups, organisms in group (a) and group (b): organisms that show hostile behavior when facing extreme conditions and organisms that possess specialized mechanisms to tolerate extreme conditions do not necessarily require extreme environmental conditions to complete their life cycles. These organisms are evolved in such a way that they could protect themselves from extreme conditions or repair the damages caused by extreme environmental conditions. Therefore, these organisms can be truly described as "extremotolerant" organisms. However, extremotolerant organisms and true extremophiles organisms that desperately need extreme conditions for their growth are generally described as extremophiles.

Although the general perception is that extreme environments are rare, a more significant fraction of the earth is covered with extreme environments. It is reported that nearly 80% of the biosphere is permanently under low temperature, mainly represented by glaciers and deep-sea environments [2]. Furthermore, nearly 70% of the earth surface is under high pressure mainly represented by deep seas [3]. Therefore, extreme environments provide habitats for a considerable fraction of the biomass and biodiversity on the earth.

The history of extremophile studies goes back to the discovery of novel bacterial strains from thermal vents at Yellowstone National Park in the United States in1965 by Thomas Brock. He discovered some bacteria that could withstand temperatures up to 88°C, which was way above the temperature that was thought to be the upper limit for life at that time. This discovery laid the foundation for the revolutionary invention of the *Taq* DNA polymerase from *Thermus aquaticus YT-1* strain isolated from the exact location in 1967 [4]. Until that period, biologists thought that life could exist only within a narrow range of environmental conditions. However, these discoveries revolutionized that idea and laid the foundation stone for a novel branch of microbiology: extremophile studies. The term "extremophiles" was first introduced by R. D. Mac Elroy in 1974 [5]. The initial focus of extremophile studies was on thermophiles. However, scientists gradually began to explore organisms that inhabit various other extreme environments. These studies focused on the diversity of extremophiles in different environments and the taxonomy, genetics, physiology, biochemistry, and evolution of those organisms. The scientists soon realized that these extremophiles are very useful in various applied aspects such as biotechnology, industrial application, bioremediation, and fundamental research areas such as astrobiological studies, origin, evolution, and future of life on earth. However, extremophile studies have some challenges: it is sometimes difficult to reach those habitats for sample collection compared to other environments, such as deep-sea hydrothermal vents, and it is challenging to provide optimum growth for some extremophiles under laboratory conditions. With the recent advances in molecular techniques, particularly 2nd and 3rd generation DNA sequencing technologies and computational biology, extremophiles research has expanded rapidly and significantly. In 2002 the scientific community working on extremophiles has also established a professional society - the International Society for Extremophiles (https://extremophiles.org/index.html) as a common platform for the advancement of extremophile research. However, despite these advances, a more significant fraction of extremophiles in diverse extreme environments is not yet explored. These new extremophiles may have a considerable potential to contribute to the industrial and biotechnological applications to achieve sustainable development and provide directions to find answers to queries related to the origin and evolution of life on earth and potential life in extra-terrestrial environments. Extremophiles belong to both microorganisms and some macroscopic plants and animals. However, the majority of extremophiles are microorganisms. Therefore, the main focus of this chapter is given to extremophilic microbial groups: bacteria, archaea and eukaryotic microorganisms. The remainder of this chapter discusses the biodiversity of extremophiles, adaptations of extremophiles, applied aspects of extremophiles, extremozymes and the future with extremophiles.

Biodiversity of Extremophiles: Distribution and Ecological Niches

Extremophiles are distributed across various extreme environments showing adaptation mechanisms to their environment (Figure 1). Some of the critical environmental parameters of which extreme ends have extensively been studied for extremophiles are temperature, pH, salinity, pressure, radiation, desiccation and oxygen stress. It is important to note that although these individual extreme environmental parameters are individually studied, many microorganisms inhabit environments characterized by more than one extreme condition. For example, most deep-sea environments experience high hydrostatic pressure and low temperatures. Therefore, microorganisms in such environments should show adaptations for high hydrostatic pressure and low temperature. Interestingly, deep sea hydrothermal vent ecosystems show high temperature and high hydrostatic pressure demanding the microorganisms inhabit such environments to develop adaptations to both extreme conditions. Microorganisms that can withstand more than one extreme environmental parameter are known as "polyextremophiles" [6]. However, this section mainly discusses microorganisms that inhabit individual extreme conditions.

Figure 1. Different habitats of extremophiles: (a) Carbonate deposits in Svalbard Islands; (b) Gypsum dunes in White Sands National Monument, New Mexico, USA; (c) Volcanic deposits in Italy; and (d) Antarctic Sea Ice.

Extremophiles in Temperature Extremes

It is estimated that nearly 75% of the earth's biosphere is covered with permanently cold environments [7]. These environments are mainly represented by Polar Regions, glaciers, deep-ocean ecosystems, and permafrost. Furthermore, nearly 70% of the earth's surface has a temperature between 1 to 5°C. There are also some high-temperature ecosystems across the earth's biosphere. Notably, the temperature is one of the most crucial factors for the survival of life. In general, when the temperature increases, the biochemical reactions in cells will speed up to a specific temperature and rapidly decline above that temperature. At low temperatures, cellular, biochemical and enzymatic activities continue more slowly. Most microorganisms studied live at an average temperature between 15 to 45°C, whereas they show the optimum growth between 25 to 37°C. These microorganisms are known as "mesophiles." However, microorganisms with extreme ends of temperatures show different temperature ranges for survival and optimum growth.

High Temperatures: Thermophiles

Any organism that could survive above 45 to 50°C is considered a thermophile. Generally, archaea can withstand higher temperatures than bacteria. There are three broad categories of thermophiles: (a) facultative thermophiles - these microorganisms are mesophiles who show the optimum growth at 25 to 37°C, however, they could tolerate temperatures nearly up to 60°C; (b) strict thermophiles - these microorganisms could not grow below 40°C, while showing the optimum growth around 60 to 65°C; and (c) extreme thermophiles or hyperthemophiles - these microorganisms prefer extremely high temperatures, hey usually cannot grow below 55°C and show the optimum growth between 80 to 115°C.

Thermophiles inhabit various high temperature environments, including terrestrial and marine geothermal ecosystems, and certain habitats associated with anthropogenic activities. Geothermal ecosystems are formed when the groundwater is heated up subsurface magma, resulting in hot water and steam that will come out from the earth surface through different types of openings. These openings include hot springs, mudpots, geysers and fumaroles [8].

Low Temperatures: Psychrophiles

Some microorganisms show the capability of surviving under low temperatures. These microorganisms are generally known as Psychrophiles. There are two main groups of Psychrophiles: (a) Strict Psychrophilic microorganisms - microorganisms permanently live in cold environments; and (b) Psychrotrophic microorganisms or facultative psychrophiles - mesophiles in general that show the optimum growth between 25 to 37°C [9]. However, they can tolerate low temperatures up to 5 to 7°C with a slower growth rate.

Psychrophiles can be found in various cold environments on earth: oceanic environments of polar regions where the temperature ranged from – 1.9 to 5°C; deep oceans where the temperature is less than 4°C; semisolid sea ice habitats where the temperature can go below – 35°C; glaciers that are formed by the compression of snow; permafrost which occupy more than 20%

of the earth's land surface; and Antarctic sub glacial lakes where the lakes are formed within glaciers.

Extremophiles in Extreme Ends Of pH

Environmental pH level is also one of the crucial factors that influence the occurrence and distribution of microorganisms. The microorganisms that show the optimum growth around pH 7 are known as Neutrophilic organisms. However, there are a large number of natural habitats that have extreme pH conditions that result in a large number of microorganisms that can withstand extreme pH conditions.

Extremophiles in Low pH: Acidophiles

Microorganisms that thrive in acidic environments are known as acidophiles. These organisms usually show the optimal growth rate at pH below 3. There are some other microorganisms that show the optimal growth around neutral pH, but they have the ability to tolerate acidic conditions to a certain extent. These organisms are known as acid tolerant organisms who can develop similar adaptation mechanisms to acidophiles to maintain suboptimal growth under slight acidic conditions. Acidophiles can be seen in diverse environments such as acid mining drainages, solfataric fields that emit sulfurous gases and acidothermal hot springs [10].

Extremophiles in High pH: Alkaliphiles

Microorganisms that thrive in alkaline environments are known as alkaliphiles. These organisms show the optimal growth around pH above 9. Furthermore, there are some organisms that show the optimal growth around neutral pH but possess the ability to tolerate alkaline conditions to a certain level. These organisms are known as alkali tolerant organisms. Alkaliphiles can be seen in environments such as hypersaline soda lakes, alkali thermal hot springs and shallow hydrothermal systems [11].

Extremophiles in High Salinity: Halophiles

Microorganisms that can tolerate and thrive in saline and hypersaline environments are generally known as halophiles. There are several categories of halophiles: (a) halotolerant organisms that shows the optimal growth under trace amounts of salts (NaCl), but have the ability to tolerate saline conditions up to 5% (w/v) NaCl. The average ocean salinity is around 3.5% NaCl; (b) slight halophiles that show the optimal growth at higher salinity levels that halo tolerant organisms, but still they are not able to survive above 5% (w/v) salinity; (c) moderate halophiles that have the capability of surviving up to 20% salinity levels; and (d) extreme halophiles that require extremely high saline conditions even up to 20-30% near saturated salinity for their growth. Halophilic microorganisms inhabit various hypersaline conditions such as the Dead sea, ancient rock salts and saline marshes. However, it is important to know that saline and hypersaline environments are not always of marine origin. Hypersaline environments that have a marine origin are known as thalassohaline environments of which NaCl is the predominant salt and locate closer to the seas and oceans. There are some other hypersaline environments that are not having a marine origin. These environments are known as a thalassaohaline environments where K, Mg, Na and CO_3^- ions mainly contribute to the hypersaline conditions of these environments [12].

Extremophiles in High and Low Pressure: Piezophiles and Hypopiezophiles

A larger fraction of the biosphere is at pressures greater than 1 atm. Therefore, such environments provide habitats for a huge biodiversity. Microorganisms that thrive at high pressures are known as piezophiles. Most of the piezophiles inhabit deep oceanic environments where the average hydrostatic pressure is around 38 MPa and sometimes that can reach up to 110 MPa. There are some microorganisms that show the optimal growth under atmospheric pressure (1 atm = 0.1 MPa), but can tolerate pressures greater than, allowing suboptimal growth. These organisms are known as piezotolerant organisms. Although high pressure environments are prevalent in the biosphere, there are also some low-pressure environments from where microorganisms have been reported. For example, microorganisms have been reported from the top of the Mount Everest where the pressure is around 0.03 MPa. Microorganisms that show the optimal growth under low pressure are known as hypopiezophiles [13].

Extremophiles in High Radiation: Radiophiles

Ionizing radiation could cause serious damages to organisms. However, the terrestrial organisms are protected from these radiations by the Ozone layer and the planetary magnetic field. Due to this protection, terrestrial organisms usually expose to 2 mGy of ionizing radiation on average per year. However, some microorganisms that inhabit environments such as deep oceanic hydrothermal vents are exposed to more than 200 mGy radiation per year. These organisms have developed mechanisms to withstand these high radiations. Microorganisms that thrive under high doses of ionizing radiations are known as radiophiles [14]. It is reported that the bacterium *Deinococcus radiodurans* can withstand an acute dose of 5000 Gy of ionizing radiation, where 5 Gy would be lethal for human.

Extremophiles in High Desiccation: Xerophiles

Desiccation could also be a serious stress for all organisms as water plays a crucial role in all cellular activities and ultimately to protect the life. However, there are some microorganisms that are adapted to thrive near complete dehydrating environments. These organisms enter into a state of metabolism which is known as an hydrobiosis. The organisms will reactivate when the water is available. These organisms are known as xerophiles [14]. There are some other organisms that show the optimal growth when the dehydration is minimal. However, they have mechanisms to tolerate dehydration conditions up to a certain extent and will die if the tolerance level is exceeded. These organisms are known as xerotolerant organisms.

Adaptations of Extremophiles

Extremophiles thrive in extreme environments and retain different structural and metabolic characteristics, which allow them to thrive and survive in these environmental conditions [15, 16]. Modifications of the cell membrane, cellular organelles and genomic composition are considered major structural modifications shown by extremophiles. Regulation of osmotic balance is considered a main physiological modification, whereas protein modifications are considered a major metabolic adaptation [17]. The extremophiles inhabit different habitats and show various unique adaptations within these broad modification types to thrive and survive in their environments.

Thermophiles

Increasing membrane fluidity, protein misfolding and nucleic acid degradation are the common challenges encountered by the microorganisms living in high-temperature regions [17, 18]. Thermophiles that thrive in such environments have developed mechanisms to support their survival [19-21].

Saturated fatty acid and alicyclic fatty acids content of the cell membrane of thermophile provides more crystalline and ordered structure to the membranes, which supports maintaining the membrane fluidity. Ether lipids or glycerol ether can maintain a liquid crystalline state even above 100°C; presence in members of domain Archaea conquers the control of membrane fluidity [18]. Besides that the isoprenoid chain's presence in the cell membrane of archaea maintains the crystalline state and acts as a high permeability barrier facilitating maintaining the membrane fluidity [18, 22].

The stability of the three-dimensional structures of the proteins should be maintained to overcome the protein misfolding in high temperatures. Since proteins from hyper/ thermophiles require sufficient structural rigidity to resist unfolding, amino acid content and structure of the protein chains in thermophiles are amended to aid the adaptations [19, 23]. Though the stable thermal proteins have similar three-dimensional structures as mesophiles, the amino acid content and the number of charged residues on the surface are much more significant than thermo liable organisms [17]. Generally, thermophiles express heat-shock chaperonins to aid protein folding, which eliminates protein denaturation. Consequently, thermophiles produce promptly crystalline thermostable proteins, which act as enzymes/ catalysts. Furthermore, the shorter loops, interior salts, disulfide bridges and densely packed nature of the thermostable proteins habitually improve the thermostability [24, 25].

Further, high temperatures encourage the degradation of nucleic acids. Guanine (G) and Cytosine (C) content of the DNA determines the temperature stability of the genome. So that the G.:.C content of the thermophile genome is comparatively higher and strongly correlates with the optimal temperature [25]. It has been recorded as 52.3% for *Geobacillus thermoleovorans CCB US3UF5*; thermophilic bacteria isolated from Malaysian hot springs [23].

Moreover, an abundance of thermophiles up regulates DNA repair pathways and preserves genetic stability [17].

Psychrophiles

Psychrophiles: microorganisms that thrive in temperatures lower than 15°C have several adaptations and mechanisms to overcome the challenges in those relaxed environments [17]. Unsaturated fatty acids, Glycerol ether lipids and exopolysaccharides support the maintenance of the semi-fluid nature of the cytoplasmic membrane. The unsaturated fatty acid content of psychrophilic cell membrane has been recorded as 52% for mesophiles, and thermophiles were 37% and 10%, respectively [23].

Further, the cold-adapted microorganisms produce secondary cold-active metabolites, antifreeze proteins, and pigments in psychrophiles as adaptations to cold environments [17, 23]. Higher amount of α-helix compared to the β-sheets in protein structure in cold adapted microorganisms is noted as a crucial determinant of protein flexibility at lower temperatures [19, 20].

The psychrophiles' genome structure also has some alterations to survive in the low temperature. Cold shock RNA chaperones: cold shock proteins bind to the RNA, which conserves the conformation of RNA, and cold shock domains in proteins that can attach to the genome are some of those adaptations in the cold-adapted microbial genomes [17].

Acidophiles

The major challenge for the acidophiles who live in high H^+ concentrations is cytoplasmic acidification. That can be maintained by moderating the H^+ influx to drive ATP (Adenosine Tri Phosphate) synthesis [17]. Limitation of the H^+ influx via nonspecific avenues by reducing the cell membrane permeability is a common adaptation of acidophilic microbes.

Microbes belong to domain Archaea, which have glycerol ethers in the cell membrane and are more resistant to acidic environments than mesophilic bacteria with glycerol esters in the cell membrane since glycerol ethers are less susceptible to acid hydrolysis [22, 23].

Alicyclic fatty acids' presence in acidophiles cell membrane also provides additional resistance against the acidic conditions. However, acid-resistant lipids may be degraded at neutral pH levels [17]. Further, the acidophilic microorganisms have diverse significant cell membrane synthesis enzymes, which are considered supportive of altering the membrane structure for fluctuating acidic pH conditions. Ferrous co-factors are used steady the enzymes in acidophiles at low pH levels.

Transmembrane channels with smaller pore sizes and positively charged channels support maintaining H+ influx in cytoplasm. In addition, acidophiles create Donnan potential in the cytoplasm via aggregation of monovalent cations [22, 23]. The high concentration of positively charged cations and ATP dependent H^+ pumps in cytoplasm prevents H^+ influx while promoting efflux of H^+, which endure acidification of cytoplasm [17].

Further, decarboxylated glutamate and arginine, phosphoric acid, and high pI (isoelectric point) proteome in the cytoplasm of acidophiles provide a buffering nature to the cytoplasm, which aid to withstand minor acidic levels. Acid-shock DNA repair mechanisms and molecular chaperones reconstruct the minor damages caused in the genome or proteome by acidic conditions [17].

Alkaliphiles

Alkaliphiles thrive in high pH conditions and have to survive in low concentrations of H^+ ions. The unique challenge of surviving in such conditions is continuous neutralization of the microbial cytoplasm and uplift of H^+influx to drive ATP synthesis [17].

The foremost adaptation of alkaliphiles to survive in low H^+ concentration is the usage of Na^+ and K^+ antiporters to the movement of H^+, as well as alkaliphiles have a diverse set of transport proteins which have a different range of optimum pH levels for the activity [23]. When considering the structure of the cell wall and cytoplasm of alkaliphiles, the acidic nature of the cell wall and low pI proteins in cytoplasm encourage the preferable H^+ influx, which allows them to maintain lower intracellular pH than the environment [17].

Halophiles

Microbes live in the high salt concentration required to tolerate the osmotic pressure influenced by high NaCl concentrations in the ambient environment [22]. Halophiles have several adaptations in cell membrane structure, cytoplasm, and proteome to overcome the osmotic pressure in the surrounding environment [17]. The surface layer surrounding the cells of halophile contains sulfated glucuronic acid residues and a higher degree of glycosylation which can increase the charge of the surface. Some of the glycoproteins in the S-layer are enriched in acidic residues. As cytoplasm alterations in halophiles, accumulation of inorganic ions such as K^+, Na^+, Cl^-, accumulation of high amounts of specific organic osmolytes such as ectoine, hydroxyectoine [23, 26, 27], and other neutral glycerols can be identified. The aggregation of high levels of inorganic cations in the cytoplasm is considered a "salt-in" strategy to balance the osmotic pressure in the surrounding environment. The intracellular K^+ ions concentration of halophilic bacteria has been recorded at extremely high levels compared to the non-halophilic bacteria. Further, the halophilic and halotolerant bacteria cytoplasm consists of some active and stable proteins in a range of salt concentrations. Those bacteria have enzymes that can remain active even at high salt concentrations [22, 23].

The surface of the three-dimension structure of proteins in halophilic bacteria contains a comparatively high amount of acidic amino acids. That acidic nature prevents denaturing of proteins in high salt concentrations [20]. The electrostatic interactions of the proteins play a significant role in the halo adaptation of microbes [17].

Thus, elevated amounts of acidic amino acid, aspartic and glutamic acids, and a small amount of lysine facilitate the halophilic nature of these proteins. Structural analyses between halophilic and mesophilic proteins reveal that the significant differences are concentrated on the protein's surface [19]. These characteristics allow cooperation with electrostatic interactions and the presence of a higher number of salt bridges. The stability of the enzymes depends on the negative charge on the surface of the protein due to acidic amino acids, the hydrophobic groups in the presence of high salt concentrations and the hydration of the protein surface due to carboxylic groups present in aspartic and glutamic acids [27]. In addition, negative surface charges are important for the solvation of halophilic proteins to prevent denaturation, aggregation and precipitation [23].

Piezophiles

Extreme pressures stress an organism through the inverse pressure-volume thermodynamic relationship. This phenomenon would influence membrane fluidity, protein stability and reaction kinetics [17]. The physiological adaptations required for growth under these extreme conditions are substantial and involve a combination of modifications of gene structure and regulation [23].

Monosaturated and polyunsaturated in the cell membranes of piezophiles provide the support to maintain the membrane fluidity under high-pressure conditions [17]. The pressure favoured ordering in the cell membrane is prevented by the cis-unsaturated phospholipids in the membrane. Other adaptation mechanisms against the high pressures include reducing cell division, modification of membrane and transport proteins and accumulation of piesolytes viz. hydroxybutyrate and EPS, which stabilize the proteins [19, 23].

In addition, dissociation of voluminous multimeric complexes like ribosomes into their monomeric constituents may also result from high-pressure conditions [17]. The elongated helices in the 16S rRNA genes increased adaptation to growth at elevated pressure and are described as a piezostable characteristic of ribosome [20, 22]. The adaptation mechanisms of low-pressure extremophiles are not well known.

Radioresistant Microorganisms

The significant threats caused by radiation (Ultraviolet) which engenders reactive radicals for radioresistant microbes, can be enumerated as; imposed on lipids, nucleic acids and proteins [17]. Especially, proteins consisting of iron-sulfur or heme group, cysteine residues and cation binding sites are highly susceptible to short wavelength UV radiations which generate superoxides and reactive oxygen species. It was documented that proteomic damage is more lethal for the living cells than genomic damage. The radio resistant microorganisms with the ability to overcome the pose of free radicals can survive in environments with radiation.

Proteomic and genetic modifications that help grow and survive in radiation environments can be observed in these microorganisms. The small molecules like Mn^{2+}, trehalose accumulation in some radio resistant microbes viz. *Lactobacillus plantarum*, *Deinococcus radiodurans*, *Synechocystis* sp. is present proteomic damages caused by radiations [17]. Those molecules preserved the enzyme activity and stabilized the proteins in the presence of radiations and in the absence of Mn^{2+} or trehalose microbes susceptible to radiations.

Further, a number of DNA cleavage preventive and repair mechanisms which help to overcome the single and double-stranded DNA cleavage posed by the radiations can be observed in radioresistant microorganisms [19]. Maintaining multiple copies of chromosomes (polyploidy), ReCA-independent double-strand break repair, and synthesis-depended single-strand annealing are the common strategies found in the radioresistant species.

Xerophiles

Water content is the limiting factor for xerophiles, and dehydration is the major problem they need to overcome to survive in such environments [17]. The xerophytic conditions lead to the disintegration of biochemical reactions, protein and cell membrane and metabolism. Thus, xerophilic microorganisms have strategies to prevent water loss from the cells and repair the damages posed by dehydration. Decrease of surface area by entering cryptobiotic "tun" state, accumulation of lyoprotectants, and extensive single and double-stranded repair mechanisms can be stated as distinctive adaptations of xerophiles for their environmental conditions [17].

Polyextremophiles

Thermoacidophiles

High temperatures increase the permeability to H^+, resulting in lethal cytoplasmic acidification. Modifications in RNA codon thermostability and a neutral surface charge in proteins prevent the occurrence of acid hydrolysis [24]. The existence of thermoacidophiles over 100°C has not yet been observed. The survival of thermoalkaliphiles is related more to the buffering capacity to maintain a stable intracellular cytoplasm than the maintenance of a bioenergetic gradient [23, 28].

Thermohalophiles

High salt concentrations allied to temperature extremes have been observed in psychrohalophiles and thermohalophiles. Many environments in polar sea ice are cold brine solutions. Thus, some halophily is typical in most psychrophiles [28]. Indeed, cold adaptation and salt adaptation have common approaches. Thermohalophiles are rare [23]. Their uncharged proteins became extremely unstable in hypersaline solutions and are denatured by solvents at high temperatures and by decreasing the electrostatic interactions required to maintain the native folding. *Thermococcus waiotapuensis* is the most hyperthermohalophilic organism discovered to date. However, the basis of its biochemical stability is still unclear.

Thermopiezophile

Another factor in extremophilic survival is the relationship between temperature and pressure. At high temperatures, volume increases are desirable, but volume increases are undesirable at high pressures. Protein adaptations in thermopiezophiles are synergistic, combining a low surface charge with a highly hydrophobic core. Other biochemical responses are the induction of both heat-shock and cold-shock response pathways reinforcing the synergistic reaction [29]. The cold-shock response is critical in psychropiezophiles, since they do not benefit from the synergistic temperature and pressure approach. They also incorporate monosaturated and polysaturated fatty acids to prevent membrane crystallization, which is caused by temperature extremes, a characteristic already incorporated in piezophiles [23].

Haloacidophiles and Haloalkaliphiles

Halo-acidophiles and halo-alkaliphiles are both found in nature. For halo-acidophiles, high extracellular concentrations allow a more favorable efflux of H^+ [30]. Halo-alkaliphiles are more commonly found because the monovalent cations of salts are essential for pH homeostasis and energetic coupling [23]. On the other hand, for halo-alkaliphiles, the less favorable influx of H^+, caused by high salt concentrations, leads to a lethal alkylation of the cytoplasm and becomes a limitation to growth. Halo-alkaliphiles can maintain a gradient of about 1.0–1.5 units [22, 28].

Piezoacidophiles

High pressures promote negative volume changes by dissociating acids and the protonation of amine groups in proteins. This leads to an acidification of the solution, creating an environment where piezoacidophiles can be found. On the other hand, the alkalinification of the environment associated with high pressures creates a propitious site for piezo-alkaliphiles. However, these organisms still need to be further investigated for systematic identification and characterization [23].

Applied Aspects of Extremophiles

Since the discovery of extremophile, they have attracted a special attention not only because of their amazing adaptations to harsh environments, but also their capability to contribute processes that are important for human. With the beginning of isolating *Taq* DNA polymerase from a thermophilic bacterium, scientists continued to explore the potential areas where extremophiles and their enzymes and metabolites can be used to achieve sustainable development. Currently, extremophiles are very useful in astrobiological studies and biotechnological and industrial applications.

Extremophiles and Astrobiology

Astrobiology is the study of the origin, evolution, distribution, and future of life in the universe. This includes both potential extraterrestrial lives: life outside the earth and life on earth. Furthermore, they also study the origin of life on earth. According to the current knowledge Mars and Europa: one of Jupiter's moons have emerged as two of the most likely locations in the Solar System for potential habitability. However, it is not easy to analyze samples from extra-terrestrial locations due to extremely high costs, and the necessity of advanced technologies associated with those studies. Interestingly, there are terrestrial environments that have analogous conditions (may not be 100% identical) to certain extra-terrestrial environments. Such terrestrial environments are known as extra-terrestrial analog terrestrial environments. These environments are mostly extreme environments as most of the extra-terrestrial environments possess extremely harsh conditions. For example, sulfate-rich evaporites in White Sand National Monument and Lake Lucero in New Mexico, United States have been identified as terrestrial analog environments to certain Martian conditions [31, 32]. Earth's deep-ocean hydrothermal vents have been identified as terrestrial analog to Europa's under-ice oceans. Therefore, extreme environments on earth are very useful to find answers to astrobiology related questions. If the origin/formation of such extreme condition on Earth and other extra-terrestrial environments are similar, exploration of the microbial communities (their origin, physiology and ecology) in terrestrial environments would enables us to extrapolate those findings to analogous extra-terrestrial environments.

Extremophiles, Biotechnology and Industrial Applications

With the growing population, industrial activities are rapidly expanding to fulfil the needs of the mankind. It is important to conduct these industrial activities in a cost effective, efficient and eco-friendly manner to achieve sustainable development. However, the physicochemical conditions used in most industrial practices are harsh and analogous to natural extreme environments: high substrate concentrations, non-aqueous conditions and extremes of pH and temperature. Chemical catalysts are generally used to speed up industrial transformation processes and achieve efficient industrial output. However, one of the major drawbacks in this approach is that these chemical catalysts could cause serious damages to the environment by accumulating as chemical wastes. Another important aspect in sustainable industrial applications is to minimize excess usage of raw materials and usage of renewable raw materials. Different enzymes from mesophilic micro-organisms can be used in industrial processes as an eco-friendly alternative to chemical catalysts [33]. However, most of the industrial processes that use mesophilic enzymes are not optimized, because these enzymes are used under extreme ends of temperature, pH, pressure, salinity, and anhydrous conditions and the enzymes will not function with best capacity. Although it is possible to enhance the efficiency of these mesophilic enzymes through genetic or chemical modification and immobilization strategies, those approaches are time consuming and costly. For example, it is possible to improve the thermostability of mesophilic α-amylase by making genetic changes to the gene encoded for the enzyme using computational biology tools and recombinant DNA technologies, but the process is time consuming and costly. Extremophiles provide an excellent source for enzymes that are already adapted to work under the extreme conditions of many industrial processes. These enzymes are known as extremozymes [34].

Extremophiles and Bioremediation

Bioremediation has emerged as an effective solution to these problems. Most bioremediation research has focused on the processes performed by the Bacteria domain. Recently, extremophiles have been the focus of growing interest in bioremediation [35]. They can tolerate very harsh environmental conditions due to their ability to produce an array of molecules or extremozymes capable of functioning in the environment without denaturing.

These extremozymes from extremophilic microorganisms have unique characteristics such as stability to elevated temperature, pH extremes, organic solvents and high ion strength. Due to the stability and persistence of these extremophilic microorganisms under adverse environmental conditions, they can be explored by finding new species for use in the bioremediation of environments contaminated with extremely recalcitrant pollutants [36, 37]. Table 1 summarizes the details of extremophilic microbes for the biodegradation of pollutants from various environments.

Table 1. Extremophiles for the biodegradation of pollutants

Pollutants	Microbial Species/ Strain
Crude oil	*Streptomyces albiaxialis*
Benzoic acid, cinamic acid and 3-phenylpropionic acid	*Haloferax* sp. strain D1227
Uranium	*Shewanella putrefaciens MR-1*
Paraffin and other petroleum derivatives	*Dietzia maris*
Petroleum hydrocarbons	*Bacterial consortia*, *Marinobacter* ssp., *Erwinia ananas* and *Bacillus* spp.
4-hydroxybenzoic Acid	*Haloarcula* sp. D1
Diesel	*Cellulomonas* spp., *Bacillus marisflavi, Dietzia maris* and *Halomonas eurihalina*
Phenol and catechol	*Halomonas campisalis*
Benzoic, *p* -hydroxybenzoic, cinnamic, salicylic, phenylacetic, phenylpropionic, *p* -coumaric, ferulic and *p* -aminosalicylic acids	*Halomonas organivorans*

Extremozymes

General Description of Extremozymes

Extremozymes are enzymes generated from extremophilic microbes that can tolerate harsh industrial processes previously assumed to be protein-destructive [38]. Extremozymes with distinctive features, such as lignocellulolytic, amylolytic, and other biomass processing extremozymes, are extensively dispersed in thermophilic prokaryotes and have a lot of promise in industrial processes [39].

Extremophile enzymes have evolved to resist harsh circumstances on their own in many cases. These adaptations correspond to significant changes in amino acid sequence, resulting in changes in enzymes' structure, flexibility, charge, and hydrophobicity [40]. Extremozymes' varied modifications enable enzymatic biocatalysis in enzymatically unfavorable circumstances prevalent in industrial processes, extending the ranges of optimal enzyme activity [39, 40].

Extremozymes can expand due to their many adaptations. Extremozymes' adaptations enable enzymatic biocatalysts to operate in enzymatically unfavorable circumstances prevalent in industrial processes, extending the ranges of optimal enzyme activity [41]. It is currently conceivable to uncover and build extremozymes that can accommodate existing industrial processes or products by actively bioprospecting harsh conditions and/or using genetic engineering [42].

Enzyme Modifications

The extremophilic microorganisms live in extraordinary conditions and additionally adjust to a range of natural factors, such as temperature (55°C to 121°C and -2°C to 20°C), pressure (> 500 atmospheres), alkalinity or acidity pH (pH > 8, pH < 4), salinity (2-5 M NaCl or KCl), geographical scale/barriers, radiation (UVR resistance > 600 J/m), chemical extremes of heavy metals (arsenic, cadmium, copper, and zinc), need of supplements (e.g., water, ice, discuss, shake, or soil), osmotic barriers, or polyextremity. Within the last decades, studies about extremophilic microorganisms have expanded [43]. Furthermore, thermophilic, piezophilic, acidophilic, and halophilic proteins have been accepting more consideration for their biotechnological and mechanical applications. They can be classified as acidophile, alkaliphile, endolith, hyperthermophile, hypolith, metalotolerant, oligotroph, piezophile, psychrophile, radioresistant, thermophile, toxitolerant, and xerophile. The significance of enzymes and their roles in many processes have been studied during the last years, especially enzymes from extremophiles [44].

Thermophiles and Applications

Thermophiles have been the most researched extremophiles in recent decades. Cellulases, amylases, pullulanase, xylanases, pectinases, chitinases, proteases, lipases, esterases, and phytases are just a few of the thermostable enzymes that have been studied [45]. Enzymes from thermophilic bacteria may tolerate proteolysis and extreme conditions such as denaturing agents, organic solvents, and high salinity. The thermozymes have physical properties and electrostatic interactions that allow them to maintain activity at high temperatures. They have several adaptations, including maintaining their shape and operating under severe temperatures [46]. They can also enhance the number of hydrophobic deposits by generating a bisulfide *liaison* between two oppositely charged ions [45, 46].

Lipases are thermostable enzymes used in various biotechnological and commercial processes, including grease hydrolysis, esterification, interesterification, transesterification, and organic biosynthesis [47]. Thermozymes have also been used in the development of optical nanosensors and analytes. Lipase is also utilized in the paper business, the milk industry, the dyeing industry, the leather industry, and pharmaceuticals [48]. Proteases employed in synthesizing dipeptides, starch processing, and DNA are examples of thermozymes. Cellulase, hemicellulases, and xylanases have all been used in the bleaching of paper and polluting the environment [47].

Bio detergents now contain enzymes resistant to extreme circumstances, such as amylase, protease, cellulase, and lipase. Thermozymes from *Pyrococcus furiosus*, such as amylase, have been used in mutational investigations [48]. The pancreatic fistula amylase mutation resulted in an increase in the production of maltoheptaose from cycloamyloses. In the food, cosmetics, and pharmaceutical industries, maltoheptaose is employed as a carrier. Thermophilic enzymes exploit not only their activity at high temperatures but cannot also demonstrate catalytic activity at low temperatures. Thermozymes that can withstand high temperatures offer huge biotechnological promises.

Psychrophiles and Applications

Psychrophile enzymes (cold-adapted enzymes) have flexible configurations that adjust to low kinetic energy in cold conditions. Cold adapted enzymes have a lower activation enthalpy (H#) and higher negative activation entropy (S#) than their mesophilic and thermophilic homologs due to their intrinsic flexible structure [49]. As a result, as the temperature is lowered, the reaction rate of psychrophile enzymes tends to decline more slowly than that of thermophile enzymes. Compared to enzymes from mesophiles or thermophiles, this balance of thermodynamic activation characteristics results in comparatively high catalytic activity (kcat) at low temperatures and relatively low structural stability.

Thermolabile cold-adapted enzymes' compositional and structural characteristics that provide great flexibility are often the opposite of those of their more rigid and stable mesophilic and thermophilic homologues [49]. Psychrophilic enzymes, for example, have a lower core hydrophobicity, higher surface hydrophobicity, lower arginine/lysine ratio, weaker interdomain and intersubunit interactions, more and longer loops, lower secondary structure content, more glycine residues, less proline residues in loops, more proline residues in helices, fewer and weaker metal binding sites, fewer and weaker disulfide bridges, and fewer and an increase in conformational entropy of the unfolded state.

Cold adapted enzymes can transform substrates that require enzyme reactions to be done at low temperatures because substrates are heat sensitive [50]. They can also be utilized to avoid undesired chemical reactions at higher temperatures. These properties are of particular relevance to the food and feed industry (As a material for removal of hemicellulose from feed, Starch hydrolysis, cheese ripening, removal of lactose from milk, wine and beverage stabilization, etc.), where it is important to avoid spoilage and change in nutritional value and flavor of the original heat-sensitive substrates and products. Furthermore, it is used in environmental biotechnology (Degradation and removal of xenobiotics and toxic compounds), tanning and hide industry, bio- bleaching in paper industry, biofuels and energy production, biodiesel production, bioethanol production, pharmaceuticals, medical and industry, an antifungal drug, etc. [51].

Acidophiles

While acidophiles' tolerance for stress generated by high concentrations of protons is a defining trait, it is far from the sole factor that has severe consequences for living in acidic settings [52]. Low pH affects the solubility of many chemicals that can harm bacteria, weighty metals, but also other compounds that hinder biological processes at high quantities. Furthermore, in many acidic situations, reactive oxygen species (ROS) are easily generated and are a source of severe oxidative damage to biological tissue. Acidophiles have evolved a variety of adaptations to cope with these conditions.

Acidophile environment pH levels fluctuate, but there is a lower limit around pH 0 below which the physical circumstances are too harsh for any bacteria to multiply (see section Diversity of acidophilic prokaryotes). Because most extreme acidophiles prefer pH 1.5-3 for growth, they have many DNA and protein repair systems to deal with acid-related cell structural damage [53]. The preservation of this gradient, termed pH homeostasis, can be achieved via three simple principles: (i) restricting proton influx, (ii) buffering of intracellular protons, and (iii) proton efflux [54, 55]. In addition to structural changes, acidophiles also exhibit a reverse, inside-positive membrane potential.

As a result, acidic waters drain enormous amounts of metal ions from the surrounding environment, exposing native acidophiles to far higher metal concentrations than neutrophils [53, 55]. Many acidophiles encourage metal dissolution, because of the nature of their surroundings, regularly come into contact with metals such as iron, copper, arsenic, aluminium, nickel, and zinc [55]. Both acidophiles and neutrophils use similar defence mechanisms to combat heavy metal toxicity. Heavy metal resistance can be genetically traced in various ways, but like other toxic substance resistances, it can be split into three categories: efflux, inactivation, and conversion.

Acid stable enzymes are used in various industries, including baking, fruit juice processing, animal feed, and pharmaceuticals, and several have already been commercialized. In the bioleaching of metals from low-grade ores, acidophiles are commonly utilized.

Alkaliphiles

Alkaliphiles' ability to proliferate abundantly at punishingly high pH, where neutralophilic bacteria cannot even live for a short time, demonstrates their distinct adaptation tactics. All alkaliphiles have a crucial adaption called intracellular pH homeostasis [56]. For decades, scientists have been trying to determine the secret of alkaliphiles' high pH adaptability. The essential adaptation mechanisms involved in pH homeostasis are detailed below, including gaining H+ from the extracellular environment, minimizing H+ leakage from the cytoplasm, producing organic acid, and preventing OH diffusion from the extracellular environment [57].

Alkaliphiles have a cytoplasmic pH that is near neutral. Alkaliphilic cells achieve this by maintaining a relatively high concentration of H+ in their cytoplasm. Translating H+ from the extracellular environment into the cell and closely regulating it is one way to accomplish this [58].

Because the H+ concentration outside the cell is larger than that in the cytoplasm, the uptake and translocation of H+ by F1F0-ATP synthases of cells suited to neutral or acidic habitats is energetically favored. Alkaliphiles, on the other hand, do not have the luxury of a high H+ concentration in their surroundings. Because H+ is limited in alkaline habitats, organisms adapted to these conditions require an effective method for capturing and transporting H+ [57, 58].

Alkaline enzymes have a dominant position in the global enzyme market as constituents of detergents. Commercially, alkaline enzymes are successfully used in detergent formulations, silk degumming, food and feed industry, photographic gelatin hydrolysis, leather dehairing, cosmetics, and pharmaceuticals.

Halophiles

Because of their polyextremophilic properties, halophilic enzymes are more stable than non-halophilic enzymes. These enzymes are thermotolerant, alkaliphilic, and active in high-salt conditions, which explains their widespread use in biotechnological applications in the food and textile industries and laundry detergents [59].

Early studies of halophilic enzymes revealed two key characteristics: (1) they have a large number of protein charges, and (2) they have a high amount of hydrophobicity, which is not present in nonhalophilic enzymes [60].

Microorganisms that are not acclimated to very saline conditions lose water, causing the cells to shrivel and eventually die due to a lack of cellular structure and function. Under such conditions, halophile enzymes have evolved two distinct strategies to increase the osmotic activity of their cytoplasm with the external environment: either producing compatible organic solutes (osmoprotectants) or reaching an equilibrium state in which the overall salt concentration within cells matches that of the environment by accumulating large salt concentrations in their cytoplasm [58-60]. Dehydration causes unfavorable interactions that alter internal microbial proteins, which can be avoided by changing their net charge. Halophile proteins contain a higher proportion of glutamate and aspartate on their surfaces than nonhalophile proteins, which is a notable difference [61]. Early microbiologists hypothesized that the molecular haloadaptation of halophilic enzymes was caused by two main characteristics: a large number of protein charges and increased hydrophobicity [60, 61]. The negativity of the halophile's proteomes was first revealed by the genome sequencing of *Halobacterium* sp. NRC-1 revealed an acidic proteome (pI * 4.5) unlike that of non-halophilic microorganisms, in which proteomes have average pIs' very close to neutrality [60]. Apart from the lack of enzymes to synthesize at least eight amino acids, encoding a group of signal transducers and transcriptional regulators may also be pertinent in their halo adaption. They contribute to most of the geochemical processes and at the industrial level as critical ingredients in most food, agricultural, and pharmaceutical industries.

Future Prospects of Extremophiles

Extremophiles, their enzymes, and metabolites have a huge potential to provide sustainable solutions to current and future problems associated with environmental pollution, depletion of natural non-renewable resources, energy crisis, human health implications, current agricultural practices and many more. Extremophiles could make new products, improve the quality of existing products and improve the cost effectiveness of the current production strategies, for example, alternative renewable raw materials can be used produce biodiesel using extreme reaction condition. However, there are some drawbacks that hinder the usage of full potential of extremophiles in industrial applications. The most significant barrier is the current lack of ability to produce most extremophiles/extremozymes on the large scales required for industrial applications. Therefore, new innovative methods should be developed produce these microorganisms and microbial products. New recombinant extremozymes can be produced in large quantities by mesophilic organisms like *Escherichia coli.* But new expression systems should be developed with extremophilic organisms as the host to achieve high expression of soluble proteins. Another challenge with harnessing the full potential of extremophiles is the lack of partnerships among academia, industry, and government. Therefore, it is crucial to coordinate all these sectors to achieve sustainable development using extremophiles.

Conclusion

Extremophiles are organisms that can survive and thrive in extreme environments: environments that comprised of extreme ends of various environmental parameters such as temperature, salinity, pH, atmospheric pressure, oxygen stress, humidity and radiation. Extreme environments are widely distributed on the earth. Therefore, extreme environments provide habitats for a considerable fraction of the biomass and biodiversity on the earth. Most extremophiles are microorganisms: bacteria, archaea and eukaryotic microorganisms. Extremophiles show various structural, physiological, and metabolic adaptations to the environments they inhabit. Modifications of the cell membrane, cellular organelles and genomic composition are considered major structural modifications, whereas regulation of osmotic balance is considered a main physiological modification, and protein modifications are considered a major metabolic adaptation. All these adaptations are governed by extremozymes: enzymes produced by extremophilic microbes that can tolerate harsh conditions. Therefore, extremophiles, their enzymes and metabolites have a huge potential to provide sustainable solutions to current and future problems associated with environmental pollution, depletion of natural non-renewable resources, energy crisis, human health implications, current agricultural practices and many more.

References

[1] Seckbach, J., Oren, A., and Stan-Lotter, H. (Eds.). 2013. *Polyextremophiles: life under multiple forms of stress* (Vol. 27). Springer Science & Business Media.

[2] Cavicchioli, R., Thomas, T., and Curmi, P. M. 2000. Cold stress response in Archaea. *Extremophiles*, *4*(6), 321-331.

[3] Mozhaev, V. V., Heremans, K., Frank, J., Masson, P., and Balny, C. 1996. High pressure effects on protein structure and function. *Proteins: Structure, Function, and Bioinformatics*, *24*(1), 81-91.

[4] Brock, T. D. 1967. The value of basic research: discovery of Thermus aquaticus and other extreme thermophiles. *Genetics*, *146*(4), 1207.

[5] MacElroy, R. D. 1974. Some comments on the evolution of extremophiles. *Biosystems*, *6*(1), 74-75.

[6] Seckbach, J., and Rampelotto, P. H. 2015. Polyextremophiles. *Microbial Evolution under Extreme Conditions. DeGruyter*, *8*, 153-170.

[7] Cavicchioli, R. 2006. Cold-adapted archaea. *Nature Reviews Microbiology*, *4*(5), 331-343.

[8] Urbieta, M. S., Donati, E. R., Chan, K. G., Shahar, S., Sin, L. L., and Goh, K. M. 2015. Thermophiles in the genomic era: Biodiversity, science, and applications. *Biotechnology Advances*, *33*(6), 633-647.

[9] Adapa, V., Ramya, L. N., Pulicherla, K. K., and Rao, K. R. S. 2014. Cold active pectinases: advancing the food industry to the next generation. *Applied biochemistry and biotechnology*, *172*(5), 2324-2337.

[10] Monnet, C., Landaud-Liautaud, S., Bonnarme, P., and Swennen, D. 2015. Growth and adaptation of microorganisms on the cheese surface. *FEMS microbiology letters*, *362*(1), 1-9.

[11] Horikoshi, K. 1999. Alkaliphiles: some applications of their products for biotechnology. *Microbiology and molecular biology reviews*, *63*(4), 735-750.

[12] Naghoni, A., Emtiazi, G., Amoozegar, M. A., Cretoiu, M. S., Stal, L. J., Etemadifar, Z., Shahzadeh Fazeli, S. A., & Bolhuis, H. 2017. Microbial diversity in the hypersaline Lake Meyghan, Iran. *Scientific Reports*, *7*(1), 1-13.

[13] Schwendner, P., and Schuerger, A. C. 2020. Exploring microbial activity in low-pressure environments. *Current Issues in Molecular Biology*, *38*(1), 163-196.

[14] Gupta, G. N., Srivastava, S., Khare, S. K., and Prakash, V. 2014. Extremophiles: an overview of microorganism from extreme environment. *International Journal of Agriculture, Environment and Biotechnology*, *7*(2), 371.

[15] Fuldeore, K. V., Patel, N., bhai Bavarva, A. H. and Wagh, V. A. 2022. Biochemistry Behind Protein Adaptations in Extremophiles. In *Physiology, Genomics, and Biotechnological Applications of Extremophiles* (pp. 124-135). IGI Global.

[16] Shah, H., Panchal, K. and Panchal, A. 2022. Major Compatible Solutes and Structural Adaptation of Proteins in Extremophiles. In *Physiology, Genomics, and Biotechnological Applications of Extremophiles* (pp. 161-186). IGI Global.

[17] Seckbach, J., Oren, A. and Stan-Lotter, H. eds. 2013. *Polyextremophiles: life under multiple forms of stress* (Vol. 27). Springer Science & Business Media.

[18] Koga, Y. 2012. Thermal adaptation of the archaeal and bacterial lipid membranes. *Archaea, 2012.*

[19] Kumar, A., Alam, A., Tripathi, D., Rani, M., Khatoon, H., Pandey, S., Ehtesham, N. Z., & Hasnain, S. E. 2018, December. Protein adaptations in extremophiles: an insight into extremophilic connection of mycobacterial proteome. In *Seminars in Cell & Developmental Biology* (Vol. 84, pp. 147-157). Academic Press.

[20] Brininger, C., Spradlin, S., Cobani, L. and Evilia, C. 2018, December. The more adaptive to change, the more likely you are to survive: Protein adaptation in extremophiles. In *Seminars in cell & developmental biology* (Vol. 84, pp. 158-169). Academic Press.

[21] Fuchida, S., Mizuno, Y., Masuda, H., Toki, T., & Makita, H. 2014. Concentrations and distributions of amino acids in black and white smoker fluids at temperatures over 200°C. *Organic Geochemistry,* 66, 98-106.

[22] Reed, C. J., Lewis, H., Trejo, E., Winston, V. and Evilia, C. 2013. Protein adaptations in archaeal extremophiles. *Archaea*, 2013.

[23] Dalmaso, G. Z. L., Ferreira, D. and Vermelho, A. B. 2015. Marine extremophiles: a source of hydrolases for biotechnological applications. *Marine drugs*, *13*(4), pp. 1925-1965.

[24] Boyd, E., Hamilton, T., Wang, J., He, L. and Zhang, C. 2013. The role of tetraether lipid composition in the adaptation of thermophilic archaea to acidity. *Frontiers in Microbiology*, *4*, p. 62.

[25] Colletier, J.-P., Aleksandrov, A., Coquelle, N., Mraihi, S., Mendoza-Barbera, E., Field, M., & Madern, D. 2012. Sampling the conformational energy landscape of a hyperthermophilic protein by engineering key substitutions. *Mol. Biol. Evol. 29*, 1683–1694

[26] Moreno, M. L., Pérez, D., García, M. T., Mellado, E. 2013. Halophilic bacteria as a source of novel hydrolytic enzymes. *Life*, *3*, 38–51

[27] DasSarma, S. and DasSarma, P. 2015. Halophiles and their enzymes: negativity put to good use. *Current opinion in microbiology*, *25*, pp. 120-126.

[28] Capece, M. C., Clark, E., Saleh, J. K., Halford, D T. 2013. Polyextremophiles and the constraints for terrestrial habitability. In *Polyextremophiles* (pp. 3-59). Springer, Dordrecht.

[29] Vijayalaxmi, S., Prakash, P., Jayalakshmi, S. K., Mulimani, V. H., & Sreeramulu, K. 2013. Production of extremely alkaliphilic, halotolerent, detergent, and

thermostable mannanase by the free and immobilized cells of Bacillus halodurans PPKS-2. Purification and characterization. *Applied biochemistry and biotechnology*, *171*(2), pp. 382-395.

[30] Poli, A., Finore, I., Romano, I., Gioiello, A., Lama, L., & Nicolaus, B. 2017. Microbial diversity in extreme marine habitats and their biomolecules. *Microorganisms*, *5*(2), p. 25.

[31] Sirisena, K. A., Ramirez, S., Steele, A., and Glamoclija, M. 2018. Microbial diversity of hypersaline sediments from lake Lucero playa in white sands national monument, New Mexico, USA. *Microbial ecology*, *76*(2), 404-418.

[32] Glamoclija, M., Ramirez, S., Sirisena, K., and Widanagamage, I. 2019. Subsurface Microbial Ecology at Sediment-Groundwater Interface in Sulfate-Rich Playa; White Sands National Monument, New Mexico. *Frontiers in Microbiology*, *10*, 2595.

[33] Chapman, J., Ismail, A. E., and Dinu, C. Z. 2018. Industrial applications of enzymes: Recent advances, techniques, and outlooks. *Catalysts*, *8*(6), 238.

[34] Nadaroglu, H., and Polat, M. S. 2022. Microbial extremozymes: Novel sources and industrial applications. In *Microbial Extremozymes* (pp. 67-88). Academic Press.

[35] Peeples, T. L. 2014. Bioremediation using extremophiles. In *Microbial biodegradation and bioremediation* (pp. 251-268). Elsevier.

[36] Giovanella, P., Vieira, G. A. L., Ramos Otero, I. V., Pais Pellizzer, E., De Jesus Fontes, B., & Sette, L. D. 2020. Metal and organic pollutants bioremediation by extremophile microorganisms. *Journal of hazardous materials*, *382*, 121024.

[37] Alavi, S., Rafieyan, S., Yavari-Bafghi, M., and Amoozegar, M. A. 2020. Extremophiles: A powerful choice for bioremediation of toxic oxyanions. In *Microbial bioremediation & biodegradation* (pp. 203-249). Springer, Singapore.

[38] Khan, N. F., Zehra, U., Reshi, Z. A., Shah, M. A., Baba, Y. R. 2022. Seasonal Dynamics of Bacterial and Fungal Lineages in Extreme Environments. In *Climate Change and Microbial Diversity*. Apple Academic Press. 63-67.

[39] Elleuche, S., Schröder, C., Sahm, K., and Antranikian, G. 2014. Extremozymes—biocatalysts with unique properties from extremophilic microorganisms. *Current opinion in biotechnology*, *29*, 116-123.

[40] DasSarma, S., Capes, M. D., Karan, R., and DasSarma, P. 2013. Amino acid substitutions in cold-adapted proteins from Halorubrum lacusprofundi, an extremely halophilic microbe from Antarctica. *PLoS One*, *8*(3), e58587.

[41] Sarmiento, F., Peralta, R., and Blamey, J. M. 2015. Cold and hot extremozymes: industrial relevance and current trends. *Frontiers in bioengineering and biotechnology*, *3*, 148.

[42] Trincone, A. 2011. Marine biocatalysts: enzymatic features and applications. *Marine drugs*, *9*, 478-499.

[43] Tiquia-Arashiro, S., and Rodrigues, D. F. 2016. *Extremophiles: applications in nanotechnology* (p. 193). New York, NY, USA: Springer International Publishing.

[44] Rasuk, M. C., Ferre, G. M., Moreno, J. R., Farias, M. E. 2016. The diversity of microbial extremophiles. *Molecular diversity of environmental prokaryotes*, 87-126.

[45] Gomes, J., and Steiner, W. 2004. The biocatalytic potential of extremophiles and extremozymes. *Food technology and Biotechnology*, *42*, 223-225.

[46] Fitter, J., Herrmann, R., Dencher, N. A., Blume, A., & Hauss, T. 2001. Activity and stability of a thermostable α-amylase compared to its mesophilic homologue: mechanisms of thermal adaptation. *Biochemistry*, *40*, 10723-10731.

[47] Hasan, F., Shah, A. A., and Hameed, A. 2006. Industrial applications of microbial lipases. *Enzyme and Microbial technology*, *39*, 235-251.

[48] Nadaroglu, H., and Polat, M. S. 2022. Microbial extremozymes: Novel sources and industrial applications. In *Microbial Extremozymes* (pp. 67-88). Academic Press.

[49] Siddiqui, K. S., and Cavicchioli, R. 2006. Cold-adapted enzymes. *Annu. Rev. Biochem.*, *75*, 403-433.

[50] Jeon, O., Bouhadir, K. H., Mansour, J. M., and Alsberg, E. 2009. Photocrosslinked alginate hydrogels with tunable biodegradation rates and mechanical properties. *Biomaterials*, *30*, 2724-2734.

[51] Chukwuma, O. B., Rafatullah, M., Tajarudin, H. A., and Ismail, N. 2020. Lignocellulolytic enzymes in biotechnological and industrial processes: a review. *Sustainability*, *12*, 7282.

[52] Canganella, F., and Wiegel, J. 2011. Extremophiles: from abyssal to terrestrial ecosystems and possibly beyond. *Naturwissenschaften*, *98*, 253-279.

[53] Crossman, L., and Dow, J. M. 2004. Biofilm formation and dispersal in Xanthomonas campestris. *Microbes and infection*, *6*, 623-629.

[54] Ram, R. J., Verberkmoes, N. C., Thelen, M. P., Tyson, G. W., Baker, B. J., Blake, R. C., Shah, M., Hettich, R. L., Banfield, J. F. 2005. Community proteomics of a natural microbial biofilm. *Science*, *308*, 1915-1920.

[55] Baker-Austin, C., and Dopson, M. 2007. Life in acid: pH homeostasis in acidophiles. *Trends in microbiology*, *15*, 165-171.

[56] Mamo, G. 2019. Challenges and adaptations of life in alkaline habitats. *Alkaliphiles in Biotechnology*, 85-133.

[57] Casey, J. R., Grinstein, S., and Orlowski, J. 2010. Sensors and regulators of intracellular pH. *Nature reviews Molecular cell biology*, *11*, 50-61.

[58] Chiquet, M., Renedo, A. S., Huber, F., and Flück, M. 2003. How do fibroblasts translate mechanical signals into changes in extracellular matrix production? *Matrix biology*, *22*, 73-80.

[59] de Lourdes Moreno, M., García, M. T., Ventosa, A., and Mellado, E. 2009. Characterization of Salicola sp. IC10, a lipase-and protease-producing extreme halophile. *FEMS microbiology ecology*, *68*, 59-71.

[60] DasSarma, S., and DasSarma, P. 2015. Halophiles and their enzymes: negativity put to good use. *Current opinion in microbiology*, *25*, 120-126.

[61] Britton, K. L., Baker, P. J., Fisher, M., Ruzheinikov, S., Gilmour, D. J., Bonete, M. J., Ferrer, J., Pire, C., Esclapez, J., & Rice, D. W. 2006. Analysis of protein solvent interactions in glucose dehydrogenase from the extreme halophile Haloferax mediterranei. *Proceedings of the National Academy of Sciences*, *103*, 4846-4851.

Chapter 3

Extremophiles and Extremozyme

Aditya Upadhyay[1]
Dharm Pal[2]
and Awanish Kumar[1,*]
[1]Department of Biotechnology, National Institute of Technology, Raipur (CG), India
[2]Department of Chemical Engineering, National Institute of Technology, Raipur (CG), India

Abstract

The majority of enzymes are used for catalyzing a biological reaction with standard conditions. Extremophilic enzymes showed as a relatively undefined and understudied class of enzymes, characterized by complex adaptation mechanisms. Extremophilic enzymes can catalyze reactions under unusual conditions, such as high and low temperatures, high salinity, exposure to radiation, high ionic strength, extreme pH values, and exposure to toxins. Thus, extremophile enzymes are a novel unexplored, and undiscovered class of enzymes due to their specific survival capacity in extreme conditions. Extremophile enzymes are produced from the extremophile microbes that belong to the archaea and bacteria groups. Thus, an extremophile bacterium can serve as a novel source of industrial enzymes. Extremophile enzymes have great potential for industrial applications as well as some therapeutic applications like anti-cancer molecules and anti-cell signaling molecules. Currently, studies are focusing on extremophilic microbes and enzymes due to their unique and unexplored characteristics. As a result, this topic has become a highly relevant and promising area of research with industrial and therapeutic applications. In this chapter, we are discussing the initial

* Corresponding Author's Email: awanik.bt@nitrr.ac.in.

In: Microbial Technologies and Their Applications
Editors: Santosh Kumar Karn and Anne Bhambri
ISBN: 979-8-88697-913-8

growing information about extremophile microbes and extremophile enzymes and also summarize up-to-date knowledge on the genomic level with various application domains.

Keywords: extremophile microbes, extremophile enzymes, industrial, therapeutic application

Introduction

Enzymes are biological products produced by microorganisms that offer effective solutions for enhancing human life. They improve food quality and can help overcome disease burdens through enzymatic medicines and disease diagnostic processes [1]. Therefore, the field of enzymes is highly diverse, as microorganisms can produce enzymes not only under normal temperature and pressure conditions, but also in extreme environments, such as high temperatures, high pressures, and extreme pH levels. These types of enzymes are known as extremozymes, and the microorganisms that produce them are called extremophiles. This shows that microorganisms are capable of surviving in a wide range of conditions, including extreme ones. However, the mechanisms by which they survive in such conditions are not yet fully understood. Therefore, there is still a vast number of extremozymes that remain undiscovered, which could potentially revolutionize the field of enzymes. Extremozymes are typically categorized into six major classes based on their ability to survive under different conditions: Thermophiles, Psychrophiles, Halophiles, Alkaliphiles, Acidophiles, and Piezophiles [2]. Thus, the six major classes of extremozymes contained a limited number of enzymes that are specific to certain microorganisms. However, these enzymes have great industrial applications. For instance, protease enzymes can work under extreme temperatures of 600-800°C and are used as detergent enzymes. Dehydrogenase enzymes are utilized for biocatalysis in organic media under high salt conditions. Amylases are used for starch processing under low pH conditions, while oxidase enzymes are used for coal desulfurization under low pH conditions of approximately 2. Despite their unpredictable nature, extremozymes are highly effective at performing their enzymatic activity. However, the actual molecular-level mechanisms underlying their survival in extreme conditions remain undiscovered. The research gap in both the fields of extremozymes and extremophiles is mainly due to the limited research work. Hence, field of extremozyme has vacant for research and innovations.

Hence, this field reflects as hot spot to research. In this chapter we are discussed the evolutionary adaptive strategy of extremozymes and extremophiles, classification with examples and industrial applications and future opportunity in this field.

Evolutionary Factor Responsible for the Development of Extremophiles

Life generally thrives under normal conditions, such as those involving normal temperatures and pressures within the ecosystem. However, microbial life is also capable of surviving under extreme conditions, including high temperatures and pressures. As such, studying the differences between extremophiles and mesophiles is a fascinating area of research. Various studies suggest that differences in nucleic acid level, protein level, and cell structural level account for the enhanced stability of extremophiles compared to mesophiles. These differences include changes in hydrophobicity, covalent bonding, G+C contents, amino acid residues, and tRNA compositions [3] (Figure 1).

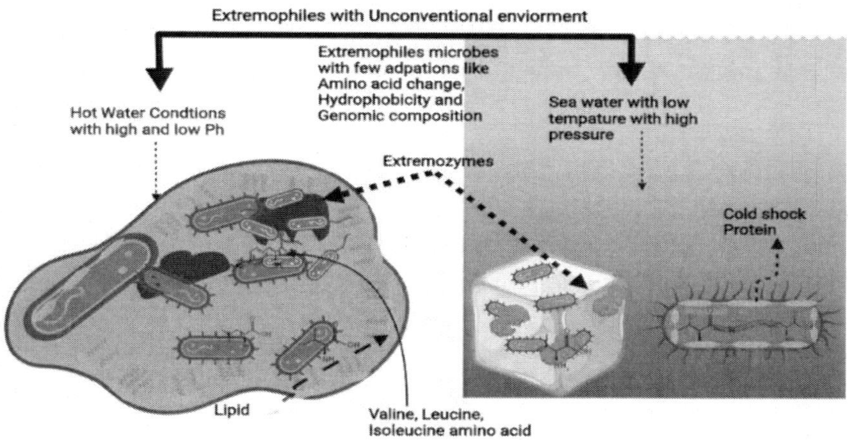

Figure 1. Evolutionary mechanism to develop the extremophiles and extremozyme.

Hydrophobicity

Hydrophobicity is a well-known factor that enhances the stability of extremophilic microorganisms through protein stability. Hydrophobicity is a key feature of biological compounds that allows them to be repelled by water molecules. Extremophile proteins exhibit high stability in extreme environments due to the hydrophobic nature of the protein. Hydrophobicity has a direct relationship with protein stability, which provides strength for the microorganisms to survive in extreme conditions. According to a study, the hydrophobic environment is responsible for the high protein stability in extremophilic microorganisms compared to mesophilic microorganisms. This study analyzed the hydrophobicity of proteins in extremophiles and mesophiles, and the results showed that 80% of extremophilic microorganism proteins exhibited the hydrophobicity feature, as compared to mesophiles [4]. Therefore, hydrophobicity plays a significant role in improving the thermostability of extremophile proteins. A recent study has demonstrated that hydrophobic interactions greatly impact protein stability. According to the study, high hydrophobic interactions improve protein stability, while lower hydrophobic interactions result in less stable proteins [5]. Another study suggests that transmembrane helices showed an important factor to enhance the hydrophobicity in the extremophilic microorganism as compared to mesophilic microorganisms. Thus, high hydrophobicity major feature finds in extremophilic proteins to and stabilize in extreme conditions like high temperatures [6].

Compositions of Protein

Protein compositions also contribute to the stability of extremophile life. Studies have shown that specific amino acids can enhance protein and cellular stability in extreme conditions, such as high temperatures above 100°C and extremely low temperatures. A previous study identified several amino acids that demonstrated greater stability in high temperatures, including Valine, Leucine, Isoleucine, Tyrosine, Lysine, Histidine, Methionine, Threonine, Serine, and Tryptophan. Valine, Leucine, and Isoleucine, in particular, are highly stable amino acids at temperatures exceeding 100°C compared to other amino acids [7]. Another study has indicated that the type of amino acid residue also plays a significant role in protein stability. This study analyzed the proteomes of extremophiles and mesophiles at temperatures ranging from

60^0 C to 100^0C. The results showed that the percentage of lysine residues increased at high temperatures in extremophiles as compared to mesophiles [8]. An additional study on the proteome analysis of extremophiles provides evidence supporting the functional role of Proteomic compositions instability factors. The study analyzed 269 proteins from extremophiles at temperatures ranging from 60^0C to 80^0C and found that lysine, arginine, and glutamic amino acids were present in the highest quantities, while cysteine was present in lower quantities [9]. Another special type of protein also responsible for the survival of extremophilic microorganisms is known as a cold-acclimations protein which provides survival capability at a very low temperature. In this study, the protein profile was analyzed after cold shock at 25^0C to 4^0C, and found the specific protein over-expressed after cold shock that protein known as cold cold-acclimations protein which provides survival strength at cold temperatures [10]. But these types of adaptions not occurred in mesophilic microbial life. Thus, amino acid compositions showed a great difference between extremophile life and mesophilic microbial life. Extremophile microorganisms showed as the specific type of microorganism due to specific amino acid compositions. This type of adaption is not sufficient for survival in extreme conditions other mechanisms are also involved in this process. However, the research volume is very less regarding extremophilic molecular life. Hence, more comprehensive work is required for the development of new adaptations.

Genomic Compositions

The genomic composition is also considered an essential factor in enhancing the stability of extremophile microorganisms. Genomic adaptation played a crucial role in the development of extremophile life. Several studies have shown that certain genomic components exhibit differences between extremophiles and mesophiles, such as high G+C content, tRNA, and DNA compositions. A recent study has indicated that extremophiles have a higher G+C content compared to mesophilic microorganisms. The study found that the G+C content was approximately 54% in extremophile microorganisms, while another study reported that mesophilic microorganisms had a G+C content of approximately 36% [11, 12]. Therefore G+C content has a direct impact on the stability of extremophile microorganisms. If G+C contents are high in DNA that means stability is also high another factor also generates the impact on extremophiles' life like tRNA. Several studies suggest that tRNA is also a different point between the extremophiles and mesophiles microorganism mesophilic microorganism has high tRNA contents as compared to the extremophiles (Especially in thermophiles microorganism). A study suggests that tRNA components are less in extremophile microorganisms because it had hydrophilic amino acid-like Arginine, Asparagine, Aspartic acid, Glutamine, Glutamic acid, and Valine. Thus, Hydrophobicity increases due to a decrease in the hydrophilic amino acid in the extremophilic life. This study also suggests that correlations between the G+C content with low tRNA contents mean high G+C content provides the stability and tRNA decreases the hydrophilic property in the extremophilic microorganism. Thus, the Stability of extremophilic microorganisms (especially thermophilic ones) increases at high temperature due to this correlation phenomenon. These adaptation processes provide more strength to extremophile microorganisms to survive under unpredictable conditions. However, the actual evolutionary mechanisms and various types of other adaptations are still unclear, and this area requires further research.

Covering System

This unique system is found in extremophilic microorganisms as an outer covering. The outer membrane mainly consists of lipids, which provide stability in stressful environments. This adaptation process is highly specific to extremophilic microorganisms. Lipid molecules do not function individually but instead form a collective structure, such as the membrane, which is referred to as the thermostable membrane. Thus, thermostable lipids are present in the membrane which is called the thermostable membrane. It protected outer stress environments. Hence, this type of feature is presented only in the extremophilic microorganism. A study suggests that a few specific lipids increase during high temperature like so fatty acid-lipid also increase at high temperatures in extremophilic bacterial species [13]. Another study also suggests that cyclic-based lipids also increased at high temperatures and provide more strength to extremophilic microorganisms but both lipids are not seem increasing conditions in mesophilic microorganisms [14]. Therefore, these specific types of lipids create a favorable environment inside extremophilic microorganisms during times of stress, allowing them to survive in unpredictable conditions. Although various types of adaptations may occur in extremophilic microorganisms, not all adaptations have been discovered yet, and the formation process of thermostable lipids and different genomic conditions in extremophiles remains unclear. These microorganisms exist in various extreme niches on Earth, with temperatures reaching approximately 100^0C and below freezing. Due to these harsh conditions, the isolation of extremophilic microorganisms is difficult. Nevertheless, their potential uses for human populations make them an important area of research. Further in-depth research is required to fully understand the capabilities and potential benefits of extremophilic microorganisms.

Classifications of Extremozymes

Extremophilic microorganisms produce a unique type of biological product known as extremozymes, which are capable of performing reactions under uncommon and unconventional conditions. These special enzymes have a greater capacity to tolerate extreme stress conditions, such as very high and low temperatures, high pH, and high pressure, compared to regular enzymes. Extremozymes are classified into six different classes based on the extreme environment in which they operate (Table 1).

Table 1. Classification of Extremophiles extremozyme with source extremophiles microorganism

S. No	Class of Extremophiles	Name of extremozyme	Mechanism	Source	References
1	Thermophiles	Taq polymerase Protease	High temperature	*Thermus aquaticus* *Anoxybacillus kamchatkensis*	40 32
2	Psychrophiles	Protease	Low temperature	*Pseudomonas prosekii*	34
3	Halophiles	Amylase	High Salt	*Pseudoalteromonas* species	36
4	Alkaliphiles	Amylase	High Alkali condition	*Luteimonasabyssi*	37
5	Acidophiles	Protease	High Acidic condition	*Bacillus licheniformis*	33
6	Piezophiles	Protease, Amylase	High pressure	Whole Extremophiles microorganism	26, 27

Thermophiles

Thermophilic extremozymes are a highly specialized and valuable product due to their ability to function at high temperatures, up to around 100^0C. The stability of thermophilic extremozymes at high temperatures is due to lipid and amino acid modifications, strong hydrogen and covalent bond modifications, and tRNA compositions. In contrast to normal enzymes, which denature at high temperatures, thermophilic extremozymes can function effectively due to these modifications. However, thermophilic extremozymes do not function well at normal temperatures, around 37^0-40^0C. One well-known example of a thermophilic extremozyme is Taq polymerase, which is used in polymerase chain reaction technology and works optimally at around 72^0C. Thermophilic extremozymes have been isolated from a temperature range of 80^0C to 115^0C, and are also found in deep sea environments [15]. Thermophile extremozymes are very similar to the normal enzyme in case sequences homologues and 3D protein structure only a few modifications occurred in thermophiles extremozyme [16]. Various examples are available for thermophile extremozyme like Proteases, amylases, glucoamylases, DNA polymerases and esterase growth characteristics between 60 to 100°C occurred with a wide range of industrial applications.

Psychrophiles

The conventional environment is not the only type of ecosystem on earth, as there are also glaciers that host microorganisms capable of surviving in such harsh conditions. Among the microorganisms found in these extremely cold environments are mostly bacteria, fungi, and archaea, such as *Pseudomonas, Bacillus, Penicillium,* and *Halorubrum,* which produce a unique type of extremozyme called Psychrophilic extremozyme [17, 18]. Psychrophiles extremozyme enzyme is a unique type of enzyme that can work with a very low temperature lower than 15^0 C. Psychrophiles microorganism have a few evolutionary mechanisms for survival in cold environments like cold shock protein and ice-binding molecules. Psychrophiles extremozyme also have a few factors for performing better activity at lower temperature like high flexibility through reducing the hydrogen bonding, and salt bridges and increasing the surface loaded residues at the molecular levels [19]. These types of evolutionary adaptions provide survival strength in extremely lower temperatures. This type of enzyme is more attractive due to the decrease in the energy constipation process. Hence food industry and the paper industry mostly used the Psychrophiles extremozyme like cellulases, lipases, dehydrogenases, etc. Thus, this type of extremozyme is very unique due to specific features as compared to other enzymes.

Halophiles

Halophiles are a unique type of microorganism that can survive in environments with high salt concentrations by maintaining balanced osmotic conditions, unlike other microorganisms that cannot survive in such conditions. They are mainly found in environments like the Dead Sea. Halophiles produce extremozymes that can perform activities in high saline environments due to the presence of highly negatively charged amino acids, which prevent the precipitation process. These types of enzymes are used for peptide syntheses, such as protease and dehydrogenase [18, 20]. This type of enzyme requires more compressive work to detailed information but large part information is unclear yet.

Alkaliphiles/Acidophiles

The classes of alkaliphiles and acidophiles differ in their ability to survive in environments with varying pH levels. Alkaliphiles can survive at pH 9 or higher due to specific adaptations, while acidophiles are capable of surviving at pH 3 or lower due to their adaptations [21, 22]. However, the adaption process common in both classes maintains the inner side pH naturally by an active proton pump. Thus, the outside environment does not generate an effect on the microorganism's inner side environment [19, 23]. However, only the proton pump is not responsible for the survival mechanism in Alkaliphiles and acidophiles but other mechanisms are also included like surface charge mechanism, high internal buffer capacity, and proton entry by antiporter membranal process, etc. Thus, alkaliphiles and acidophile extremozyme play the maximum activity on their own optimum pH conditions. Alkaliphiles and acidophiles have great potential use for industrial purposes. This type of enzyme is mainly used in the detergent industry, polymer degradation process, starch hydrolysis, and paper industry [24, 25]. Various examples of Alkaliphiles and acidophiles are available like xylanases, pullulanases, pectinases, chitinases and Amylases, glucoamylases respectively [2].

Acidophilic extremozyme has great industrial attenuation due high degradation capacity as compare the other enzymes. Thus, both type of extremozyme showed own importance in different- different conditions. These types of enzymes showed single limitations like it presents in very unpredictable place like deep niche etc. isolation source is a main challenge with these type of enzyme.

Piezophiles

Piezophiles microorganism very a unique type of microorganism that survives under high pressure approx. few 100 to 400 MPa [2]. This type of microorganism finds in the deep sea so it is very hard to study this type of microorganism. This type of microorganism generally belongs form the genera Shewanella, Methanococcus, Pyrococcus, and Moritella [26, 27]. At the current time, If an extremozyme belongs to piezophile that means it is highly resistant to high-pressure conation not need any new adaptations [28]. This type of enzyme is mainly used in the food industry and sterilization process because proteins can tolerate high pressure. Hence denaturation pressure does not occur. Thus, this type of enzyme is very useful but Piezophiles extremozyme classless explore due to the limitation of source (difficult to isolate from the deep sea). Thus, this class highly requires more comprehensive work to improve better understating of the actual uses.

The above classification is based on the isolation source of extremophiles microorganisms that summarized all types of extremozyme but it is not limited to summarizing various other types of extremozyme or extremophiles available which belong to new class but few special types of extremozyme are very limited information's like few enzymes can tolerate the radiations affect that type called Radiophiles extremozyme and few microorganisms have the power to grow under the extremely dry condition called xerophilic extremophiles and this type of biological product name as xerophilic xtremozyme and few microorganisms also grow under high metal concentration so this type microorganism known as Metallophiles extremophiles. Thus, microorganisms have the quality to grow in unpredictable and unconventional conditions. Hence, the classification process is not limited to this requiring more research to develop more understanding.

Example of Industrial Potential Extremophile Extremozymes

Numerous examples demonstrate the commercial applications of extremozymes, such as in the paper, food, polymer, and detergent industries. Currently, extremophile extremozymes are highly sought after for their unique abilities to survive in extreme conditions, including high and low temperatures, extreme pH levels, and high salinity. However, we will only be discussing a few of the most commonly used extremozymes in this section, including cellulase, protease, amylase, and chitinase.

Cellulase

Cellulose is a widely recognized polymer that is synthesized from thousands of D-glucose monomer units. It is predominantly used in the paper industry for products such as paperboard, paper cards, and fibers. In addition, certain bacterial strains are known to produce cellulose as a protective outer covering in response to external stresses, such as antibiotic attacks, as seen in *Salmonella* [29]. Thus, cellulose degradation is very necessary to control infections and other paper industries also require the degradation of cellulose at high temperatures. In the formation industry, cellulose degradation is also required to produce biofuel and generate energy. But normal cellulase enzyme has the capability to degrade the cellulase but it does not work at high temperatures and high pH conditions. So, we require the extremophile extremozymes cellulase which degrades the cellulose polymer in extreme conditions. A study indicates that the researcher isolated the thermostable cellulase from the bacillus species ($DUSELR_{13}$strain) and the Geobacillus species ($WSUCF_1$ strain). Both strain cellulase optimum temperature 70^0 to 75^0C and cellulase degrading up to 78 percent and 89 percent [30]. This type of cellulose product is more cost-effective as compared to normal cellulase production and is highly used in the fermentation industry for biofuel production due to its thermostability features. Another study also suggests that *Myceliophthora thermophila* (ATCC 42464) satin also produces thermostable cellulose [31]. Another study production of Alkaliphiles cellulase from the *Paenibacillus tarimensis bacterial strain which has the capability to survive alkali conditions.* Thus, thermostable and Alkaliphilic stable cellulase solves the higher temperature and pH problems. Hence, extremophile extremozymes is better than normal enzyme and that type enhances industrial uses.

Protease

Protein is an essential nutritional part of animals, humans, and plants. Protein is composed of small thousands of amino acids. Sometimes protein requires breakdown for a valuable purpose like the nutritional part (food industry). So, mesophilic protease is a good option to break down proteins but it works only under conventional conditions. This is not performed activity in unconventional conditions like high temperature and pH etc. Some required the decomposition of protein part due to industrial purposes so required the more stable protease as compared to the mesophilic protease. In this way, A study suggests the identification of a novel protease from the *Anoxybacillus kamchatkensis* M1V which showed the optimum activity at high temperatures approx. 70^0C and pH 11 Which solved the detergent industry problem due to thermostability and Alkaliphilic stability [32]. Another study also indicates that the researcher isolated the extremophilic extremozymes protease from *Bacillus licheniformis* species which showed the optimum activity at a high salt concentration which means it halophiles extremozyme protease [33]. Another recent study showed that isolated the protease from the Antarctic freshwater lakes microbes *Pseudomonas prosekii* strain A which have cold adaptions. This protease showed activity at an optimum temperature at 4^0 C. Hence this type of protease consider a psychrophile extremozyme protease [34]. Thus, Extremophilic extremozymes protease is better than mesophilic protease. For this reason, the food industry and detergent industries mostly focus on extremophilic extremozymes protease due to high tolerance working capacity temperature levels and cost-effectiveness as compared to mesophilic protease.

Amylase

Amylase enzyme well-known enzyme for breakdown starch into glucose by using the hydrolysis process. This enzyme is mainly used in the brewing process for the production of sugar. The detergent industry or bakery (food industry) showed high interest in amylase enzymes. Before a few decades, amylase enzyme was produced by a mesophilic microorganism which has a risk of contamination and not showed survival in harsh conditions like high and cold temperatures and the cost of production was also high. Thus, the Industry's desire was focusing to new amylase enzymes which have a low risk of contamination and tolerate harsh conditions. Hence extremophilic microorganism provides an effective solution in the direction of industrial requirement with amylase enzymes. The various study suggests that the numerous name of microorganism which produces the amylase enzyme with specific power against harsh conditions Yassin et al. (2021) suggest that they isolated a few isolates (M_{13}, M_8, M_2) from the Bacillus genus which capable to produce the thermostable amylase. This type of thermostable amylase showed optimum activity around 70-75^{0}C. Hence, this type of amylase considers an extremozyme [35]. Another study suggests that the researcher identified the novel amylase extremozyme from *Pseudoalteromonas* species which showed high tolerance power against high salt concentration. Hence this type of amylase enzyme belongs to the halophiles extremozyme which is highly used in the detergent industry [36]. Another study also indicates that a new type of amylase enzyme isolated from *Luteimonas abyssi* XH_{031}(T) showed survival capacity at high alkali conditions. This type of amylase showed the optimum activity of pH 9 and showed the highest activity between pH 9 to 11 which was very helpful for the detergent industry [37]. Thus, extremozyme amylase is shown as the most favourable choice for the industry as compared to mesophilic amylase because it overcomes the contamination risk and economic cost decreases, and survival under a high pH range, etc.

Chitinase

Chitin is a well-known polymeric substance which mainly finds in the insect skeleton and fungal cells as covering. Thus, chitin polymer is highly distributed on earth. It is also useful for agricultural lands as fertilizer and production of eye lenses etc. Thus, the chitinase enzyme used to break down the chitin polymer makes good biofertilizers. But mesophilic chitinase showed less degradation of chitin due to temperature limitation. Therefore, extremophilic extremozyme chitinase overcomes this problem and is responsible for high biofertilizer production. A study indicates the identification of thermostable chitinase from *Humicola grisea* strain from Rajasthan desert soil [38]. Another study also reported the novel thermostable chitinase extremozyme from the marine bacterium *Paenibacillus barengoltzii* which showed optimum activity at the 55^{0}C and pH 5.5 which is used in caricatural industry and decreases the temperature and pH problem [39]. Thus, extremophilic extremozyme chitinase created a new milestone in the agricultural industry.

DNA Polymerase

DNA polymerase enzymes are very useful enzymes for the living body like prokaryotes and eukaryotes. It is responsible for DNA replications in the body but it works only at normal temperatures. But currently days this enzyme is highly useful in disease diagnostic. The molecular diagnostic process is completely based on DNA polymerase enzyme and polymer chain reaction machine. But normal DNA polymerase showed temperature limitations because it showed the denaturation between enzymes. Hence, Molecular diagnostic required the thermostable polymer because the polymer chain reaction machine temperature raised up to 100^0C. Thus, a species reported for the production of thermostable polymerase known as *Thermus aquaticus* and DNA polymer name Taq polymerase enzymes [40, 41]. Thermostable DNA polymerase (Taq polymerase) created a new milestone in the disease diagnostic world. At the current people are suffering from various types of infection but polymer chain reaction machines and Thermostable DNA polymerase provide a new hope to detect infections with high accuracy. Thus, extremophile extremozyme is very helpful for the human race and provides more beneficial products and these types of enzymes are responsible for revolutions in the enzyme industry.

Conclusion

Enzymes are unique biological products of nature that facilitate valuable biochemical reactions. Various microorganisms are responsible for enzyme production, and they can be divided into two classes: mesophilic and extremophilic. Mesophilic microorganisms produce enzymes that work under normal temperature and pressure conditions, but these enzymes are not suitable for many industrial applications. To fulfill industry requirements, researchers have discovered a new class of microorganisms that can survive under unconventional conditions and produce enzymes capable of working under harsh conditions, such as high or low temperatures, high pressure, and high or low pH (Table 1). These enzymes are called extremophile extremozymes, and they have diverse applications in various industries, including paper, detergent, textile, food, and medical fields. However, the discovery of extremophile extremozymes is still in its early stages, and further research is needed to better understand their potential uses.

References

[1] Robinson PK. Enzymes: principles and biotechnological applications. *Essays Biochem*. 2015;59:1–41.

[2] Van den Burg B. Extremophiles as a source for novel enzymes. *Curr Opin Microbiol.* 2003;6(3):213–8.

[3] Kohli I, Joshi NC, Mohapatra S, Varma A. Extremophile – An Adaptive Strategy for Extreme Conditions and Applications. *Curr Genomics*. 2020;21(2):96–110.

[4] Gromiha MM, Pathak MC, Saraboji K, Ortlund EA, Gaucher EA. Hydrophobic environment is a key factor for the stability of thermophilic proteins. *Proteins Struct Funct Bioinforma.* 2013;81(4):715–21.

[5] Pace CN, Fu H, Fryar KL, Landua J, Trevino SR, Bret A, Hendricks, M. M., Iimura, S., Gajiwala, K., Scholtz, J. M., Grimsley, G. R., c2112-AD V.77.4 - 2007-0708 - 4dsocial - interactive design environments - *Lucy Bullivant*. 2012;408(3):514–28.

[6] Meruelo AD, Han SK, Kim S, Bowie JU. Structural differences between thermophilic and mesophilic membrane proteins. *Protein Sci.* 2012;21(11):1746–53.

[7] Robb FT, Clark DS. Adaptation of proteins from hyperthermophiles to high pressure and high temperature. *J Mol Microbiol Biotechnol*. 1999;1(1):101–5.

[8] Farias ST, Bonato MCM. Preferred amino acids and thermostability. *Genet Mol* Res. 2003;2(4):383–93.

[9] Goldstein RA. Amino-acid interactions in psychrophiles, mesophiles, thermophiles, and hyperthermophiles: Insights from the quasi-chemical approximation. *Protein Sci*. 2007;16(9):1887–95.

[10] Berger F, Morellet N, Menu F, Potier P. Cold shock and cold acclimation proteins in the psychrotrophic bacterium Arthrobacter globiformis SI55. *J Bacteriol.* 1996;178(11):2999–3007.

[11] Tõlgo M, Hüttner S, Rugbjerg P, Thuy NT, Thanh VN, Larsbrink J, Olsson, L., Genomic and transcriptomic analysis of the thermophilic lignocellulose-degrading fungus Thielavia terrestris LPH172. *Biotechnol Biofuels* [Internet]. 2021;14(1):1–16. Available from: https://doi.org/10.1186/s13068-021-01975-1.

[12] Ma, LJ, Ibrahim, AS, Skory, C, Grabherr, MG, Burger, G, Butler, M, Elias, M, Idnurm, A, Lang, BF, Sone, T, Abe, A, Calvo, SE, Corrochano, LM, Engels, R, Fu, J, Hansberg, W, Kim, JM, Kodira, CD, Koehrsen, MJ, … Wickes, BL. Genomic analysis of the basal lineage fungus Rhizopus oryzae reveals a whole-genome duplication. *PLoS Genet*. 2009 Jul;5(7):e1000549.

[13] Weerkamp A, Heinen W. Effect of temperature on the fatty acid composition of the extreme thermophiles, Bacillus caldolyticus and Bacillus caldotenax. *J Bacteriol.* 1972 Jan;109(1):443–6.

[14] Sprott GD, Meloche M, Richards JC. Proportions of diether, macrocyclic diether, and tetraether lipids in Methanococcus jannaschii grown at different temperatures. *J Bacteriol*. 1991;173:3907–10.

[15] Vieille C, Zeikus GJ. Hyperthermophilic Enzymes: Sources, Uses, and Molecular Mechanisms for Thermostability. *Microbiol Mol Biol Rev*. 2001;65(1):1–43.

[16] Davies GJ, Gamblin SJ, Littlechild JA, Watson HC. The structure of a thermally stable 3-phosphoglycerate kinase and a comparison with its mesophilic equivalent. *Proteins*. 1993 Mar;15(3):283–9.

[17] Takimura Y, Saito K, Okuda M, Kageyama Y, Saeki K, Ozaki K, Ito, S., Kobayashi, T., Alkaliphilic Bacillus sp. strain KSM-LD1 contains a record number of subtilisin-like serine proteases genes. *Appl Microbiol Biotechnol*. 2007 Aug;76(2):395–405.

[18] Sakamoto T, Ihara H, Kozaki S, Kawasaki H. A cold-adapted endo-arabinanase from Penicillium chrysogenum. *Biochim Biophys Acta*. 2003 Dec;1624(1–3):70–5.

[19] Gupta GN, Srivastava S, Khare SK, Prakash V. Extremophiles: An Overview of Microorganism from Extreme Environment. *Int J Agric Environ Biotechnol*. 2014;7(2):371.

[20] Piera-Velázquez S, Marhuenda-egea FC, Cadenas EC. The dependence of a halophilic malate dehydrogenase on \$ω\$o and surfactant concentration in reverse micelles. *J Mol Catal B-enzymatic*. 2001;13:49–55.

[21] Rothschild LJ, Mancinelli RL. Life in extreme environments. *Nature*. 2001 Feb;409(6823):1092–101.

[22] Johnson DB. Physiology and Ecology of Acidophilic Microorganisms. In: *Physiology and Biochemistry of Extremophiles* [Internet]. John Wiley & Sons, Ltd; 2007. p. 255–70. Available from: https://onlinelibrary.wiley.com/doi/abs/10.1128/9781555815813.ch20.

[23] Horikoshi K. Alkaliphiles: some applications of their products for biotechnology. *Microbiol Mol Biol Rev*. 1999 Dec;63(4):735–50, table of contents.

[24] Sarethy IP, Saxena Y, Kapoor A, Sharma M, Sharma SK, Gupta V, Gupta, S., Alkaliphilic bacteria: Applications in industrial biotechnology. *J Ind Microbiol Biotechnol*. 2011;38(7):769–90.

[25] Sharma A, Parashar D, Satyanarayana T. Acidophilic Microbes: Biology and Applications. *Gd Challenges Biol Biotechnol*. 2016;1(December 2017):215–41.

[26] Kato C, Nogi Y. Correlation between phylogenetic structure and function: examples from deep-sea Shewanella. *FEMS Microbiol Ecol*. 2001 May;35(3):223–30.

[27] Kato C, Li L, Nogi Y, Nakamura Y, Tamaoka J, Horikoshi K. Extremely Barophilic Bacteria Isolated from the Mariana Trench, Challenger Deep, at a Depth of 11,000 Meters. *Appl Environ Microbiol*. 1998;64:1510–3.

[28] Gross M, Jaenicke R. Proteins under pressure. The influence of high hydrostatic pressure on structure, function and assembly of proteins and protein complexes. *Eur J Biochem*. 1994 Apr;221(2):617–30.

[29] Jonas K, Tomenius H, Kader A, Normark S, Römling U, Belova LM, Melefors, O., Roles of curli, cellulose and BapA in Salmonella biofilm morphology studied by atomic force microscopy. *BMC Microbiol*. 2007 Jul;7:70.

[30] Rastogi G, Bhalla A, Adhikari A, Bischoff KM, Hughes SR, Christopher LP, Sani, R. K., Characterization of thermostable cellulases produced by Bacillus and Geobacillus strains. *Bioresour Technol* [Internet]. 2010;101(22):8798–806. Available from: https://www.sciencedirect.com/science/article/pii/S0960852410009594.

[31] Wang J, Gong Y, Zhao S, Liu G. A new regulator of cellulase and xylanase in the thermophilic fungus Myceliophthora thermophila strain ATCC 42464. 3 *Biotech*

[Internet]. 2018;8(3):1–6. Available from: https://doi.org/10.1007/s13205-017-1069-y.

[32] Mechri S, Bouacem K, Zaraî Jaouadi N, Rekik H, Ben Elhoul M, Omrane Benmrad M, Hacene, H., Bejar, S., Bouanane-Darenfed, A., Jaouadi, B., Identification of a novel protease from the thermophilic Anoxybacillus kamchatkensis M1V and its application as laundry detergent additive. *Extremophiles* [Internet]. 2019;23(6):687–706. Available from: https://doi.org/10.1007/s00792-019-01123-6.

[33] Suganthi C, Mageswari A, Karthikeyan S, Anbalagan M, Sivakumar A, Gothandam KM. Screening and optimization of protease production from a halotolerant Bacillus licheniformis isolated from saltern sediments. *J Genet Eng Biotechnol* [Internet]. 2013;11(1):47–52. Available from: http://dx.doi.org/10.1016/j.jgeb.2013.02.002.

[34] Matsui M, Kawamata A, Kosugi M, Imura S, Kurosawa N. Diversity of proteolytic microbes isolated from Antarctic freshwater lakes and characteristics of their cold-active proteases. *Polar Sci* [Internet]. 2017;13:82–90. Available from: http://dx.doi.org/10.1016/j.polar.2017.02.002.

[35] Nureddin Yassin S, Milkessa Jiru T, *Indracanti M. Screening and Characterization of Thermostable Amylase-Producing Bacteria Isolated from Soil Samples of Afdera, Afar Region, and Molecular Detection of Amylase-Coding Gene*. 2021; Available from: https://doi.org/10.1155/2021/5592885.

[36] Wang X, Kan G, Ren X, Yu G, Shi C, Xie Q, Wen, H., Betenbaugh, M., Molecular cloning and characterization of a novel -Amylase from antarctic sea ice bacterium pseudoalteromonas sp. M175 and its primary application in detergent. *Biomed Res* Int. 2018;2018.

[37] Song Q, Wang Y, Yin C, Zhang X-H. LaaA, a novel high-active alkalophilic alpha-amylase from deep-sea bacterium Luteimonas abyssi XH031(T). *Enzyme Microb Technol*. 2016 Aug;90:83–92.

[38] Kumar M, Brar A, Vivekanand V, Pareek N. Production of chitinase from thermophilic Humicola grisea and its application in production of bioactive chitooligosaccharides. *Int J Biol Macromol* [Internet]. 2017;104:1641–7. Available from: https://www.sciencedirect.com/science/article/pii/S0141813016327295.

[39] Yang S, Fu X, Yan Q, Guo Y, Liu Z, Jiang Z. Cloning, expression, purification and application of a novel chitinase from a thermophilic marine bacterium Paenibacillus barengoltzii. *Food Chem* [Internet]. 2016;192:1041–8. Available from: https://www.sciencedirect.com/science/article/pii/S0308814615011255.

[40] Tindall KR, Kunkel TA. Fidelity of DNA synthesis by the Thermus aquaticus DNA polymerase. *Biochemistry*. 1988 Aug;27(16):6008–13.

[41] Coker JA. *Extremophiles and biotechnology: Current uses and prospects*. F1000Research. 2016;5:1–7.

Chapter 4

Sulphur Pollution and Remediation

Diwakar Aggarwal[1]
and Chadetrik Rout[2,*]

[1]Department of Biotechnology, M. M. Engineering College,
Maharishi Markandeshwar (Deemed to be University), Mullana-Ambala, Haryana, India
[2]Department of Civil Engineering, Maharishi Markandeshwar (Deemed to be University), Mullana- Ambala, Haryana, India

Abstract

Sulphur is one of the most interesting as well as challenging elements to work with chemically; it is plentiful geochemically and is crucial biochemically. Sulphur and its compounds have been the subject of more than 100 years of scientific inquiry, resulting in a large volume of literature that is difficult to search for credible information. However, there are several credible studies in the literature that critically and comprehensively assess the chemical literature. Increased SO_2 concentrations in the atmosphere have been linked to fossil fuel combustion, moist deposition, and acid rain in the previous century. Despite recent reductions in SO_2 emissions, urban and industrial areas continue to be vulnerable to excessive sulphur contamination, necessitating further research into the deleterious effects of excess sulphur on environmental parameters and components. With increased human activities, more sulphur is introduced into the environmental system and put the ecosystem and human health in jeopardy. The sources, effects and remedial techniques of sulphur pollution are reviewed and discussed in this chapter. Moreover, a typical study on sewer deterioration has also been discussed in this context.

---

[*] Corresponding Author's Email: chadetrikrout@gmail.com.

In: Microbial Technologies and Their Applications
Editors: Santosh Kumar Karn and Anne Bhambri
ISBN: 979-8-88697-913-8

Keywords: bioleaching, inundation, AMD, surfactants, phase separation

Introduction

The intricacy of sulphur chemistry arises from the several oxidation states that sulphur may take, as well as sulphur's proclivity to catenate, generating chains and rings of astounding diversity. Over 200 years of scientific research on sulphur and its compounds has resulted in a massive body of literature that is difficult to search for credible information. Furthermore, this literature is riddled with inaccuracies and inconsistencies since older researchers, without the methodologies that are now common, frequently made statements that were not always validated afterwards. However, there are several credible studies in the literature that critically and comprehensively assess the chemical literature of S.

Sulphur is the 10th most abundant element on Earth [1] and is dynamically exchanged between the geosphere and biosphere. It exists in a variety of valence states ranging from +6 in sulphate to –2 in sulphides. The reduced states (-2 and –1) are found in metal and metalloid sulphides (Ag, Fe, Cd, Hg, Mn, Ca, Te, Se, As, Sn, Cu, Pb, Pt, Sb, Co, Ni, Mo, Rn, W) is the most prevalent of which being pyrite (FeS_2) [2, 3]. The intermediate valencies (0, +2, +4) occur only briefly in the geochemical cycle and are typically reaction intermediates (e.g., elemental sulphur in sulphide oxidation). The bulk of sulphur in nature is sulphate or sulphide in water and soil, and sulphur dioxide in the atmosphere [4], with the oceans serving as a significant reservoir. Sulphur is abundant in this area, both as dissolved sulphate and as sedimentary minerals such as gypsum [5].

Sulphur dioxide (SO_2) is the most prevalent cause of sulphur pollution. Not only is SO_2 harmful to humans, but SO_2 and its by-products (sulphate, sulfuric acid) have a regional and worldwide impact on the environment and agriculture (e.g., acid deposition, direct and indirect radiation forcing) [6]. SO_2 occurs naturally, mostly as a result of volcanic activity, biological and photochemical synthesis. To a lesser extent, it is also emitted by forest fires, sulphur springs, and sea salt [6, 7]. Although natural SO_2 is thought to account for around one third of overall SO_2 emissions, the majority of sulphur emissions are manmade. Anthropogenic sources are estimated to account for more than 70% of worldwide SO_2 emissions, half of which occur from the burning of fossil fuels, notably coal, which generally contain 1.0-3.0% sulphur in the form of pyrites and marcasites. Furthermore, sulphur pollution occurs during the refinement of petroleum, the smelting of sulfidic ores in the manufacture of heavy metals, and the manufacturing of sulfuric acid, paper, and sulphur [6, 8-10].

Sulphur is an essential nutrient for life and is abundant in all organisms, where it can be found in amino acids like methionine and cysteine, (poly-)peptides, enzyme cofactors, redox-active co-substrates like glutathione, as well as redox-active enzyme cofactors (iron sulphur clusters), sulfolipids, and carbohydrates. It is a macroelement essential for plant growth and development, with optimal concentrations in plant leaf tissues ranging from 0.1-0.5% of a plant's dry weight [11, 12]. Plants absorb sulphur mostly in the form of SO^{2-} through their aerial (or aboveground) parts and SO_4^{2-}through their roots [12]. Plants use sulphur to synthesise the amino acids cysteine, cystine, and methionine, as well as metabolites (such as biotin, thiamine, and glutathione). Sulphur shortage (now more common than excess, particularly in agricultural regions) causes a reduction in root development and leaf area, as well as declines in chlorophyll and protein concentration (chlorosis) [12].

Sulphurous and acidic soils are most common in industrialised areas, particularly in the United States, Canada, and China [13]. Areas with excessively high sulphur concentrations are frequently found around sulphur mines, sour gas processing plants, and abandoned coal mining sites [13-17]. Furthermore, coal mining regions associated with pyrites and marcasites, as well as iron and gold mines, are frequently associated with Fe, Al, or $Fe_2 (SO_4)^3$ pollution, as well as sulphur contamination [14, 18, 19]. Yet, locations with high sulphur concentrations are rather unusual; the opposite tendency in agriculture has been noted; however, a lack of control over the industry responsible for introducing sulphur into the soil may result in substantial environmental alterations. Each ecosystem should be thoroughly evaluated in order to properly repair the ecosystem (soil, water, vegetation, and fauna, as well as their components).

Sulphur Related Problems

SO_2, which is responsible for air pollution, might be one of the major causes contributing to forest decrease in the northern hemisphere [20]. Acid deposition causes a high concentration of SO_2 in the atmosphere. The US Environmental Protection Agency detected severely acidic precipitation with a pH of less than 4.3 in 2000. Acid deposition harms forests by acidifying soil (a one-point pH reduction was seen in Germany, Sweden, and China) and disrupting tree mineral management [16]. In 2000, China (one of the world's major SOx emitters) calculated that acid deposition and its consequences cost the country 1.77% of its GDP [21, 22]. Although sulphur deposition can be adequate for optimal plant yields in industrial and urbanised regions [12], SO_2 emissions have reduced in the last decade as a result of tougher controls. The EU 2017 adopted SO_2 emissions limitations for some liquid fuels (announced in their regulations). As a result, the energy sector has shifted from high sulphur solid and liquid fuels to low sulphur fuels such as natural gas and has created industrial facilities to deploy flue gas desulfurization abatement technology.

Furthermore, elevated sulphur concentrations have an impact on the root system (perhaps by altering carbon allocation) and diminish foliage. Excessive amounts also disrupt leaf stomata and promote early senescence, leading to crown thinning, distorted canopy morphologies, restricted development, and the nutritional deficiency [6, 8, 20, 23, 24]. Dust from industrial pollution accumulates on plants, interfering with photosynthesis and transpiration processes [20, 24, 25]. Excess sulphur deposits increase the amount of nutrients that seep into the soil. The displacement of the base cations Ca^{2+} and Mg^{2+} from the sorption complex, for example, results in soil acidification. Trace elements are toxic to soil organisms in acidic soil, disrupting biogeochemical cycles and vascular plant feeding [26]. Soil acidification can also result in the emission of toxic Al, which can harm fine root development and hinder Ca, Mg, and K absorption [20]. Sulphuric pollution can deplete nutrients and amplify the impacts of natural stresses such as physiological drought, frost, pests, and fungal diseases [8].

Although sewer degradation has been documented in the past, evidence of sulphide-related concerns was discovered only recently. The first evidence of concrete corrosion induced by sulphide was probably documented in 1895 for an outfall sewer in Los Angeles, USA [27]. Sewer gas issues have been recognised for ages. However, it was only around 100 years ago that it was recognised, based on strong scientific and technological understanding, that corrosion and odour issues in wastewater collecting systems are intimately tied to anaerobic conditions and the formation of hydrogen sulphide.

The process of formation of hydrogen sulphide:

$$SO_4^{2-} + \text{Organic Carbon} \rightarrow HCO_3^{-} + H_2S$$

In summary, sulphide is formed by the activities of anaerobic sulphate reducing bacteria, which are mostly found in sewer biofilms where sulphide enters the wastewater phase. might have catastrophic consequences.

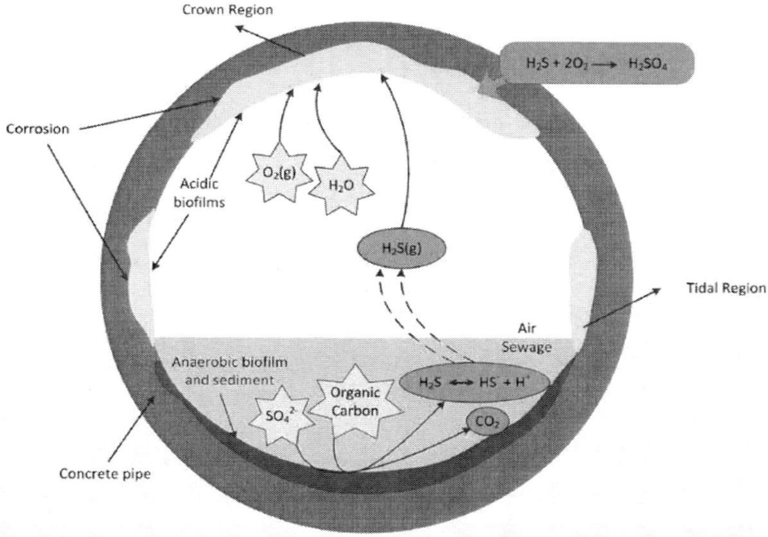

Figure 1. A cross section of a sewage concrete pipe highlights the primary mechanisms that contribute to acid production in aerobic biofilms and the commencement of sewer corrosion [28].

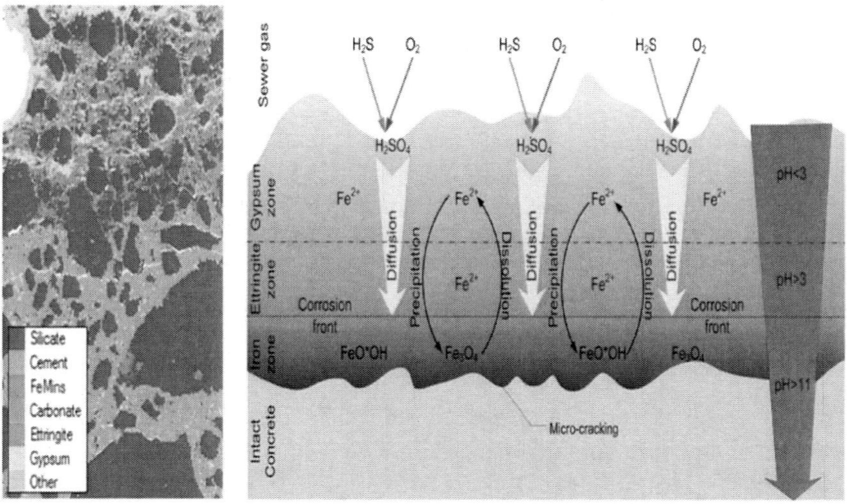

Figure 2. The sewage concrete corrosion layer conceptual model [29].

Figure 3. Corroded sewer pipe with stones and gravel from the concrete aggregate where the cement in the surface layer has decayed.

The following emission of hydrogen sulphide into the sewer atmosphere causes corrosion of concrete and metal structures in the wastewater collecting system, and ventilation causes smell nuisance in the urban environment. Although discovering this cause-effect link is relatively easy, the complicated sewage networks need to consider and investigate the whole sulphur and carbon cycles in order to predict and efficiently reduce the negative consequences of hydrogen sulphide. Furthermore, in anaerobic conditions, sulphate is the primary external electron acceptor for bacteria in sewage biofilms, leading to the production of hydrogen sulphide. Subsequently release of H_2S into the wastewater, sewer gas phase, and urban environment.

Factors Affecting Sulphide Related Problems in Sewers

The presence of sulphate is the most critical element affecting sulphate reduction and hence sulphide generation [30]. Other factors like temperature, dissolved oxygen, pH [31-37], area-to-volume ratio of sewer pipes [38], quality and quantity of biodegradable organic matter [30], anaerobic residence time in the sewer network and flow velocity [39, 40] are reported to effect sulphide generation. The aforementioned factors are critical for predicting future consequences and providing associated mitigating information. These variables determine the scope of sulphur processes and offer information on the significance of external parameters such as wastewater redox potential, DO, temperature, flow, and biodegradable organic matter concentration. It is critical to "assemble" such parameters in a way that they may be used to foresee and mitigate the sulphide problem.

Methods to Control of Sulphide Problems in Sewers

The evaluation of sulphide control strategies has received substantial attention in scholarly publications, manuals, and textbooks, both in general and in particular terms [28, 39-49]. Three strategies for control of sulphide problems have been summarised as suppression or inhibition of sulphide generation, decrease of sulphide content in the water phase, and reduction or dilution of sewer gases.

Strategy 1. Suppression or Inhibition of Sulphide Formation

Table 1. Control strategies for sulphide formation suppression or inhibition

Method	Action	Reference
Suppression of the sulphate reducing bacterial activity	• pH increase • Removal of biofilm • Use of biocides	[28, 50]
Prevention of the anaerobic activity by changing the electron acceptor	• Aerobic conditions by supplying of oxygen (air or pure oxygen) • Anoxic conditions by nitrate dosing	[44-46, 51, 52]

Strategy 2. Reduction of the Sulphide Concentration in the Water Phase

Table 2. Control methods for reduction of the sulfide concentration in the water phase

Method	Action	Reference
Reduction	• Supplying of oxygen • Dosing of nitrate	[53]
Chemical oxidation	• Addition of chemical oxidants, e.g., Cl compounds, H_2O_2, O_3 and permanganate	[43]
Precipitation	• Addition of ferrous iron and ferric iron salts	[28, 40, 41, 48]

Strategy 3. Reduction or Dilution of Sewer Gases

Table 3. Control methods for reduction of foul sewer gases

Method	Action	Reference
Treatment of sewer gas	• Dilution, e.g., via a chimney • Chemical or biological treatment	[49, 53-56]
Rising of pH	• Addition of alkaline substances	[28, 57]

In addition to using control strategies, adequate sewer network design, construction, and maintenance are critical for decreasing the impact of sulphide production and other unpleasant volatile compounds. Several handbooks developed by organisations, municipalities, and consultants give useful assistance in this regard [43, 48, 58].

Treatment of Inorganic Sulphur Compounds

Inorganic Sulphur Transformations

Different types of solid phase systems, such as ores, soils, and sediments, have large quantities of inorganic sulphurous chemicals in the environment. The important criterion in determining the environmental risk of inorganic sulphurous compounds found in these materials is whether they remain attached to the solid phase or are mobilised into the solution. The mobility of inorganic sulphur in natural water/solid biphasic systems is primarily governed by oxidative or reductive changes and sulphurous compound precipitation constants.

Reductive Processes

Microbial sulphate reduction is the most prevalent reductive mechanism impacting sulphur in the environment. Sulphate-reducing bacteria require an anaerobic environment with a pH of 5-6 or above. Even in anoxic systems with very acidic leachates (pH < 3), microbial sulphate reduction was seen. This is owing to the sulphate reduction's high alkalizing capability, which leads to the establishment of micro areas within the soil or sediment profile with conditions favourable for sulphate reducing bacteria. As seen in the following sulphate reduction process [59], the primary products are sulphide and OH^-.

$$8H_2 + 2SO_4^{2-} \rightarrow 6\ H_2S + HS^- + 5H_2O + 3OH^-$$

Oxidative Processes

In nature, oxidation of reduced inorganic sulphur compounds occurs through both spontaneous chemical and microbiological oxidation processes [60]. At neutral to alkaline pH, spontaneous chemical reactions dominate oxidation. The presence of oxygen is required for the oxidation of inorganic reduced sulphur compounds. However, oxidation can take place even in anoxic conditions. The significant oxidising ability of ferric (Fe^{3+}) iron can promote spontaneous chemical oxidation:

$$MeS + 2Fe^{3+} \rightarrow 6\ Me^{2+} + 2Fe^{2+} + S^0$$

Treatment Strategies

Prevention of Sulphur Pollution

The fundamental method to mitigate the risk associated with sulphur buildup in the environment is to prevent sulphur pollution [60, 61]. This includes enhanced flue-gas treatment efficiency, desulfurization of fossil fuels, or the use of clean technology in ore processing and smelting [62]. Systematic mining management that considers the whole mine life-cycle can significantly minimise AMD production [63, 64]. Similarly, proper and sensitive river basin management and regulations can result in considerable reductions in pollution, which leads to fewer hazards connected with riverine sediment handling [65].

Elimination of Sulphurous Compounds from Solid Materials

The removal of inorganic sulphurous chemicals from solid materials is often an expensive technique. In practice, it is used mostly to remove hazardous metals that are present with sulphur. Any prospective strategy for removing sulphur from solid materials must fulfil the special criteria imposed by the presence of sulphur in the solid form. The removal of solid impurities from a solid substance is mostly accomplished via the use of phase or molecular separation methods [66, 67].

Phase Separation

For the enrichment of sulphur-containing non-ferrous ores, several gravity-separation and flotation processes are widely used [67-69]. Metal-polluted sediments are commonly treated via phase separation. To the best of our knowledge, no example of phase separation for the removal of solid sulphurous chemicals from trash, soil, or sediment has been documented.

Molecular Separation

The extraction of molecules containing the sulphur atom referred known as molecular separation entails the solubilization of contaminants followed by their removal from the process liquid [70]. Metals are removed using a variety of extractants, such as mineral (HCl, H_2SO_4) or organic (lactate, formate) acids, complexing agents (EDTA, DTPA, NTA), or pure water. Reduced inorganic sulphurous compounds must first be dissolved because they are generally present as solid particles. As previously stated, inorganic sulphurous compounds are soluble under oxidative circumstances. As a result, sulphur oxidation is a fundamental concept used in the molecular separation of sulphurous compounds and associated hazardous metals.

Bioleaching as a Special Case of Molecular Separation

When microbial activity (growth of *thiobacilli*) dominates the oxidative process, it is referred to as "bioleaching" [71, 72]. Bioleaching is used in practice, particularly in mining operations, for example, biohydro-metallurgical mining, low-grade ore enrichment, and so on [72]. The method has also been used to treat sewage sludge, soils, and sediments [73-77].

The goal of bioleaching is to reach pH levels low enough to solubilize as many hazardous metals as possible. In other circumstances, extra elemental sulphur or ferrous iron are introduced into the system as bacterial substrates. Bioleaching has been tried in suspension reactors such as Pachuca tanks, rolling reactors, and containers agitated by propellers [74]. Treatment on heaps at the site of permanent or temporary disposal is a less-intensive option for removing sulphur and related contaminants from large amounts of garbage. Enhancing heap leaching operations, for example, by forced aeration or the inclusion of hydrophilic sulphurous substrates, may be a viable alternative to the more expensive slurry leaching techniques [76, 77].

Suppression of the Sulphur Oxidation

Another strategy to deal with sulphur and related pollutants is to keep them out of the environment. The below discussion highlights the primary solutions for preventing sulphur oxidation that is now in use or has been presented as viable methods.

Inundation

Inundation may be a low-cost and simple solution for many systems. When topped with water, oxygen cannot reach the sulphur compounds and oxidation cannot begin, as observed for sulphur rich hard coal tailings by [78]. However, this approach cannot always ensure complete leaching suppression since bacterial leaching might occur even in the absence of oxygen. Furthermore, due to ferric iron's oxidative capability, spontaneous chemical solubilization can proceed [71, 72, 79].

Hydrogeological Isolation

Another method of preventing leaching is to isolate the place hydro-geologically. Isolation membranes or a clay-soil combination can be used to inhibit vertical or horizontal water flows [80]. Proper surface treatment and restoration can lead to a reduction in contaminated discharges from mine tailings or waste banks [69]. Covering the deposit with soil and planting fast-growing trees is a widely used approach for rehabilitating enormous regions ravaged by mining and covering massive heaps of waste all over the world [62].

Use of Surfactants

The addition of surface-active chemicals (surfactants) can inhibit bacterial cells from adhering to sulphur containing minerals. [81] described pyrite treatment utilising different fatty-acid amines, which resulted in a significant decrease in the bioleaching process. [82] demonstrated the potential use of oxalic, tannic, and fulvic acids to inhibit bacterial adhesion, with fulvic chemicals yielding the greatest results. [83] Investigated pyrite cell adhesion in the presence of potassium phosphate. Their research revealed a combination impact, as potassium phosphate reduced bacterial adherence to pyrite while also keeping pH levels unfavourable for bacterial growth and activity.

Liming

Another technique for suppressing leaching is to add lime to a system containing inorganic sulphurous chemicals. It results in the acidic environment being neutralised. The pH of neutral is unfavourable for bioleaching [84]. Furthermore, neutralisation of acidic circumstances prevents hazardous cationic metal leaching. However, because the lime is constantly leached out of the system, it is typically only a temporary solution, and it is also very costly. According to [85], liming may be more costly than ferric phosphate coating.

Treatment of Acid Sulphate Soils

In practise, four fundamental tactics are used: limiting pyrite oxidation, leaching of reaction products, liming, and self-neutralization under reductive circumstances.

Curtailing the Pyrite Oxidation

Pyrite oxidation can be inhibited by reducing the availability of oxygen or altering the rate of one or more intermediary stages in oxidation, for as by decreasing the concentration of Fe^{3+}. However, in practice, the only possible method of preventing oxygen entry into the system is its persistent flooding [86].

Leaching of Reaction Products

Another treatment technique is to eliminate soil soluble or exchangeable acidity. Fresh water leaching is effective for removing free sulfuric acid as well as the Fe and Al ions that have been solubilized by acidity. Leaching with salt or brackish water causes the exchangeable Al to be replaced by Na, Ca, and Mg [87]. When decreasing circumstances prevail in the soil profile, however, leaching is inefficient. As a result, effective leaching requires highly aerated soil, particularly soil that has been subjected to an extended drying phase prior to leaching [87].

Liming of Acid Sulphate Soils

To balance the acidity of acid sulphate soil, lime or limestone can be added. A dosage of around 30 tonnes of $CaCO_3$ per 10000 m^2 and per 10 cm of soil is typically required [86]. Only after the water soluble acidity has been leached away, than liming is typically effective. Otherwise, the massive volumes of lime required would make the process uneconomical. Excessive liming would also result in significant salinization of the location, which would be detrimental to plants and fauna.

Self-Neutralizing in Reductive Conditions

Initiating microbial sulphate reduction in acid sulphate soils can efficiently reduce acidity while also stabilising sulphate content. This makes rice growing on acid sulphate soils, for example, a promising ameliorative approach since it combines the impact of flooding with activation of sulphate reductive mechanisms. Although the harmful effects of acidity and cationic metals (particularly aluminium) are inhibited under anoxic conditions, they are replaced by the deleterious effects of reduced Fe^{2+}, which is water soluble. However, [88] observed that sulphate reduction can prevent Fe^{2+} toxicity. The sulphide formed interacts with free Fe^{2+} to make pyrite.

Treatment of Acid Mine Drainage (AMD)

Mining waste, spoil, or abandoned mines are sometimes difficult to handle, and mine drainage water cannot be avoided. In these circumstances, appropriate water treatment is required. The treatment tries to remove the water's (1) acidity, (2) harmful metals, (3) sulphate, and (4) hardness. To cure AMD, several methods have been developed and are in use. However, many of them do not target all four pollutant categories indicated above in AMD. Microbial sulphate reduction and chemical neutralisation are the two most fundamental treatment concepts in use.

Treating AMD by Microbial Sulphate Reduction

The principles of employing microbial sulphate reduction for AMD therapy are thoroughly explained. Sulphate reduction can take place in a variety of technical settings, such as suspension tanks [59], anaerobic ponds [89], wetlands [62, 90, 91], or a permeable reactive wall [92]. Many installations include organic substrate in the system to boost reaction speeds. Sulphate reduction treatment is the most comprehensive: it neutralises acidity, eliminates harmful metals (through precipitation as metal sulphides), removes part of the sulphate from the water, and reduces the salt content in AMD by sulphides precipitation [91].

Chemical Neutralization

Neutralization is the second key therapy strategy for AMD. The use of alkali causes acidity to be neutralised and poisonous cations to precipitate. Lime is most commonly used for neutralisation. The procedure is often carried out in a horizontally percolated network of drains or ditches filled with limestone [93]. A significant negative impact of lime treatment is generally an increase in water hardness. Lime application might result in sulphate retention as well. When sulfuric acid combines with lime ($CaCO_3$), gypsum is formed ($CaSO_4.2H_2O$). However, the production of gypsum is pH dependent. In the presence of ambient CO_2 and a pH greater than 7 (typical for a lime-saturated environment), gypsum production is thermodynamically unfavourable. This is addressed by isolating limestone layers from the environment. The absence of free CO_2 in this environment allows gypsum production even at high pH levels. Furthermore, the usage of anoxic lime drains benefits from the presence of iron in its ferrous state. This prevents an undesirable coating of lime particles with ferric precipitates (armouring).

Conclusion

It is evident that sulphur has a wide range of environmental consequences. However, the number of sulphurous chemicals present is not always the most important indicator of environmental quality or degradation. Rather, the side consequences of sulphur cycling are of concern, particularly the coupling of sulphur transformations with environmental acidification/alkalinization and the mobilization/retention of cationic hazardous metals. The negative side effects, such as self-acidifying soils or AMD, are obvious. The treatment of sulphur pollution is highly dependent on its severity. The majority of the approaches are labour-intensive in terms of chemicals, energy, or equipment. They can, however, only be used for concentrated pollution in very tiny amounts of treated material (point pollution). Sulphur, on the other hand, is generally prevalent in the form of diffuse pollution or contamination that affects huge regions or quantities of trash (AMD, spoil banks, mine tailings, etc.). Less intensive techniques should be utilised in this case. Extensive treatment processes need fewer chemical or energy inputs. Treatment times, on the other hand, are suitably extended. Natural systems, preventative measures, and whole life management of mining sites are examples of such broad methods.

References

[1] Steudel, R. and Chivers T., 2019. The role of polysulfide dianions and radical anions in the chemical, physical and biological sciences, including sulfur-based batteries. *Chemical society reviews*, 48(12), pp. 3279-3319.

[2] Muyzer, G. and Stams A. J., 2008. The ecology and biotechnology of sulphate-reducing bacteria. *Nature reviews microbiology*, 6, pp. 441-454.

[3] Johnson, D. B., Kanao, T. and Hedrich S. 2012. Redox transformations of iron at extremely low pH: fundamental and applied aspects. *Frontiers in microbiology*, 3, p. 96.

[4] Middelburg J. J., 2000. The geochemical sulfurcycle. In: *Environmental technologies to treat sulphur pollution*, Lens., P and Pol., L. H. (eds), IWA Publishing, London, pp. 33-46.

[5] Sievert, S. M., Kiene, R. P. and Schulz-Vogt H. N., 2007. *The sulfurcycle. Oceanography*, 20(2), pp. 117-123.

[6] Lu, Z., Streets, D. G., Zhang, Q., Wang, S., Carmichael, G. R., Cheng, Y. F., Wei, C., Chin, M., Diehl, T. and Tan Q., 2010. Sulphur dioxide emissions in China and sulfur trends in East Asia since 2000. *Atmospheric chemistry and physics*, 10(13), pp. 6311-6331. https://doi.org/10.5194/acp-10-6311-2010.

[7] Elias, T., Kern, C., Horton, K.A., Sutton, A.J. and Garbeil H., 2018. Measuring SO_2 emission rates at Kīlauea Volcano, Hawaii, using an array of upward-looking uv spectrometers, 2014-2017. *Frontiers in earth science,* 6, p. 214, doi:10.3389/feart. 2018.00214.

[8] Tomlinson G. H., 1983. Air pollutants and forest decline. *Environmental science & technology,* 17(6), pp. 246A-256A.

[9] Schweinfurth P. S., 2009. An introduction to coal quality. *The national coal resource assessment overview*. U.S. geological survey professional paper, 1625-F, pp. 1-16.

[10] Dmuchowski, W., Gozdowski, D., Baczewska-Dąbrowska, A. H., Dąbrowski, P., Gworek, B. and Suwara I., 2018. Evaluation of the impact of reducing national emissions of SO_2 and metals in Poland on background pollution using a bioindication method. *PloS one,* 13(2), e0192711, https://doi.org/10.1371/journal. pone.0192711.

[11] Tisdale, S. L., Nelson, W. L. and Beaton J. D., 1985. *Soil fertility and fertilizers*. Collier Macmillan Publishers.

[12] Marcshner H., 2012. *Marschner's mineral nutrition of higher plants*. Croydon: Academic Press Elsevier.

[13] Likus-Cieślik., J. and Pietrzykoski M., 2019. High sulfur contamination effect on the environment at the afforested former Jeziórkosulfur mine. *Journal of the American society of mining & reclamation*, 8(3), pp. 66-90.

[14] Katzur, J. and Haubold-Rosar M., 1996. Amelioration and reforestation of sulfurous mine soils in Lusatia (Eastern Germany). *Water air and soil pollution*, 91, pp. 17-32, https://doi.org/10.1007/BF00280920.

[15] Larssen, T. and Carmichael G.R., 2000. Acid rain and acidification in China: the importance of base cation deposition. *Environmental pollution*, 110(1), pp. 89-102.

[16] Liu, K., Liu, K., Fang, Y., Yu, F., Liu, Q., Li, F. and Peng S., 2010. Soil Acidification in Response to Acid Deposition in Three Subtropical Forests of Subtropical China. *Pedosphere*, 20, pp. 399-408.

[17] Likus-Cieślik, J., Pietrzykowski, M., Szostak, M. and Szulczewski M., 2017. Spatial distribution and concentration of sulfur in relation to vegetation cover and soil properties on a reclaimed sulfur mine site (Southern Poland). *Environmental monitoring and assessment,* 189 (2), p. 87, https://doi.org/10.1007/s10661-017-5803-z.

[18] Katzur, J. and Liebner F., 1998. Effects of superficial tertiary dump substrates and recultivation variants on acid output, salt leaching and development of seepage water quality. In: Geller, W., Klapper, H., Salomons, W. (eds) Acidic Mining Lakes. *Environmental Science*. Springer, Berlin, Heidelberg. https://doi.org/10.1007/978-3-642-71954-7_13.

[19] Grande, J. A., de la Torre, M. L., Cerón, J. C., Beltránc, R. and Gómez T., 2010. Overall hydrochemical characterization of the Iberian Pyrite Belt. Main acid mine drainage-generating sources (Huelva, SW Spain). *Journal of hydrology*, 390, pp. 123-30.

[20] de Vries, W., Dobbertin, M. H., Solberg, S., Van Dobben, H. F. and Schaub M., 2014. Impacts of acid deposition, ozone exposure and weather conditions on forest ecosystems in Europe: an overview. *Plant soil*, 380, pp. 1-45, https://doi.org/10.1007/s11104-014-2056-2.

[21] Hu, B., Li, Z. and Zhang L., 2019. Long-run dynamics of sulphur dioxide emissions, economic growth and energy efficiency in China. *Journal of cleaner production*, 227, pp. 942-949.

[22] Qian, Y., Cao, H. and Huang S., 2020. Decoupling and decomposition analysis of industrial sulfur dioxide emissions from the industrial economy in 30 Chinese provinces. *Journal of environmental management*, 260, p. 110142.

[23] Rinne, K. T., Loader, N. J., Switsur, V. R., Treydte, K. S. and Waterhouse J. S., 2010. Investigating the influence of sulphur dioxide (SO_2) on the stable isotope ratios (δ13C and δ18O) of tree rings. *Geochimica et cosmochimica acta*, 74(8), pp. 2327-2339.

[24] Wei, H., Liu, W., Zhang, J. and Qin Z., 2017. Effects of simulated acid rain on soil fauna community composition and their ecological niches. *Environmental pollution*, 220, pp. 460-468.

[25] Feliciano, M. S., Pio, C. A. and Vermeulen A. T., 2001. Evaluation of SO_2 dry deposition over short vegetation in Portugal. *Atmospheric environment*, 35(21), pp. 3633-3643.

[26] Menz, F. C. & and Seip H. M., 2004. Acid rain in Europe and the United States: an update. *Environmental science and policy*, 7, pp. 253-265.

[27] Olmsted, W. and Hamlin H., 1900. Converting portions of the Los Angeles outfall sewer into a septic tank. *Engineering news*, 44(19), pp. 317-318.

[28] Hvitved-Jacobsen, T., Vollertsen, J. and Nielsen A. H., 2013. *Sewer processes: microbial and chemical process engineering of sewer networks*, 2nd edn, CRC Press, Boca Raton, FL, USA, 379.

[29] Jiang, G., Wightman, E., Donose, B. C., Yuan, Z., Bond, P. L. and Keller J., 2014. The role of iron in sulfide induced corrosion of sewer concrete. *Water research,* 49, pp. 166-174.

[30] Nielsen, P. H. and Hvitved-Jacobsen T., 1988. Effect of sulfate and organic matter on the hydrogen sulfide formation in biofilms of filled sanitary sewers. *Journal of water pollutioncontrol federation, 60(5),* pp. 627-634.

[31] Hvitved-Jacobsen, T., Raunkjaer, K. and Nielsen P. H., 1995. Volatile fatty acids and sulfide in pressure mains. *Water science & technology*, 31(7), pp. 169-179.

[32] Nielsen, P. H., Raunkjaer., K. and Hvitved-Jacobsen T., 1998. Sulfide production and wastewater quality in pressure mains. *Water science & technology*, 37(1), pp. 97-104.

[33] Nielsen, A. H., Vollertsen., J. and Hvitved-Jacobsen T., 2003. Determination of kinetics and stoichiometry of chemical sulphide oxidation in wastewater of sewer networks. *Environmental science & technology*, 37(17), pp. 3853-3858.

[34] Nielsen, A. H., Vollertsen., J. and Hvitved-Jacobsen T., 2004. Chemical sulphide oxidation of wastewater-effects of pH and temperature. *Water science & technology, 50*(4), pp. 185-192.

[35] Nielsen, A. H., Lens., P., Vollertsen., J. and Hvitved-Jacobsen T., 2005a. Sulfide-iron interactions in domestic wastewater from a gravity sewer. *Water research*, 39, pp. 2747-2755.

[36] Nielsen, A. H., Hvitved-Jacobsen., T. and Vollertsen J., 2005b. Kinetics and stoichiometry of sulphide oxidation by sewer biofilms. *Water research*, 39, pp. 4119-4125.

[37] Nielsen, A. H., Vollertsen., J. and Hvitved-Jacobsen T., 2006. Kinetics and stoichiometry of aerobic sulphide oxidation in wastewater from sewers-effects of pH and temperature. *Water environment research,* 78(3), pp. 275-283.

[38] Liu, Y., Ni, B., Ganigué, R., Werner, U., Sharma, K. and Yuan Z., 2015. Sulfide and methane production in sewer sediments. *Water research*, 70, pp. 350-359.

[39] USEPA., 1974. *Process design manual for sulfide control in sanitary sewerage systems.* USEPA (US Environmental Protection Agency), 625/1-74-005, Technology transfer, Washington, DC.

[40] Kiilerich, B., Wagner, M., Nielsen, A. H. and Vollertsen J., 2018. Apparent diffusion coefficients in sewer force main biofilms treated with iron salts. *Environmental science: water research and technology*, 4(10), pp. 1501-1510.

[41] Pomeroy, R. and Bowlus F., 1946. Progress report on sulfide control research. *Sewage works journal*, 18(4), pp. 597-640.

[42] ASCE., 1989. Sulfide in wastewater collection and treatment systems. ASCE (American Society of Civil Engineers) manuals and reports on engineering practice No. 69, p. 324.

[43] ASCE and WEF., 2007. Gravity sanitary sewer design and construction, 2nd edn, ASCE (American Society of Civil Engineers) Manuals and reports on engineering practice No. 60. WEF (Water Environment Federation), *Manual of practice* No. FD-5, p. 422.

[44] Boon, A. G., 1995. Septicity in sewers: causes, consequences and containment. *Water science & technology,* 31(7), pp. 237-253.

[45] Gutierrez, O., Mohanakrishnan, J., Sharma, K. R., Meyer, R. L., Keller, J. and Yuan Z., 2008. Evaluation of oxygen injection as a means of controlling sulfide production in a sewer system. *Water research*, 42(17), pp. 4549-4561, https://doi.org/10.1016/j.watres.2008.07.042.

[46] Thistlethwayte D. K. B., 1972. *The control of sulphides in sewerage systems*, Butterworths, Sydney, Australia, 173.

[47] USEPA., 1985. *Odor and corrosion control in sanitary sewerage systems and treatment plants.* USEPA 625/1-85/018, Washington, DC.

[48] WERF., 2007. *Minimization of odors and corrosion in collection systems-Phase 1*, WERF (Water Environment Research Foundation) Co-published by IWA (International Water Association) Publishing, Final Report.

[49] Zhang, L., De Schryver, P., De Gusseme, B., De Muynck, W., Boon, N. and Verstraete W., 2008. Chemical and biological technologies for hydrogen sulfide emission control in sewer systems: a review. *Water research*, 42(1-2), pp. 1-12.

[50] Lin, H. W., Lu, Y., Ganigué R., Sharma K. R., Rabaey K., Yuan Z. and Pikaar I., 2017. Simultaneous use of caustic and oxygen for efficient sulfide control in sewers. *Science of the total environment*, 601-602, pp. 776-783.

[51] Yang, W., Vollertsen, J. and Hvitved-Jacobsen T., 2005. Anoxic sulfide oxidation in wastewater of sewer networks. *Water science and technology: A journal of the international association on water pollution research*, 52(3), pp. 191-199.

[52] Jiang, G., Sharma, K. R., Guisasola, A., Keller, J. and Yuan Z. 2009. Sulfur transformation in rising main sewers receiving nitrate dosage. *Water research*, 43(17), pp. 4430-4440.

[53] WEF., 2004. Control of odors and emissions from wastewater treatment plants. WEF (Water Environment Federation) *Manual of Practice No. 25*, Alexandria, VA, USA, p. 537.

[54] Stuetz, R. M. and Frechen F. B., 2001. *Odours in wastewater treatment – measurement, modelling and control*. IWA Publishing, London, UK, 437.

[55] WERF., 2009. *Collection system ventilation research report*, WERF (Water Environment Research Foundation), Report No. 04-CTS-1A.

[56] Shammay, A., Sivret, E. C., Le-Minh, N., Fernandez, R. L., Evanson, I. and Stuetz R. M., 2016. Review of odour abatement in sewer networks. *Journal of environmental chemical engineering*, 4(4), pp. 3866-3881.

[57] Talaiekhozani, A., Bagheri, M., Goli, A. and Khoozani M. R. T., 2016. An overview of principles of odor production, emission, and control methods in wastewater collection and treatment systems. *Journal of environmental management*, 170, pp. 186-206.

[58] WEF., 1999. Wastewater collections systems management, WEF (Water Environment Federation) Manual of Practice, *MOP No. 7*, 5th Edition, p. 244.

[59] Van Houten, R. T., Pol, L. and Lettinga G., 1994. Biological sulphate reduction using gas-lift reactors fed with hydrogen and carbon dioxide as energy and carbon source. *Biotechnology and bioengineering*, 44(5), pp. 586-594.

[60] Ivanov M. V., 1983. Major fluxes of the global biogeochemical cycle of sulphur. *The Global Biogeochemical Sulphur Cycle*, Wiley, Chichester, pp. 449-463.

[61] Brimblecombe, P., Hammer, C., Rodhe, H., Ryaboshapko, A. and Boutron C. F., 1989. Human influence on the sulphur cycle. *Evolution of the global biogeochemical sulphur cycle*, In. P. Brimblecombe and A. Y. Lein, Eds. (Wiley, New York), pp. 77-121.

[62] Tichý, R. and Mejstrík V., 1996. Heavy metal contamination from open-pit coal mining in Europe's Black Triangle and possible remediation. *Environment reviews*, 4(4), pp. 321-341.

[63] Orava, D. A. and Swider R. C., 1996. Inhibiting acid mine drainage throughout the mine life cycle. *CIM bulletin*, 89(999), pp. 52-56.

[64] Gray, N. F., 1997. Environmental impact and remediation of acid mine drainage: a management problem. *Environmental geology*, 30(1), pp. 62-71.

[65] Salomons W., 1995. Long-term strategies for handling contaminated sites and large-scale areas. In *Biogeodynamics of pollutants in soils and sediments*, Springer, Berlin, Heidelberg, pp. 1-30.

[66] Rulkens, W. H., Tichy, R. and Grotenhuis J. T. C., 1995. Sites polluted with heavy metals: current techniques for clean-up and desirable future developments. In *Proc. Int. Conf. on Heavy metals in the environment.* Duitsland, 1, pp. 10-19.

[67] Rulkens, W. H., Tichy, R. and Grotenhuis J. T. C., 1998. Remediation of polluted soil and sediment: perspectives and failures. *Water science & technology*, 37(8), pp. 27-35.

[68] Spottiswood, E. G., 1982. *Introduction to mineral processing*. John Wiley & Sons Inc., New York.

[69] Richards, I. G., Palmer, J. P. and Barratt P. A., 1993. *The reclamation of former coal mines and steelworks*. Elsevier.

[70] Hinsenveld, M., 1993. *Physical/chemical extraction technologies. Demonstration of remedial action technologies for contaminated land and groundwater.* Ed. Olfenbuttel, R. F., NATO/CCMS, Brussel.

[71] Karavaiko G. I., 1985. *Microbiological processes for the leaching of metals from ores*. Centre of International Projects GKNT, Moscow.

[72] Bruynesteyn, A., 1989. Mineral biotechnology. *Journal of biotechnology*, 11(1), pp. 1-10.

[73] Iske, U. and Glombitza F., (1988). Possibilities of using biotechnological processes for the extraction of metals from ores, minerals and waste products of the chemical industry. *Chemical engineering*, 40(11), pp. 464-468.

[74] Tyagi, R. D., Couillard, D. and Tran F. T., 1991. Comparative study of bacterial leaching of metals from sewage sludge in continuous stirred tank and air-lift reactors. *Process biochemistry*, 26, pp. 47-54.

[75] Blais, J. F., Tyagi, R. D. and Auclair J. C., 1993. Bioleaching of metals from sewage sludge: microorganisms and growth kinetics. *Water research*, 27(1), pp. 101-110.

[76] Tichý, R., Lens, P., Grotenhuis, J. T. C. and Bos P., 1998a. Solid-state reduced sulfur compounds: environmental aspects and bio-remediation. *Critical reviews in environmental science and technology*, 28(1), pp. 1-40.

[77] Tichý, R., Rulkens, W. H., Grotenhuis, J. T. C., Nýdl, V., Cuypers, C. and Fajtl J., 1998b. Bioleaching of metals from soils or sediments. *Water science & technology*, 37(8), pp. 119- 128.

[78] Schuring, J., Kolling M. and Schulz, H. D., 1997. The potential formation of acid-mine drainage in pyrite-bearing hard-coal tailings under water-saturated conditions-an experimental approach. *Environmental geology*, 31(1-2), pp. 59-65.

[79] Van Breemen, N., 1988. Effects of seasonal redox processes involving iron on the chemistry of periodically reduced soils. In *Iron in soils and clay minerals*, Springer, Dordrecht, pp. 797-809.

[80] Bell, A. V., Riley, M. D. and Yanful E. K., 1995. Evaluation of a composite soil cover to control acid waste rock pile drainage. *CIM bulletin*, 88(995), pp. 41-46.

[81] Thomson, B. M. and Turney W. R., 1996. Minerals and mine drainage. *Water environment research*, 68(4), pp. 542-546.

[82] Sasaki, K., 1997. A study on the dissolution of pyrite and its suppression in acidic environments. *Bunsekikagaku*, 46(7), p. 612.

[83] Takeuchi, T. L. and Suzuki I., 1997. Cell hydrophobicity and sulfur adhesion of *Thiobacillus thiooxidans*. *Applied and environmental microbiology*, 63(5), pp. 2058-2061.

[84] Evangelou, V. P., 1995. Acid mine drainage, surface chemistry, molecular oxidation mechanisms, microbial role, kinetics, control, ameliorates, limitations, microencapsulation. *Pyrite Oxidation and its Control*. CRC/Lewis Press, Boca Raton, FL, https://doi.org/10.1201/9780203741641.

[85] Georgopoulou, Z. J., Fytas, K., Soto, H. and Evangelou B., 1996. Feasibility and cost of creating an iron-phosphate coating on pyrrhotite to prevent oxidation. *Environmental geology*, 28(2), pp. 61-69.

[86] Van Breemen, N., 1975. Acidification and de-acidification of coastal plain soils as a result of periodic flooding. *Soil science society of America Journal*, 39(6), pp. 1153-1157.

[87] van Mensvoort, M. E. F., Duong van, Ni. and van der Schans, J., 1991. Improvement of acid sulphate soils by leaching with salt or brackish water. In P. Deturck, & F. N. Ponnamperuma (Eds.), *Rice production on acid soils of the tropics*, pp. 219-224.

[88] Park, N. J., Park, Y. S. and Kim Y. S., 1971. Effect of Lime on Growth of Rice and Changes in pH, Eh, Fe^{2+} and Al in an Acid Sulfate Soil. *Korean journal of soil science and fertilizer*, 4(2), pp. 167-175.

[89] Rose, P. D., Maart, B. A., Dunn, K. M., Rowswell, R. A. and Britz P., 1996. High rate algal oxidation ponding for the treatment of tannery effluents. *Water science & technology*, 33(7), pp. 219-227.

[90] Wildeman, T. R., and Laudon L. S. 1989. Use of wetlands for treatment of environmental problems in mining: non-coal-mining applications. *Constructed wetlands for wastewater treatment*, pp. 221-231.

[91] Eger, P., 1994. Wetland treatment for trace metal removal from mine drainage: the importance of aerobic and anaerobic processes. *Water science & technology*, 29(4), pp. 249-256.

[92] Benner, S. G., Blowes, D. W. and Ptacek C. J., 1997. Sulfate reduction in a permeable reactive wall for prevention of acid mine drainage. In *Abstracts of papers of the American chemical society*, 213, pp. 45-46.

[93] Ziemkiewicz, P. F., Skousen, J. G., Brant, D. L., Sterner, P. L. and Lovett R. J., 1997. Acid mine drainage treatment with armored limestone in open limestone channels. *Journal of environmental quality*, 26(4), pp. 1017-1024.

Chapter 5

Oil Pollution Remediation Technologies: Novel Technologies with Future Challenges

G. Y. Liyanage[1,3]
M. S. Ekanayake[1,4]
and P. M. Manage[1,2,*]
[1]Centre for Water Quality and Algae Research, Department of Zoology, University of Sri Jayewardenepura, Sri Lanka
[2]Faculty of Graduate Studies, University of Sri Jayewardenepura, Sri Lanka
[3]Department of Aquatic Bioresources, Faculty of Urban and Aquatic Bioresources, University of Sri Jayewardenepura, Sri Lanka
[4]Department of Aquaculture and Fisheries, Faculty of Livestock, Fisheries and Nutrition, Wayamba University of Sri Lanka, Sri Lanka

Abstract

Pollution has increased due to urbanization, industrialization, and population pressure. Around 1.7 to 8.8 million metric tons of oil are released into the world's water annually, wreaking havoc on many ecosystems. Diesel, crude oil, petroleum, lubricating oil, bitumen, bunker oil, and other types of oil pollute the environment. Oil spills, urban runoff, wastewater treatment plants, and other sources account for more than 90% of oil pollution. Oil pollution harms the environment, economy, and society. Contaminants must be dealt with in three different ways:

Identify the issue, determine the nature and amount of the pollutant, and select the most appropriate remediation method. As a result of the need to clean up polluted sites, new technologies have emerged that focus on the detoxification and destruction of contaminants rather than the

* Corresponding Author's Email: pathmalal@sjp.ac.lk.

In: Microbial Technologies and Their Applications
Editors: Santosh Kumar Karn and Anne Bhambri
ISBN: 979-8-88697-913-8

traditional disposal method. Oil remediation can be accomplished using a variety of chemicals, biological, and mechanical methods. Solidification, soil washing, dehalogenation, thermal technologies, flotation, and chemical reduction are among the physicochemical remediation methods available. There are technologies for bioaccumulation, bioaugmentation, biotransformation, bioaccumulation, biosorption, bioleaching, attenuation, and other biological remediation. Micro/nanotechnologies have sparked tremendous interest in oil remediation as a novel technology due to the synergistic effect of unique motion characteristics and inherited physicochemical properties of nanomaterials. Another novel challenge is integrating electrochemical methods with biological routes, microbial fuel cells, enzyme technologies, and bio-electrochemical systems. Integrated remediation techniques have recently received more attention than single-technology remediation. On the other hand, bioremediation is a green, appealing, and promising option for oil remediation at a contaminated site, as it is more cost-effective and selectively degrades pollutants without harming the site or its native flora and fauna. The development of remediation processes must integrate various disciplines such as microbial ecology, biochemistry, microbial physiology, and biochemical and bioprocess engineering in order to be commercially successful.

Keywords: oil pollution, remediation, bioremediation, microbes

Introduction

Oil can be identified as a significant non-renewable natural resource mainly used as a source of energy [1, 2]. Diverse petrochemicals, including synthetic materials like plastics, were manufactured using oil [3]. The majority of the crude oil and its refined products are transported around the world by pipelines, ships, and trains. Spills are always a possibility, and oil spills can be devastating to the environment [4]. Different types of oil are crude oil, oily garbage, gasoline and diesel like petroleum products, ship bunkers, etc. [2, 5]. Oil contained residential waste and some industrial operations such as cutting oils, lubrication, hydraulic, etc., contributes to oil contamination in the marine environment [1, 5].

Petroleum is mined in huge quantities. Petroleum contributes more than 42% of the commercial energy production in the world. Furthermore, global petroleum use increased by 11% between 2004 and 2019 [6, 7]. Rapidly expanding economies, such as India (75%) and China (59%), show high utilization of petroleum. Further, relatively wealthy, developed countries account for 48% of the world petroleum utilization, though they support about 20% of the human population [8].

Oceanic tankers and overland pipelines transport oil and refined products globally, while smaller tankers, barges, pipelines, railroads, and trucks are involved in local distribution systems [6, 7]. All of these modes of transportation have the potential for spillage. Those spillages lead to substantial environmental damage. In addition, a considerable amount of oil is released into the environment via a variety of smaller sources [6, 7].

Under natural conditions, marine environment recovery is quick. However, the internal mechanisms of the natural environment are not endless [8]. Therefore, proper treatment and protection are needed for the marine environment. More and more attention to the marine environment is needed because of the rise of many new oil related development projects.

Naval technologies have resulted in removing the pollutants from polluted sites, focusing on the degradation and the detoxification of pollutants rather than the traditional disposal method [1]. Oil remediation could be accomplished using a variety of chemicals, biological, and mechanical methods. To find the solution, prevention, partial measures, compensation, and remedial action have all been considered.

However, to be commercially successful, remediation procedures must incorporate a variety of disciplines, including microbial ecology, microbial physiology, biochemistry, and biochemical and bioprocess engineering.

Types of Hydrocarbons

Oil is a mixture of various hydrocarbon compounds. Hydrocarbons can be classified into the following three groups [9]:

1. Aromatic hydrocarbons
2. Alicyclic hydrocarbons
3. Aliphatic hydrocarbons

The specific mixtures of hydrocarbons and other chemicals such as sulphur (1-2%), vanadium (0.15%), and nitrogen (1%) in crude petroleum vary significantly [10]. Different hydrocarbon fractions: gasoline, kerosene, lubricating oils, heating oil, waxes, jet fuel, and residual fuel oil are separated by distillation at different temperatures while petroleum is handled in an industrial processing plant [9]. In addition, catalytic cracking converts some of the heavier fractions into lighter, more valuable hydrocarbons like some of those found in gasoline. Gas, naphtha, kerosene, light gas, and residuals make up crude oil [10].

Scenarios of Global Oil Pollution

Any kind of spillage related to the petroleum or its refined products results in oil pollution. Discharges of oil or bunker fuel can happen via disabled tankers, ships, barge, well blowout, broken pipelines, etc. [11].

Oil spills into the world's ocean account for nearly 1.4 million tonnes annually (Figure 1). Furthermore, there is enormous natural emission of hydrocarbons to the sea [12]. Phytoplankton synthesized and released these chemicals at an estimated 26-million tonnes/year (Figure 1) [12]. These massive biological releases contribute to the background concentration of hydrocarbons of about 1 ppb (1 μg/L) in seawater [13]. These biogenic emissions fall from natural contamination which has no records of significant biological harm. Though these biogenic emitted hydrocarbons have not been recorded in known biological harm, the natural emission of oil through underwater seepage, which is about 0.6-million tonnes/year, may sometimes cause local ecological damages [13].

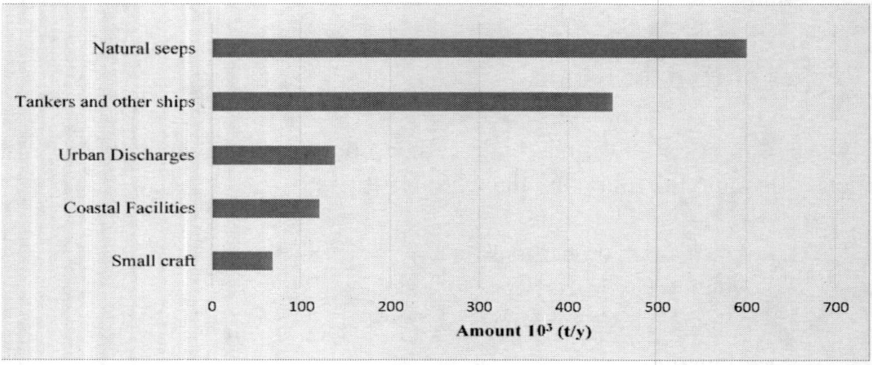

Figure 1. Sources of oil into sea. The data are in 10^3 tonnes per year over the period 1988 to 2017. Sources: [14].

Some of the most disastrous marine spills of petroleum are:

- 1967-Torrey Canyon (117 thousand tons)- England;
- 1973-Metula spilt (53 thousand tons)- Magellan;
- 1989-Exxonvaldez (36 thousand tons)- South Alaska;
- 1999-Erica (20 thousand tons)- France and Spain;
- 2002-Prestige (63 thousand tons)- Atlantic off France and Spain;
- 2003-Tasman Spirit (30 thousand tons)- Pakistan;
- 2009-Montera (30 thousand tons)- Australia;
- In 2012, MV Thermopylae Sierra (40-50 thousand) Panadura, Sri Lanka;
- In 2021, the X-Press Pearl ship: the vessel contained approximately 350 tonnes of oil.

Ruptured pipelines are the most common source of terrestrial oil pollution. Pipeline breaks may be caused by pump malfunctions, corrosion or faulty welding, as well as by earthquakes, erosion slumps and even armed vandals engaged in target practice [15].

Furthermore, oil is released into the ocean due to washing petroleum storage tanks in the ships that transport petroleum and liquid fuels [16]. After delivering the petroleum load for the refinery process, tanks fill with seawater to balance the ballast water [16]. Ballast may be discharged into the ocean as the tanker approaches its destination. This wastewater could contain hydrocarbon residues equal to about 1% of the tanker's capacity [3].

The Fate of Spilled Oil

After petroleum and refined products spill into the environment, they are affected by a variety of natural processes (Figure 2). Different hydrocarbon fractions will selectively evaporate, spread across the surface, dissolve in water, collect as persistent residues, or be destroyed by microbes and sun UV radiation, depending on their chemical and physical features [17].

When petroleum enters the natural environment by accident or due to human activity, its soluble component becomes very fast portioning in water, air, and sediments. The insoluble fraction forms a layer of 0.01-3.00 mm thickness on the water layer [16]. When hydrocarbon contamination occurs in the marine environment, some hydrocarbons are evaporated while the others remain in the water [17]. The evaporation can change the hydrocarbon composition by about 0.1% for heavier oil in water and 17.3% for lighter oils [17]. Evaporation of lighter oil fractions facilitates the higher biodegradation process of heavier oil as more delicate oils show higher toxicity towards oil degrading microorganisms [18].

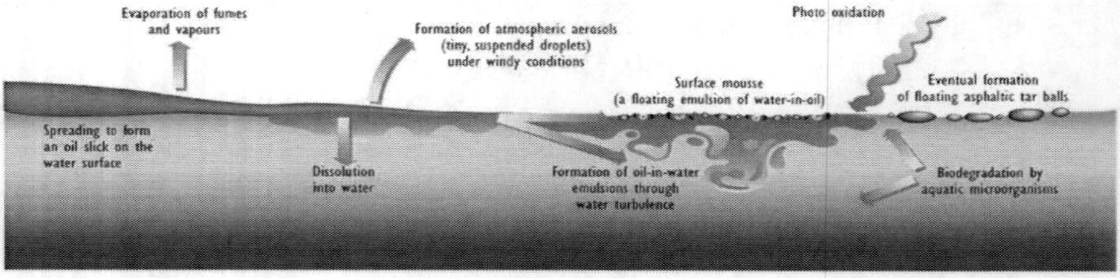

Figure 2. Fate of oil in water.

The crude oil droplets can easily interact with planktonic organisms as droplets are frequently in the food size spectra of zooplanktons [19]. PAHs bioaccumulate in fish and other animals' lipid-rich tissues, liver and pancreas. The hydrophilic PAHs uptake via ventilated water, food and sediments [20]. Furthermore, some hydrocarbons like PAH may cause progeny's death or mutations in aquatic living beings [21].

More often, the hydrocarbon-contaminated soils hinder the natural biodegradation processes due to its low water-soluble nature and the large molecular weight. Further, most of these may enter the soil macro and micropores, limiting the water and air movement required for organic matter conversion [22]. The sorption of oil pollutants to the sediments depends on oil properties, weather conditions, pH of the environment, etc. [23].

The weather and the composition of crude oil spilt at sea determine its fate. Once it reaches coastal environments, such as lagoons and salt marshes, its environmental impact is amplified. Hydrocarbon contaminated fur and feathers of marine mammals and birds, affect their hydrophobic properties and leads to hypothermia or fatal doses due to oil absorption during preening [24].

However, human activities such as dredging and bioturbation can cause buried and adsorbed oil particles in the sediment to be released into the environment [25].

Oil Toxicity

Ingestion of petroleum, refined products, or pure hydrocarbons causes acute toxicity, followed by cellular membrane damage, leading to tissue death. Some crude oil components, such as PAH, benzene and its substitutes, and cycloalkane rings, are highly toxic to plants and animals and accumulate in high concentrations through food chains [26]. Especially the polyaromatic hydrocarbons (PAHs) with 4 or 5 rings are known as carcinogens, and usual intake takes place via the inhalation, ingestion, or dermal contact would have adverse impacts on the aquatic organisms and human beings [27].

Despite the fact that this is a crucial factor when considering oil spills in an ecological context, few research have looked at the influence of oil exposure on exposed species' ability to endure commonly encountered environmental stressors. If take fish as an example, the toxicity of oil negatively affects temperature balance, osmoregulatory, hypoxia condition and giving social stress (Figure 3).

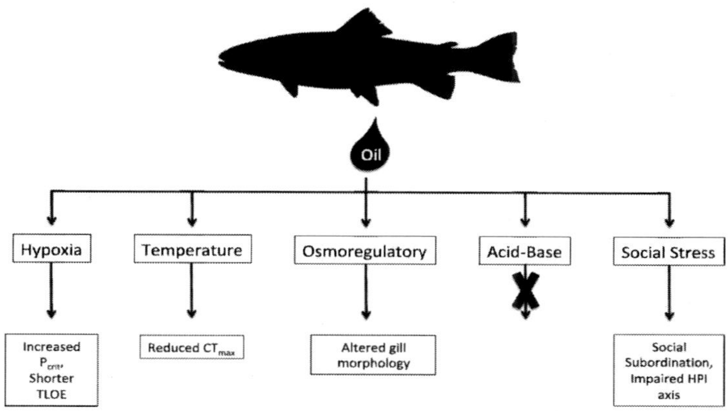

Figure 3. Oil toxicity in fish [Critical oxygen tension (*Pcrit*); Time to Loss of Equilibrium (*TLOE*); Critical Thermal maximum (*CTmax*), the Hypothalamus-Pituitary-Internal (*HPI*) axis].

PAHs are rapidly absorbed by the gastrointestinal tract owing to their high lipid solubility. The production of reactive epoxide and quinine intermediate can happen during the metabolism of PAH in aquatic organisms, which leads to the formation of mutations in cells that ultimately lead to cancers [28]. Other toxicity effects of PAHs are associated with lung, liver, intestine, pancreas and skin cancers, and the degree of toxicity varies with the molecular weight of the PAHs [3].

The use of dispersants and emulsifiers, burning, removal of oiled substrates, hot-water washing, and tilling of oiled soil to improve aeration and decomposition during the conventional cleanup process may result in more severe environmental damage.

Further, some negative ecological consequences are also stimulated by way of damage triggered to keystone species inside the food web.

Effect of Oil on Aquatic Flora and Fauna

Oil impacts aquatic life through physical contact, ingestion, inhalation, and absorption [29]. Floating oil can affect the plankton, algae, larvae stages and coral. Fish can be contaminated due to exposure directly to oil or as a result of eating contaminated food items in the aquatic environment. Larger organisms in the food chain, such as humans, may consume harmful organisms when they eat these fish [30].

Fur-bearing mammals like sea otters and birds' feathers lose their water repellency due to the oil, exposing them to the environment. Hypothermia will kill birds, turtles, and mammals if they cannot repel water and insulate themselves [30].

Oil can enter the body of juvenile sea turtles who ingest these oils mistakenly as food, dolphins and whales during their breathing and finally affect their body functions, especially in respiratory, immunological and reproduction systems [22]. Further, these oils can be ingested by birds and animals while trying to clean themselves, leading to several poisonous issues [22].

Though fish and some other aquatic organisms, including shellfish and corals, may not be exposed to the oil spills directly, they may contact these oils later, especially due to the water movements [23]. The alteration of heart and respiration rates, stunted growth, enlarged livers, fine erosion, and reproduction problems are the main issues visible in adult fish after exposure to oil [23]. Further, more sensitive stages of the fish life cycle: eggs and larvae, show particular vulnerability towards fatal and sublethal effects. Even though some fish and shellfish do not show any harmful effects after exposure to an oil spill, biomagnification in the food chain can occur. Finally, some toxic effects can be seen at higher levels in the food chains, including humans [22].

Furthermore, oil can protect atmospheric gases from diffusing into the water. As a result, the amount of dissolved oxygen in the water decreases, resulting in the death of planktonic organisms. Even oil prevents solar radiation from reaching the seawater. It directly impacts plankton and algae's photosynthesis [31].

Mangroves, extremely productive and beneficial ecosystems in the coastal belts, are highly susceptible to severe damage by various oil pollution activities, including oil and gas exploration, petroleum production and accidental spills. For instance, oil pollution on the coast of Sao Paulo, Brazil, resulted in the defoliation of *Rhizophora mangle*, *Laguncularíara cemosa* and *Avicennias chaueriana* by 25.9%, 43.9%, and 64.5%, respectively [32].

Methods of Oil Remediation

Multiple oil remediation technologies under physical remediation, chemical remediation and biological remediations were developed and applied for the recover contaminated environments.

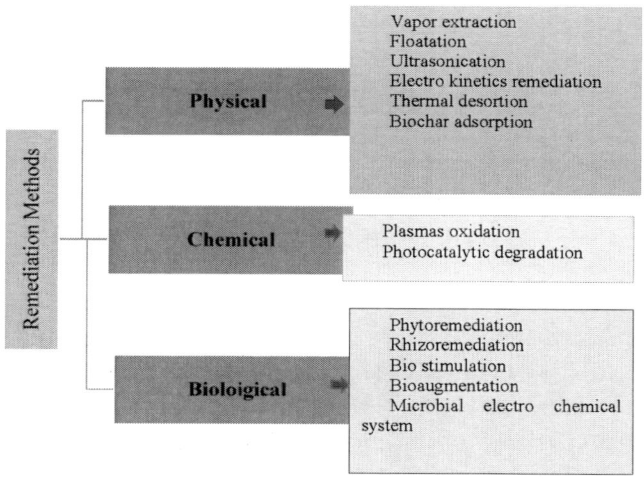

Figure 4. Oil remediation methods.

Chemical Remediation

Thermal desorption as heat to increase volatility, burn [29, 33], plasma oxidation, and photocatalytic degradation are some of the methods used in the remediation of oil-polluted sites [34].

Chemical oxidation can rapidly be deposing or preprocessing oil contaminants. Here, contaminants chemically convert to non-hazardous or less toxic, less mobile or inert compounds and enhance bioavailability [35].

Plasmas oxidation is another novel technology to recover the contaminated oil site. Several active compounds such as ozone, hydrogen peroxide, hydroxyl radicals and high energy electrons were generated in plasma generation by ionization [35].

Photocatalytic degradation is one of the most effective decomposition methods for PAH in the environment. This method degraded organic compounds to a small, nontoxic compound using semiconductor metal oxide as a catalyst [31].

Physical Remediation

To destroy or separate contaminants, physical remediation techniques rely on the physical properties of the selected contaminants or the medium that it dissolved. Identified physical remediation methods are soil vapour extraction, electrokinetics remediation, biochar adsorption, floatation, ultrasonication, etc. [29, 33].

In soil vapour extraction, inducing transport of oil contaminant from subsurface to the surface for a treat with other methods. The efficiency of this method depends on soil properties, operational conditions and contaminant properties [36].

Floatation is characterized as a simple, low-cost, and highly efficient method for remediating oil. It can separate tiny or lightweight particles which have low settling velocities [36].

In the ultrasonication form strong oxidants and hydroxyl radicals, which are increased efficiency of remediation process. Ultrasonication can remove pollutants entirely without using any chemical [37].

In electro, kinetics remediation employs a direct electric current between appropriately distributed electrodes embedded in an oil-contaminated area to form an electric field. The disadvantage of this method is that it is ineffective in low contaminant concentrations [33].

Due to high surface adsorption and partitioning properties, biochar was used as an amendment to implement organically contaminated water remediation [33]. However, chemical and physical remediation methods have been labour intensive, highly expensive and want expertise and knowledge to get better results. Another important factor is most practices do not obliterate contaminants from the site. Also, some mathematical models have been developed to explain pollutant behaviour in the water and sediments [38]. Thus, researchers are being investigated low-cost and environmentally friendly methods to remove oil contaminants from the contaminated sites using harnessing microbes in recent years.

Biological Remediation

Because of its low cost and ability to prevent the accumulation of contaminants, bioremediation is one of the most successful and eco-friendly technologies which can be applied for the treatment of oil-contaminated sites [39-41]. Bioremediation is breaking or degrading hazardous substances into less toxic or nontoxic substances using living organisms such as plants, yeast, fungi, and bacteria. Phytoremediation, bioremediation, bioventing, biopolymer shields, and other biological processes can be identified [42, 43].

In phytoremediation, plants and associated microorganisms are used to degrade the contaminant. Plant growth-promoting bacteria, called rhizobacteria, play a significant role in phytoremediation [43].

The microbial electrochemical system was considered an emerging technique for bioremediation. Through that technique, pollutants convert into value-added and nontoxic products.

The primary method of bioremediation is using microorganisms. Microbes (yeast, fungi or bacteria) can be used to recover the contaminated oil site as some specific microorganisms consume organic substances in crude oil as the sole carbon source for their energy [1, 27, 44]. Therefore, bacteria convert hazardous hydrocarbon in the soil to less harmful compounds such as carbon dioxide, methane and water by aerobic and anaerobic reaction mechanisms [45]. The bioremediation process is outlined in Figure 5. Bioaugmentation and bio-stimulation are the two main approaches to oil spill bioremediation. In bioaugmentation, oil-degrading bacteria are added to enhancement of the prevailing microbial population where the growth of indigenous oil degraders is stimulated by the addition of nutrients or other growth-limiting co-substrates in bio-stimulation.

Prior to the 1989 Exxon Valdez oil spill, there was little information on the performance of commercial bio-augmentation products. After the Exxon Valdez spilt 41 million litres of oil in Alaska in 1989, bioremediation techniques were improved [36, 46, 47].

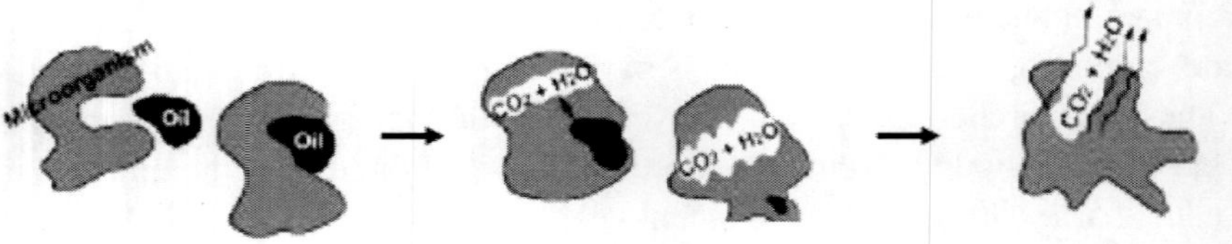

Figure 5. Schematic diagram of aerobic biodegradation of oil contaminants.

Microorganisms Work to Clean Up the Environment

Various microorganisms use oil as a source of carbon and energy. As a result, they clean up the oil on the site. Previous researchers discovered that Acinetobacter and Enterobacter species could degrade crude oil [38]. Kostka et al., (2011) and colleagues (2011) found that the majority of the bacteria isolated were Gammaproteobacteria, including members of genera known to degrade oil (Alcanivorax, Marinobacter, Pseudomonas, and Acinetobacter) [48]. As the oil degrades, Omatayo (2012) recorded the bacteria *Micrococcus varians*, *C. amycolatum, Corynebacterium ulcerans,* and *Bacillus badius*. Further, *Sphinogobacterium* spp., *Yokenella* spp., *Moraxella* spp. *Roseomonas* spp., *Alcaligenes* spp., and *Flavobacter* spp., were also identified as bacteria that degrade petroleum hydrocarbons [48]. Pseudomonas, Aeromonas, Moraxella, Beijerinckia, Flavobacteria, Chrobacteria, Nocardia, Corynebacteria, Acinetobacter, Mycobacteria, Modococci, Streptomyces, Bacilli, Arthrobacter, Aeromonas, Cyanobacteria, Arthrobacter, Aeromonas, Cyanobacteria, Arthrobacter, Aeromon [24]. *Bacillus* sp., *Enterococcus* sp., and *Micrococcus varians* have recently been identified as oil degraders by Liyanage and Manage (2016). Thus, as a result, bioremediation is a more convenient approach because it does not require a large amount of input, and bacteria consume oil till no residue remains [49, 50].

Bacterial Oil Degradation Pathways

It was found that different bacteria perform different steps in the degradation pathway. The enzymatic activation of hydrocarbons by oxygen has long been identified as a crucial step in biodegradation. Further, several mechanisms for aromatic and aliphatic compounds have been suggested [26], including aerobic and anaerobic pathways with some nutrients such as nitrate, sulfate, or iron [35]. In aerobic degradation, microbes degrade organic compounds into water, carbon dioxide and minerals such as nitrate and sulfate (Figure 5). Aerobic degradation does not produce any harmful byproducts such as hydrogen sulfide or methane. Moreover, aerobic reactions need less energy to remediate contaminants (Figure 5) [35]. Aerobic degradation in the marine environment is associated with various microorganisms, including bacteria, actinomycetes and fungi. Actinobacteria, Rhodococcus, and Pseudomonas are a few bacterial strains often associated with petroleum contaminated habitats [51]. It was recorded that the bacterium *Azoarcus* sp., *Desulfavilona* sp., and *Brevibacillus* sp. degraded petroleum hydrocarbons under anaerobic conditions [37, 52]. Other available biological techniques, bio-venting, water flushing, and biopolymer shield techniques, are much more expensive than bioremediation. It is the best method for deep soil layer cleanup and is susceptible to different conditions.

Bioremediation Technologies

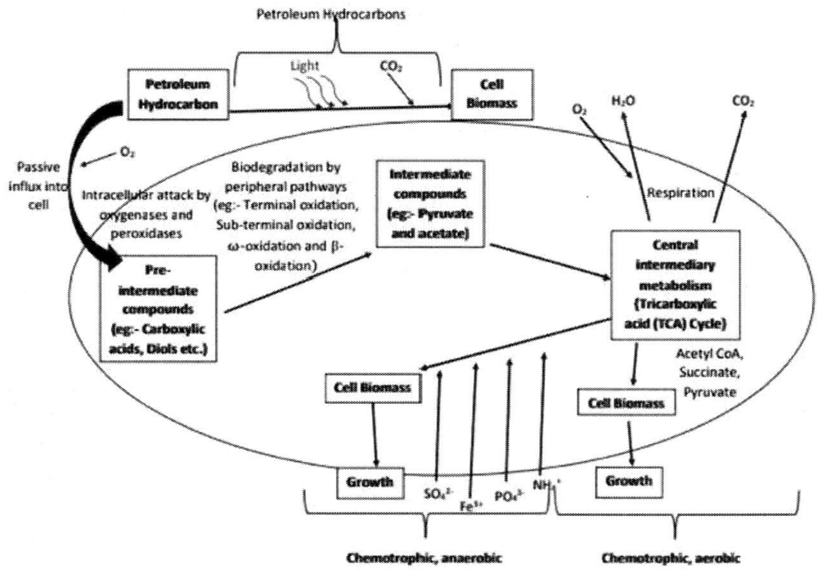

Figure 6. Mechanism for oil remediation.

There are three distinct groups of bioremediation technology for responding to oil spills:

1. Biostimulation (Nutrient enrichment);
2. Bioaugmentation (Seeding with naturally occurring microorganisms);
3. Bioaugmentation with Genetically Engineering Microorganism (GEMs).

Biostimulation (Nutrient Enrichment)

The process of providing nutrients to indigenous bacteria that digest hydrocarbons to destroy hazardous chemicals is known as biostimulation. In some cases, the most important and most easily modifiable factor slowing down petroleum biodegradation in marine environments is a lack of appropriate nutrients such as nitrogen and phosphorus. Nutrient enrichment (also known as nutrition) has received more attention than the other two options, especially now that the EPA, Exxon, and the State of Alaska have done extensive nutrient enrichment studies on beaches damaged by the Exxon Valdez oil spill.

As a result of these factors, many scientists now consider nutrient enrichment to be the most promising of the three techniques for oil spill situations where bioremediation is appropriate. This method, conceptually similar to fertilizing a lawn, involves adding nutrients to a spill site that reduce biodegradation rates (but do not introduce new microbes).

Bioaugmentation (Seeding with Naturally Occurring Microorganisms)

Seeding is a technique for increasing biodegradation rates by introducing microorganisms into a contaminated environment. The inoculums could be a mix of oil-degrading bacteria collected from the remediation site and mass-cultured in the lab or on-site bioreactors, or a mix of non-native microbes collected from various degraded habitats and specifically selected and cultivated for their oil-degrading abilities. Seed cultures are commonly supplemented with nutrients. Microorganisms are introduced to a spill site because native microbial communities may lack the diversity and density of oil degraders required to digest the spill's various components efficiently.

The advantages of introducing nonindigenous microbes to marine ecosystems are currently being studied. With a few exceptions, the scientific community has been skeptical of the potential of seeding marine oil spills. There is no data on the effectiveness of this strategy because no controlled trials have been conducted in such circumstances. Many experts question whether adding microbes to a spill site is necessary because most locations already have good oil-degrading microbes, and biodegradation in most ecosystems is hampered more by a lack of nutrients than a lack of germs.

Bioaugmentation with GEMs (Genetically Engineered Microorganisms)

Even though it was not superior to indigenous organisms and had never been tested in the field, the first frost organism to be patented was a bacterium genetically modified to break down oil. The rationale for developing such organisms is that they can be either more efficient than naturally occurring species or capable of breaking down portions of petroleum that naturally occurring species cannot. To be effective, such microorganisms would have to overcome all of the problems associated with seeding a spill with nonindigenous germs.

Bioengineered microbes for environmental remediation are unlikely to happen soon due to scientific, economic, regulatory, and public perception barriers. A lack of robust research infrastructure, the prevalence of small businesses in the bioremediation industry, a lack of data sharing, and regulatory challenges impede the commercial use of genetically altered organisms.

Integrated Remediation Methods

The microorganism can be used with physical and chemical methods as the integrated remediation methods. Most of the integrated strategies increase the microbes' activity and aeration of contaminated sites due to the hydrophobicity and fluidity of the oil.

Nanomaterials on Bioremediation

As conventional remediation methods have several drawbacks like long remediation period, survival of microbes under natural wave actions and other environmental fluctuations, those methods need to be advanced in a way that is more efficient and effective. Use of nanomaterials is one of the better solutions for addressing those issues in an environmental safe manner as it can improve the bioremediation effort. Wu et al., (2015) illustrates how different nanoparticles can enhance the bioavailability of heavy crude oil, using a Molecular Dynamic (MD) simulation model [53]. Immobilization of potential oil degrading microbes by entrapping, adsorbing or encapsuling on different surfaces is another way to reduce movements of microbes and make the degradation process more efficient. Bestawy et al. (2020) recorded the potential applicability of the exogenous bacterial consortium consists with *Pseudomonas otitidis* and *Enterobacter cloacae* which were decorated (immobilized) on Fe_3O_4 nanoparticles for the removal of petroleum hydrocarbon-contaminated wastewater [54]. Similarly, combination of polyvinylpyrrolidone (PVP) coated magnetite nanoparticles and selected oil-degrading bacteria have showed 100% oil removal after 48h of incubation [21].

Microorganism and Biochar

Biochar is enriched with high carbon content, adsorption capacity, stability and nutrient immobilization capacity. Mineralization, biodegradation and adsorption or a combination of these three are the major modes of this integrated method. Zhang et al., (2019) put forward that potential applicability of immobilized hydrocarbon degrading bacteria on to a mushroom substrate-based biochar, suggesting that the immobilized microbes enhanced the biodegradation rate of petroleum [55]. Similarly, Li et al., (2013) recorded that modified wheat biochar and peanut shell biochar are two low-cost biochar that can be applied for remediation along with different oil degrading microbes [23].

Microorganisms and Nutrients

The rapid depletion of the primary inorganic nutrients such as nitrogen and phosphorous, happens due to a significant carbon source (oil) input. It caused a reduction in microbial metabolism. Therefore, the addition of nutrients stimulates the biodegradation of the pollutant [30]. When adding N to the low nutrient sample, it accelerates the microorganisms' cell growth in that site and increases the hydrocarbon degradation. Hassanshahian et al. (2013) recorded around 95% of oil removal at the end of the bioaugmentation with *A. borkumensis,* and 80% for the biostimulation treatment with the same bacteria species [56]. Therefore, the combination of bioaugmentation and biostimulation are used to enhance and speed up the treatment processes [30].

Future Challenges

Petroleum extraction has led to a significant increase in harmful chemical contaminants. Currently, bioremediation is a research hotspot for the remediation of oil due to its ease of use, economic feasibility, and lack of secondary contaminants. According to previous research, most micro-organisms and plants are better suited to remediation when there is less than 5% oil pollution. When oil concentrations at a site rise, the ability of microorganisms and plants to remediate is reduced.

Extreme climatic changes, such as changes in site temperature (below 10°C) and pH (below 4 or greater than 9), will limit microorganism and plant activity and slow the remediation process. Furthermore, due to high costs and technological factors, the applicability of chemical and physical methods is a challenge for many countries. Finally, despite the low cost and minimal secondary contamination, microbial remediation faces significant challenges.

Conclusion

Oil is a valuable natural resource that must be transported over long distances from extraction to the point of use. Accidental spills and deliberate spills such as those caused by acts of war or terrorism are always a possibility. If they occur, they can cause significant environmental damage. Though massive cleanup initiatives and wildlife rehabilitation can sometimes help the natural recovery process, evading oil spills and other ecological disasters is the best option than spending a lot of money on cleanup processes. Oil can be removed from contaminated locations using chemical, physical, and biological cleanup approaches. Among these different methods, bioremediation is an eco-friendly green technique with lower processing cost. Producing different bacterial consortiums, genetically modified microbes, and integrated processes are novel technologies for oil remediation. Among all the methods, bioremediation and integrated methods are the most effective and eco-friendly methods. However, developing and improving novel bioremediation technologies is still long.

References

[1] Manage, P.M., Kannangara, S.D. and Liyanage, G.Y., 2015. Role of native bacteria as a bioremediating agent for naphthalene and phenanthrene. *Proceedings of Academics World 12th International Conference, Singapore.* 20th December 2015.

[2] Liyanage, G.Y. and Manage, P.M., 2016. *Isolation and characterization of oil degrading bacteria from coastal waters and sediments from three locations in Sri Lanka,* 44(4) 351-358.

[3] Hylland, K., 2006. Polycyclic aromatic hydrocarbon (PAH) ecotoxicology in marine ecosystems. *Journal of Toxicology and Environmental Health, Part A, 69*(1-2), pp.109-123.

[4] Beyer, J., Trannum, H.C., Bakke, T., Hodson, P.V. and Collier, T.K., 2016. Environmental effects of the Deepwater Horizon oil spill: a review. *Marine pollution bulletin,* 110(1), pp.28-51.

[5] Damodharan, D., Rajesh Kumar, B., Gopal, K., De Poures, M.V. and Sethuramasamyraja, B., 2019. Utilization of waste plastic oil in diesel engines: A review. *Reviews in Environmental Science and Bio/Technology, 18*(4), pp.681-697.

[6] Dessbesell, L., Paleologou, M., Leitch, M., Pulkki, R. and Xu, C.C., 2020. Global lignin supply overview and kraft lignin potential as an alternative for petroleum-based polymers. *Renewable and Sustainable Energy Reviews, 123*, p.109768.

[7] Williams, K.N., Kephart, J.L., Fandiño-Del-Rio, M., Simkovich, S.M., Koehler, K., Harvey, S.A., Checkley, W. and CHAP Trial Investigators, 2020. Exploring the impact of a liquefied petroleum gas intervention on time use in rural Peru: a mixed methods study on perceptions, use, and implications of time savings. *Environment international,* 145, p.105932.

[8] Venosa, A.D., Campo, P. and Suidan, M.T., 2010. Biodegradability of lingering crude oil 19 years after the Exxon Valdez oil spill. *Environmental science & technology, 44*(19), pp.7613-7621.

[9] Crone, T.J. and Tolstoy, M., 2010. Magnitude of the 2010 Gulf of Mexico oil leak. *Science, 330*(6004), pp.634-634.

[10] Atlas, R.M. and Hazen, T.C., 2011. Oil biodegradation and bioremediation: a tale of the two worst spills in US history. *Environmental Science and Technology,* 45, 6709–6715.

[11] Franco, M.N., Galeano-Díaz, T., López, Ó., Fernández-Bolaños, J.G., Sánchez, J., De Miguel, C., Gil, M.V. and Martín-Vertedor, D., 2014. Phenolic compounds and antioxidant capacity of virgin olive oil. *Food chemistry*, *163*, pp.289-298.

[12] Brezna, B., Khan, A.A. and Cerniglia, C.E., 2003. Molecular characterization of dioxygenases from polycyclic aromatic hydrocarbon-degrading Mycobacterium spp. *FEMS microbiology letters*, *223*(2), pp.177-183.

[13] Evelyn, D.E. and Dickson, P., 2008. *On this Spot: Pinpointing the Past in Washington, DC*. Capital Books.

[14] *Group of Experts on the Scientific Aspects of Marine Environmental Protection* (GESAMP), available at: http://www.gesamp.org/, access: May, 2021.

[15] Hoegh-Guldberg, O., Cai, R., Poloczanska, E.S., Brewer, P.G., Sundby, S., Hilmi, K., Fabry, V.J., Jung, S., Skirving, W., Stone, D.A. and Burrows, M.T., 2014. *The ocean.*

[16] Carrera-Martinez, D., Mateos-Sanz, A., Lopez-Rodas, V. and Costas, E., 2011. Adaptation of microalgae to a gradient of continuous petroleum contamination. *Aquatic Toxicology,* 101(2), pp.342-350.

[17] Fingas, M., 2010. Review of the North Slope Oil properties relevant to environmental assessment and prediction. *Spill Science,* Edmonton, Alberta, Canada.

[18] Li, X., Li, P., Lin, X., Zhang, C., Li, Q. and Gong, Z., 2008. Biodegradation of aged polycyclic aromatic hydrocarbons (PAHs) by microbial consortia in soil and slurry phases. *Journal of hazardous materials*, *150*(1), pp.21-26.

[19] Ribeiro, H., Mucha, A.P., Almeida, C.M.R. and Bordalo, A.A., 2013. Bacterial community response to petroleum contamination and nutrient addition in sediments from a temperate salt marsh. *Science of the Total Environment*, *458*, pp.568-576.

[20] Lam, S.S., Mahari, W.A.W., Ok, Y.S., Peng, W., Chong, C.T., Ma, N.L., Chase, H.A., Liew, Z., Yusup, S., Kwon, E.E. and Tsang, D.C., 2019. Microwave vacuum pyrolysis of waste plastic and used cooking oil for simultaneous waste reduction and sustainable energy conversion: Recovery of cleaner liquid fuel and techno-economic analysis. *Renewable and Sustainable Energy Reviews*, *115*, p.109359.

[21] Alabresm, A., Chen, Y.P., Decho, A.W. and Lead, J., 2018. A novel method for the synergistic remediation of oil-water mixtures using nanoparticles and oil-degrading bacteria. *Science of the Total environment*, *630*, pp.1292-1297.

[22] Caravaca, F. and Roldan, A., 2003. Assessing changes in physical and biological properties in a soil contaminated by oil sludges under semiarid Mediterranean conditions. *Geoderma*, *117*(1-2), pp.53-61.

[23] Li, H., Peng, Z., Yang, X., Wang, W., Fu, J., Wang, J., Han, Y., Chai, Y., Guo, T., Yang, N. and Liu, J., 2013. Genome-wide association study dissects the genetic architecture of oil biosynthesis in maize kernels. *Nature genetics*, *45*(1), pp.43-50.

[24] Thapa, B., Kc, A.K. and Ghimire, A., 2012. A review on bioremediation of petroleum hydrocarbon contaminants in soil. *Kathmandu university journal of science, engineering and technology*, *8*(1), pp.164-170.

[25] Widdel, F. and Rabus, R., 2001. Anaerobic biodegradation of saturated and aromatic hydrocarbons. *Current opinion in biotechnology*, *12*(3), pp.259-276.

[26] Fuchs, G., Boll, M. and Heider, J., 2011. Microbial degradation of aromatic compounds—from one strategy to four. *Nature Reviews Microbiology*, *9*(11), pp.803-816.

[27] Bouwer, E.J. and Zehnder, A.J., 1993. Bioremediation of organic compounds—putting microbial metabolism to work. *Trends in biotechnology*, *11*(8), pp.360-367.

[28] Enríquez-Fernández, B.E., Álvarez de la Cadena y Yañez, L. and Sosa-Morales, M.E., 2011. Comparison of the stability of palm olein and a palm olein/canola oil blend during deep-fat frying of chicken nuggets and French fries. *International journal of food science & technology*, *46*(6), pp.1231-1237.

[29] Aggarwal, S., Schnabel, W., Buist, I., Garron, J., Bullock, R., Perkins, R., Potter, S. and Cooper, D., 2017. Aerial application of herding agents to advance in-situ burning for oil spill response in the Arctic: A pilot study. *Cold Regions Science and Technology*, *135*, pp.97-104.

[30] Luo, Q., Hou, D., Jiang, D. and Chen, W., 2021. Bioremediation of marine oil spills by immobilized oil-degrading bacteria and nutrition emulsion. *Biodegradation*, *32*(2), pp.165-177.

[31] Rodríguez-Rodríguez, A.A., Moreno-Trejo, M.B., Meléndez-Zaragoza, M.J., Collins-Martínez, V., López-Ortiz, A., Martínez-Guerra, E. and Sánchez-Domínguez, M., 2019. Spinel-type ferrite nanoparticles: synthesis by the oil-in-water microemulsion reaction method and photocatalytic water-splitting evaluation. *International journal of hydrogen energy*, *44*(24), pp.12421-12429.

[32] Ke, L., Zhang, C., Wong, Y.S. and Tam, N.F.Y., 2011. Dose and accumulative effects of spent lubricating oil on four common mangrove plants in South China. *Ecotoxicology and environmental safety*, *74*(1), pp.55-66.

[33] Nomack, M. and Cleveland, C.J., 2010. *Oil Spill Control Technologies. The Encyclopedia of Earth.*

[34] Dave, D.A.E.G. and Ghaly, A.E., 2011. Remediation technologies for marine oil spills: A critical review and comparative analysis. *American Journal of Environmental Sciences*, *7*(5), p.423.

[35] Widdel, F. and Rabus, R., 2001. Anaerobic biodegradation of saturated and aromatic hydrocarbons. *Current opinion in biotechnology*, *12*(3), pp.259-276.

[36] Martins, R., Seabra, J., Brito, A., Seyfert, C., Luther, R. and Igartua, A., 2006. Friction coefficient in FZG gears lubricated with industrial gear oils: biodegradable ester vs. mineral oil. *Tribology international*, *39*(6), pp.512-521.

[37] Rocchetti, L., Beolchini, F., Ciani, M. and Dell'Anno, A., 2011. Improvement of bioremediation performance for the degradation of petroleum hydrocarbons in contaminated sediments. *Applied and Environmental Soil Science, 2011*. https://doi.org/10.1155/2011/319657

[38] Yin, H., Qiang, J., Jia, Y., Ye, J., Peng, H., Qin, H., Zhang, N. and He, B., 2009. Characteristics of biosurfactant produced by Pseudomonas aeruginosa S6 isolated from oil-containing wastewater. *Process Biochemistry*, *44*(3), pp.302-308.

[39] El-Nawawy, A.S., El-Bagouri, I.H., Abdal, M. and Khalafawi, M.S., 1992. Biodegradation of oily sludge in Kuwait soil. *World Journal of Microbiology and Biotechnology*, *8*(6), pp.618-620.

[40] Manage, P.M. and Liyanage, G.Y., 2019. Antibiotics induced antibacterial resistance. In *Pharmaceuticals and personal care products: waste management and treatment technology* (pp. 429-448). Butterworth-Heinemann.

[41] Ekanayake, M. and Manage, P.M. 2020. Green Approach for Decolorization and Detoxification of Textile Dye - CI Direct Blue 201 using Native Bacterial Strains. *Environment and Natural Resource Journal,* Vol. 18(1), pp. 1-8.

[42] Ekanayake, M., Udayanga, D., Wijesekara, I. and Manage., P.M. 2021. Phytoremediation of synthetic textile dyes: Bio-sorption and enzymatic degradation involved in efficient dye decolorization by *Eichhornia crassispes* and *Pistia* sp. *Environmental Science and Pollution Research*, Vol. 28(16), pp.20476-20486.

[43] Idroos, F.S., De Silva, B.G.D.N.K. and Manage, P.M., 2017. *Biodegradation of microcystin analogues by Stenotrophomonas maltophilia isolated from Beira Lake Sri Lanka.*

[44] Idroos, F.S., Hettiarachchi, I.U. and Manage, P.M., 2016. Optimization of a standard method for enumeration of total cell counts of colonial Microcystis aeruginosa in environmental samples collected from Boralasgamuwa Lake, Sri Lanka. *Vidyodaya Journal of Science*, *18*.

[45] Kafilzadeh, F., Sahragard, P., Jamali, H. and Tahery, Y., 2011. Isolation and identification of hydrocarbons degrading bacteria in soil around Shiraz Refinery. *African Journal of Microbiology Research*, *5*(19), pp.3084-3089.

[46] Santos, H.F., Carmo, F.L., Paes, J.E., Rosado, A.S. and Peixoto, R.S., 2011. Bioremediation of mangroves impacted by petroleum. *Water, Air, & Soil Pollution*, *216*(1), pp.329-350.

[47] Kostka, J.E., Prakash, O., Overholt, W.A., Green, S.J., Freyer, G., Canion, A., Delgardio, J., Norton, N., Hazen, T.C. and Huettel, M., 2011. Hydrocarbon-degrading bacteria and the bacterial community response in Gulf of Mexico beach sands impacted by the Deepwater Horizon oil spill. *Applied and environmental microbiology*, *77*(22), pp.7962-7974.

[48] Omotayo, A.E., Ojo, O.Y. and Amund, O.O., 2012. Crude Oil Degradation by Microorganisms in Soil Composts. *Research Journal of Microbiology* 7 (4) 209-218.

[49] Kim, S.J., Choi, D.H., Sim, D.S. and Oh, Y.S., 2005. Evaluation of bioremediation effectiveness on crude oil-contaminated sand. *Chemosphere*, *59*(6), pp.845-852.

[50] Maiti, A., Das, S. and Bhattacharyya, N., 2012. Isolation and characterization of a new bacterial strain from petroleum oil contaminated soil, India. *J Sci,* 2, pp.103-108.

[51] Koma, D., Sakashita, Y., Kubota, K., Fujii, Y., Hasumi, F., Chung, S.Y. and Kubo, M., 2003. Degradation of car engine base oil by Rhodococcus sp. NDKK48 and Gordonia sp. NDKY76A. *Bioscience, biotechnology, and biochemistry*, *67*(7), pp.1590-1593.

[52] Grishchenkov, V.G., Townsend, R.T., McDonald, T.J., Autenrieth, R.L., Bonner, J.S. and Boronin, A.M., 2000. Degradation of petroleum hydrocarbons by

facultative anaerobic bacteria under aerobic and anaerobic conditions. *Process Biochemistry*, *35*(9), pp.889-896.
[53] Wu, G., Zhu, X., Ji, H. and Chen, D., 2015. Molecular modeling of interactions between heavy crude oil and the soil organic matter coated quartz surface. *Chemosphere*, *119*, pp.242-249.
[54] Bestawy, E.E., El-Shatby, B.F. and Eltaweil, A.S., 2020. Integration between bacterial consortium and magnetite (Fe3O4) nanoparticles for the treatment of oily industrial wastewater. *World Journal of Microbiology and Biotechnology*, *36*(9), pp.1-16.
[55] Zhang, B., Zhang, L. and Zhang, X., 2019. Bioremediation of petroleum hydrocarbon-contaminated soil by petroleum-degrading bacteria immobilized on biochar. *RSC advances*, *9*(60), pp.35304-35311.
[56] Hassanshahian, M., Emtiazi, G., Caruso, G. and Cappello, S., 2014. Bioremediation (bioaugmentation/biostimulation) trials of oil polluted seawater: a mesocosm simulation study. *Marine environmental research*, *95*, pp.28-38.

Chapter 6

Hydrocarbon Remediation: A Green Solution to the Problem of Hydrocarbon Pollutants

K. R. V. Bandara[1,2]
G. Y. Liyanage[1,4]
K. A. Sirisena[1,3]
and P. M. Manage[1,2,*]
[1]Centre for Water Quality and Algae Research, Department of Zoology, University of Sri Jayewardenepura, Gangodawila, Nugegoda, Sri Lanka
[2]Faculty of Graduate Studies, University of Sri Jayewardenepura, Gangodawila, Nugegoda, Sri Lanka
[3]Department of Environmental Technology, University of Colombo, Sri Lanka
[4]Department of Aquatic Bioresources, Faculty of Urban and Aquatic Bioresources, University of Sri Jayewardenepura, Sri Lanka

Abstract

Hydrocarbons are organic chemicals based on carbon (C) and hydrogen (H) compounds. Pollution of hydrocarbons (primarily petroleum hydrocarbons) in the environment is becoming a major issue worldwide. This is most likely due to the world's heavy reliance on petroleum as a primary energy source, population growth, rapid industrialization and a total disregard for environmental health. Polycyclic Aromatic Hydrocarbons (PAHs) and related compounds such as acridine, benzopyrene, fluoranthene, naphthalene, quinolone and phenanthridine have been found in high concentrations in the environment all over the

---

* Corresponding Author's Email: pathmalal@sjp.ac.lk.

In: Microbial Technologies and Their Applications
Editors: Santosh Kumar Karn and Anne Bhambri
ISBN: 979-8-88697-913-8

world. The unintentional or intentional release of hydrocarbons into the environment is a major source of pollution in the water, soil, and air. The natural equilibrium that exists between living species and their surroundings is usually disrupted by these hydrocarbon pollutants. Carcinogens and neurotoxic organic pollutants have been identified in hydrocarbon components. Hydrocarbons in the environment enter the human food web via plants due to their proclivity for bioaccumulation and transfer from one food chain to another, posing a threat to the ecosystem. Hydrocarbons have been found in various food chains, with disastrous results for both wildlife and humans. Man-made hydrocarbons are resistant to degradation and have a long half-life in the environment. Because hydrolysis and photolysis are ineffective for all hydrocarbon types, different degradation techniques of physical and chemical treatment methods are used. However, most of these methods have limitations when it comes to entirely remediating hydrocarbon contaminated sites. Some of these methods result in daughter compounds that are more harmful to the environment than the parental compounds. Bioremediation is the most environmentally friendly way for removing hydrocarbon contamination as it relies on natural microorganisms' ability to break down hydrocarbons into harmless substances. Contaminants from the environment are transformed or degraded using microbial enzymatic activities. Because it is a natural procedure that does not often produce harmful by-products, it is a cost-effective remediation strategy compared to other treatments. Due to the complete mineralization of toxins from the environment, it also provides a long-term solution. Biological remediation has several advantages over other treatment methods, including the ability to destroy pollutants rather than transfer them to another medium, low worker exposure to contaminants, long-term public health protection and the possibility of a shorter remediation time.

Keywords: hydrocarbon, pollution, neurotoxic, bioremediation, long-term solution

Introduction

Anthropogenic activities produce hydrocarbons into the environment and rehabilitation of polluted settings with classical physicochemical technologies is a challenging and expensive operation [1]. To remove toxic hydrocarbons from polluted areas, ecotoxicological consequences and human health hazards necessitated the use of "green technology". It is critical to transition away from costly old physicochemical treatments and toward more ecologically friendly biological treatment technologies that consume less energy, chemicals, and money to operate and allow for more long-term risk-based approaches to environmental cleanup [2]. The chapter provides an overview of the numerous biological treatment technologies applied to the cleanup of hazardous hydrocarbon-polluted areas. Biological treatment technologies include bioremediation, biostimulation, bioaugmentation, bioattenuation, bioventing, biosparging, bioslurry, biopiling, biotransformation, landfarming, composting, windrow, phytoremediation, mycoremediation, phyto-remediation, electro bioremediation, nanoremediation, and trichoremediation [1]. In contrast to the damaging and expensive physicochemical treatments technologies, they are widely embraced green technology techniques that are scientifically defensible, sustainable, non-invasive, eco-friendly, and cost-effective in the remediation of petroleum hydrocarbon-polluted settings. The benefits and downsides of various biological treatment technologies are illustrated in-depth in this chapter to aid incomplete knowledge and successful execution of their benefits and drawbacks.

Hydrocarbons in Nature and Their Significance

Hydrocarbons are colorless gases with extremely mild scents entirely made up of only two kinds of atoms, carbon and hydrogen [3]. Hydrocarbons contain carbon atoms from the periodic table's carbon 14th group, which contains carbon, silicon, germanium, tin, and lead. Aromatic and aliphatic hydrocarbons are the two basic types of hydrocarbons (Figure 1). Aromatic hydrocarbons contain at least one benzene ring, while alkanes, alkenes, and alkyne-based compounds are examples of aliphatic compounds. Carbon atom only has 4 electrons in the outermost shell, which means a single carbon atom can make a maximum of 4 covalent bonds to be stable. Hydrocarbons can form more complex long chains such as decane and nonane. Cyclic compounds like benzene can be formed by making bonds between themselves. This feature is known as catenation and it has the ability of hydrocarbons to form bonds with one another [4, 5].

Significance of the Hydrocarbons

- Hydrocarbons are primarily utilized for fuel.as an example: Liquefied Natural Gas (LNG) and Liquefied Petroleum Gas (LPG).

- Hydrocarbons are used for making polymers such as polystyrene, polyethene etc.
- Hydrocarbons are also used as raw materials in the production of pharmaceuticals and dyes.
- They are used as grease and lubricants.

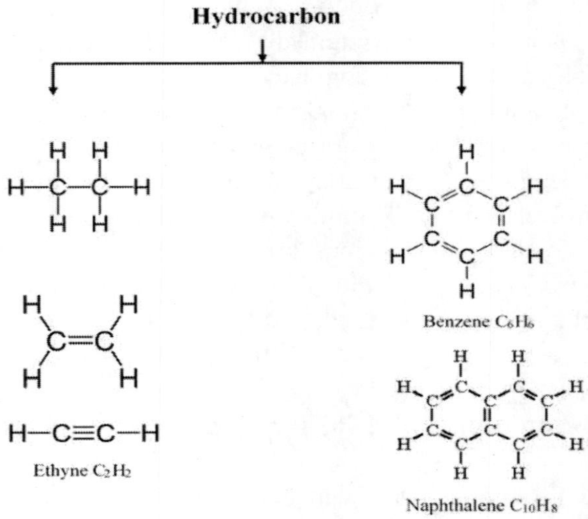

Figure 1. Major types of hydrocarbons.

Hydrocarbons are the most frequently applied organic compound globally and the primary component of the volatile organic carbons [6]. These compounds act as the driving force of modern civilization, as they are the foundation of fossil fuels. Hydrocarbons such as butane and propane are used in lighters, grills, lanterns and as internal combustion engine fuel [7]. Another common hydrocarbon is pentane and saturated pentane becomes a liquid at room temperature: this liquid is used as an organic solvent, fuel for transportation and as cleanser [8]. The larger hydrocarbon molecules referred to as kerosene and used as diesel fuel, jet fuel and heating oil and they are frequently used as lubricants and greases for engines [9]. The other hydrocarbons which are thicker than that will form substances related to wax or tar and that are commonly used in road construction and roofing [10].

Evolution of Hydrocarbons

The growth of new energy sources and related conversion technologies can be seen from the history of human culture. Human comfort, longevity, and economic success have improved as a result of these advancements [11]. Solar energy is naturally stored in the hydrocarbon bonds of plants and animals [12]. Then, the human cultural evolution has profited from stored hydrocarbon energy. The connecting of the energy in the hydrocarbon bonds of timber by fire, which allowed humans to exploit even more food resources, to smelt metals, and to bake ceramics, was one of the turning point in human history. The industrial revolution began around three hundred years ago with water-powered technologies and stationary wind which have been largely replaced by fossil hydrocarbons: coal in the nineteenth century, oil since the twentieth century, and now increasingly natural gas [13]. Since 1750, the worldwide consumption of hydrocarbons as a source of energy has increased by nearly 800 times during the 20th century [14]. Energy derived from hydrocarbons is essential for the economic, social, and environmental aspects of human development. Recently, scientific reports have published that the current world has entered a "postindustrial" era in which computers and more generally, human knowledge which have been replaced raw energy and materials in the generation of wealth [15]. However, there is still a strong correlation between energy and economic activity in the majority of industrialized and developing nations.

However, environmental hydrocarbons undergo a chemical reaction when exposed to sunlight or nitrogen oxides. Humans have produced hazardous pollutants and pollution during this industrial revolution, with hydrocarbons accounting for many of these toxic substances [16]. The great majority of crude oil, natural gas, and pesticides are made up of hydrocarbons. These compounds contribute to the greenhouse effect and ozone layer depletion [17]. They also reduce plant photosynthetic ability, raise the risk of cancer in humans and animals, and increase the risk of respiratory disease. Oil spills are the most well-known and well-documented harmful hydrocarbons, destroying marine plant life and causing the extinction of thousands of endangered species every year [18].

Hydrocarbon Pollution

There are two types of hydrocarbon pollution: point sources and non-point sources [19]. A single, identifiable source of pollution, such as a pipe or a drain, is referred to as a point source. This is a common method of dumping industrial waste into rivers and the sea. Diffuse pollution refers to pollution that originates from non-point sources. They're inputs and impacts that happen all over the place and can't be attributed to a single source. In contrast to individual point source discharges, they are frequently linked to specific land uses. As the example, pesticides, petroleum, and various harmful organic materials that are discharged into natural water bodies as effluents. Hydrocarbon contamination is caused by contaminated oil spills, lands, pesticides and urban storm water wastes [20]. The contamination triggered by petroleum hydrocarbons is a cause for concern due to these substances are toxic to numerous forms of life. It is common for crude oil to contaminate water due to its extensive use, accidental spills and connected dumping process. Petroleum is a complex blend of a broad spectrum in hydrocarbons with high and low molecular weight. As a result of human action, the concentration of hydrocarbons in the soil is rising. In some regions, higher hydrocarbons and metalloid concentrations have a significant impact [21].

Accidents of ships and oil platforms which are used to transport hydrocarbon containing products and discharging water into the sea which used to wash vessels' tanks are the primary causes of hydrocarbon pollution in the coastal stretch. Petroleum products and crude oil form a water-proof film on water and avoid oxygen transition in between the water and environment. It adversely affects to marine plants, animals, and human beings [22]. "Double-hulled" tankers are now used in international shipping to prevent leaks in any accidental situation [23].

Pesticides including fungicides, herbicides and insecticides are another source of hydrocarbon contamination in natural water sources. The majority of the pesticide remains in soil, where rain can wash it away. All these substances are responsible to the greenhouse effect and depletion of the ozone layer [24]. Industrialized lands highly contribute to the contamination of water with hydrocarbons [25]. The discharge of urban wastewater is also one of the most significant factors of hydrocarbon pollution [26]. Car parks and roads in urban areas are often polluted by oil and gasoline from the vehicles [20].

Impact of Hydrocarbon on Environment and Human Health

Recent studies have identified the toxicity of hydrocarbons including carcinogenicity and mutagenicity [27]. Increasing the hydrocarbon contamination levels in the environment and aquatic media are the main environmental concern due to they are frequently used and have toxic effects on the environment [28, 29]. The World Health Organization classifies polycyclic aromatic hydrocarbons (PAHs) and total petroleum hydrocarbons (TPHs) as persistent organic pollutants (POPs) (WHO). Because of their great potential for bioaccumulation in the food chain, extended half-life, vast distribution, and possible toxicity to humans, most PAHs have been banned by health authorities. These chemicals are highly lipophilic, and they bioaccumulate in the gastrointestinal tracts of mammals due to their interaction with the environment [30]. Because of their susceptibility and vulnerability to these components, animals and humans have experienced various health problems when exposed to hydrocarbons [27]. Some hydrocarbons such as benzo[a]pyrene, pyrene, and benzo[a]anthracene, have been found to have carcinogenic and mutagenic effects in studies [31, 32].

Continuous exposure to huge quantities of lubricant may interfere to the respiratory function of animals and humans [33]. Large oil spills are not only harmful to the environment, but even small emissions from automobiles and other sources cause to cumulate adverse effects on the environment. As an example, fish ingest contaminated water through their gills during respiration, which then accumulates in their gall bladder, liver, and stomach, rendering them unfit for human consumption [34]. Benzene containing hydrocarbons that uses as a solvent and a fuel responsible to red blood cells depletion, causes cancer in mammals, and damages bone marrow [35]. Polynuclear aromatic compounds are also responsible for increasing cancer risk. Hydrocarbons which containing alkyl nitrates chemically react with molecules in the atmosphere and produce nitrous oxide and resulted to affect blood vessels, liver, kidneys and nervous system of living animals and human beings [28, 29, 36].

Hydrocarbon-contaminated wastewater has a negative impact on plants, animals and humans. The hydrocarbon deposition on plant leaves reduced the surface area available to absorb sufficient oxygen and resulted to decrease crop yields. When farmland is irrigated with water containing hydrocarbons, harvest and available food for households will be reduced due to inappropriate crop growth [37]. Soil fertility might be decreased due to oil contamination in paddy water as the important nutrients are no longer available to plant consumption results in decrease of the harvest or crop production. The decline of crop production results to reduce farmer's incomes.

Hydrocarbon Remediation

Hydrocarbon contamination is being remedied nationally and internationally, posing a hazard to the environment and public health. Polluted environments have become a significant health issue for humans. Over the last few decades, several cleanup procedures have been developed and implemented. Phytoremediation, chemical remediation, and bioremediation are the most commonly used remediation procedures [38]. Because diverse activities are required for distinct kinds of molecules, it is challenging to break down petroleum hydrocarbons physiologically or chemically. As a result, cleaning up an oil-stained environment is challenging. Remediation strategies can be created after assessing the polluted site's oil content and physicochemical nature. Chemical and physical qualities and the pH of polluted water and soil all have a role in crude oil deterioration. Because oil-producing locations are often located near the shoreline, seawater is predominantly contaminated by oil spills during oil production. Chemical and biological approaches have been used to manage oil spills throughout the previous decade [1].

Chemical and Physical Remediation Methods

Chemical Remediation

Various chemicals can be used to treat hydrocarbon contaminants during the chemical remediation process. Typically, chemicals are able to alter the chemical and physical properties of contaminants [39]. Solidifiers, dispersants and chemical oxidants are the main steps for chemical remediation.

Dispersants

Dispersant surfactants can break down oil slicks into minute droplets, which can then be quickly diluted and easily destroyed [40]. Chemical dispersants can increase the surface area of oil droplets, resulting in a faster rate of natural biodegradation; this process makes the oil less adherent to the surface by preventing the development of oil-water emulsions, allowing for faster treatment. This is accomplished by turning oil layers into droplets that bacteria can destroy [41]. Chemical dispersants include Corexit 9500, Nokomis 3-F4, Neon AB3000, Slickgone NS, Finasol OSR 52, Saf-Ron Gold, ZI-400, Corexit 8667, and SPC 1000TM [41].

Solidifiers

During this process, lubricant is extracted using a physical method involving the interaction of dry granular ingredients in oil, which transforms its liquid state into a rubber-like solid state. Solidifiers may be applied in a variety of forms, including dry particulates and semisolid substances such as pucks, balls, sponge, etc. Because seawater provides energy for mixing, solidification can be enhanced by deploying solidifiers in the ocean. However, there are some disadvantages of this method: solidifiers are hard to recover after the solidification while they are time consuming with less efficient [42].

Chemical Oxidation

Oxidation contains the application of chemical agents capable of oxidizing organic pollutants. These chemical agents are applied into the soil or water in the contaminated area with the aid of mixing equipment. The effectiveness of the oxidation is dependent on oxidant quality, reaction efficiency, geological and pollutant conditions and the reaction time [43]. This technique is quick and suitable to apply in all weather conditions of the environment.

Physiochemical Remediation

Oil Spilled on Water

There are different practices for removing oil from surface water and preventing oil reach the coastal shoreline. The most common techniques are the use of dispersants and mechanical recovery. The crude oil spreads across the surface of the water due to it's lighter than water and the oil film quickly becomes thin. The velocity of oil spreading depends on the sea tide, type of oil and the temperature of atmosphere [44].

Using booms and skimmers, which collect the oil spill to a small area, accidental oil spills can be prevented from spreading and oil layer can then be collected into a container. Oil can be solubilized using biosurfactants, which are generally non-toxic to the environment. Burning can be applied as an optional method for oil spill remediation. However, this method is only useful when: the oil spill float on the surface of water, after oil has been reduced to a narrow area by booms or the oil spill is fresh. The technique described above has the disadvantage of polluting aquatic systems with the byproducts and smoke produced by burning of oil. Weather, tides, and ocean currents are the variables that determine the effectiveness of this remediation process. If oil contaminates the shoreline, various techniques are used to remove the sand and gravel. Occasionally, oil is absorbed by sponge-like sorbents. The oiled vegetation is washed with water to remove the oil, but severely damaged plants must be removed entirely. As a remediation method, contaminated water is pumped to the processing site when its concentration is less than ex situ. The beach sand and pebbles are excavated and cleaned away from the polluted sites.

Oil Spilled on Soil

Wellheads during offshore oil production and drilling activities, leaks from pipes and leaks from underground petroleum storage tanks, overflow from gathering stations, petroleum yields, and inappropriate dumping of petroleum waste all contribute to oil pollution of soil. This procedure may offer considerable dangers during the transportation, excavation, and handling of contaminated materials. Finding new landfills for the substance's final disposal is quite tricky. Continuous monitoring and adequate maintenance of separation barriers are essential if the contaminant remains at the site. As a result, the cap and physiochemical approaches are only a short-term fix.

Bioremediation

Bioremediation has emerged as the most attractive option for cleaning up oil-contaminated soil due to its low cost and capacity to prevent the accumulation of contaminants among the technologies and methodologies that have been explored [45]. Plants, microorganisms such as yeast, fungi, and bacteria are used in bioremediation to break down or degrade hazardous chemicals into less toxic or harmless substances. Bioremediation can be utilized to clean up an oil-contaminated site since some microbes use organic chemicals found in crude oil as their primary carbon source [27]. As a result, bacteria use aerobic and anaerobic reaction pathways to convert dangerous hydrocarbons in the soil to less damaging molecules like carbon dioxide, methane, and water [46]. Various enzymes and metabolic processes break down the hydrocarbons in petroleum. Hydrocarbon breakdown is hampered by the lack of a suitable enzyme [47].

Utilizing natural microorganisms for environmental decontamination is known as bioremediation [48]. Bacteria and fungi are naturally occurring microorganisms that use this mechanism to convert contaminants into beneficial or harmless compounds. This is also a process in which microorganisms degrade and metabolize chemical compounds to improve environmental quality [49]. *Bacillus* sp., *Achromobacter* sp., *Arthrobacter* sp., *Brevibacter* sp., *Corynebacterium* sp., *Flavobacterium* sp., *Nocardia* sp., *Pseudomonas* sp., and *Vibrio* sp. are the most common bacteria used in hydrocarbon cleanup. *Aspergillus* sp., *Candida* sp., *Clasosporium* sp., *Penicilium* sp., *Trichoderma* sp. and *Fusarium* sp. are the most common fungal species [42].

Biological remediation is a more efficient and cost-effective technique than the physicochemical methods. The rate of degradation of petroleum products is increasing as a result of the development of multiple remediation techniques. With comparison to other biological methods, microorganism-based bioremediation is considered as the most effective. However, the low adsorption and solubility of hydrocarbons with high molecular weight limit their approachability to microorganisms [50].

Phytoremediation

Phytoremediation Techniques

Phytoremediation is the process of extracting, degrading, and removing pollutants from water, soil, and the air using plants [51]. It involves several mechanisms that can cause pollutants to degrade, dissipate, immobilize, or accumulate [52]. Table 1 lists many phytoremediation applications with illustrations.

Table 1. Applications of phytoremediation with examples

Application	Description	Examples
Phytovolatilization	Volatilization of pollutants; chlorinated aliphatics, herbicides, ammunition waste aromatics, nutrients,	Grasses, Legumes
Rhizosphere bioremediation	Influence of specific plant species on the biodegradation of resistant organic pollutants by root-associated bacteria and fungi.	Phenolic releasers Grasses with fibrous roots Phreatophyte trees Aquatic plants
Phytostabilization	Reduction of bioavailability of contaminants in soil. Metals- (Pb, Cd, Zn, As, Cu, Cr, Se, U), PAHs, DDT, hydrophobic organics - PCNBs, dioxins, furans, pentachlorophenol, dieldrin	Phreatophyte trees Grasses soil erosion
Phytoextraction	Concentration of contaminants from soil to plant tissues. Metals (Pb, Cd, Zn, Ni, Cu)	Sunflowers, Indian mustard Serpentine plants, Barley Rape seed plants, Nettles, Hops, Crucifers, Dandelions
Rhizofiltration	Water based technique of remediation where plants take up contaminants by its roots; it is very useful for contamination of wetlands and estuaries. hydrophobic organics, Metals (Pb, Cd, Zn, Ni, Cu),	Aquatic plants: submerges (algae, stonewort, hydrilla); emergent (cattail, coontail, pondweed, arrowroot, duckweed)

Source: [42].

Mechanisms of Phytoremediation

Plants are more feasible than physical and chemical remediation treatments for various contaminants in contaminated land and water [53]. Because of their sedentary nature, plants have evolved several defenses against hazardous chemicals. Plants take pollutants from the soil through their roots and transfer them to various plant tissues. They can be digested, volatilized, or sequestered, acting as solar-powered pumping and filtering systems [45, 54]. The plant uses a variety of ways to remove contaminants from the soil. Biophysical and biochemical activities such as adsorption, translocation, transport, mineralization and transformation by plant enzymes are all part of the phytoremediation process [55]. Dehalogenases, laccases, P450 dioxygenases, monooxygenases, nitroreductases, phosphatases, peroxidases, and nitrilases are among the contaminant-degrading enzymes found in plants [56]. Mineral nutrition, photosynthesis, transpiration, and metabolism are among them. The root of the plant is responsible for the uptake of organic and inorganic compounds from the soil and the binding and stabilization of substances on its external surfaces via interactions with rhizosphere bacteria. The uptake or release of molecules is caused by the exchange of gases between the plant's components and the atmosphere. Six phytotechnologies have been recognized by the Interstate Technology and Regulatory Cooperation for treating a variety of pollutants in a variety of substrates [57]. The mechanism of phytoremediation is depicted in Figure 2.

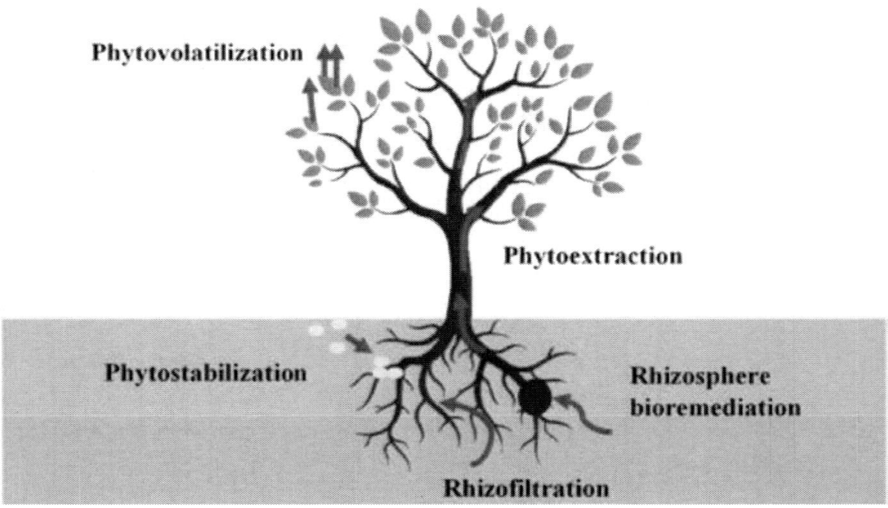

Figure 2. Illustration of the mechanism of phytoremediation.

Microorganisms

Bacteria

Bacteria play an important role in hydrocarbon degradation among these microorganisms. [58] reviewed that low molecular weight alkanes degrade rapidly. Biodegradation of petroleum is driven by the ability of microorganisms to use hydrocarbons to meet their cell growth and energy needs through hydrocarbon degradation [59]. Compared to pure cultures, mixed cultures perform a greater amount of biodegradation of petroleum. In many ecosystems, the indigenous microbial community is capable of biodegrading oil, but environmental conditions must be favorable for oil degradation metabolic activity. Hydrocarbon degradation by indigenous microorganisms has several advantages over adding microorganisms.

The genera Micrococcus, Bacillus, Streptomyces, Actinomyces, Azatobacter, Aspergillus, Penicillium, Enterobacter and Trichoderma recorded as hydrocarbon utilizers [60]. [61] recorded that the genera Micrococcus, Bacillus, Corynebacterium are hydrocarbon degrading microorganisms and the other studies have reported the same bacteria as hydrocarbon degraders as well [62]. Information regarding bioremediation of oil in Sri Lanka is not available and according to the knowledge and literature available. [63] identified Corynebacterium sp., Bacillus sp., Enterobacter sp. and Micrococcus sp. as oil degradation bacteria and that is the first report in Sri Lanka.

Fungi

In the biodegradation of hydrocarbons in soils, fungi are more critical than bacteria. Filamentous fungi are primarily found in surface films and sediments in aquatic environments. To break down polycyclic aromatic hydrocarbons, fungi use the same metabolic pathways as the human species (PAHs). In filamentous fungus, two significant kinds of cytochrome P450 monooxygenases have distinct properties. PAHs can be converted to dihydrodiols, phenols, and other metabolites and conjugates by various fungi. One rare species that can catabolize them to CO_2 is *Phanerochaete chrysosporium* [64, 65].

Types of bioremediation

Complex mixtures of hydrocarbons, such as crude oil in aquatic environments, freshwater and soil are degraded by different microorganism species with extensive enzymatic capacities [55]. There are two processes in the bioremediation as follows:

1. Bioaugmentation
2. Biostimulation

Bioaugmentation

Bioaugmentation is a process that degrades toxic substances including hydrocarbons through the addition of microorganisms to reduce the environmental pollution [66]. The applied microorganisms have ability to degrade hydrocarbon in the polluted water. It is occasionally involves the biodegradation of hydrocarbon through the addition of genetically modified microorganisms to polluted [66].

Biostimulation

Biostimulation reduces harmful compounds by providing the nutrients that indigenous bacteria need to break down hydrocarbons [67]. Increased carbon encourages the growth of microbes that degrade oil in an oil spill. The ability of microorganisms to break down hydrocarbons is boosted by adding a sufficient amount of additional nutrients. Microorganisms can achieve their maximum growth rate and, as a result, their maximum contaminant uptake rate as a result of this. Maximal biostimulation was detected at the optimal nutrient concentration required for maximum microbe growth, and this concentration was maintained for as long as possible [68].

Bioremediation by Genetically Engineered Microorganisms

In most oil-polluted areas, native microorganisms are incapable or unwilling to digest the pollutants, necessitating the employment of genetically engineered or modified microorganisms (GEMs) that can boost the breakdown yield. GEMs have shown tremendous potential in the bioremediation of hydrocarbon-contaminated sites, soil-groundwater, and activated sludge settings. When specific genes are turned on, enzymes and proteins that break down pollutants are produced [69]. GEMs offer a plethora of study options that have important ramifications. It is sometimes feasible to create and introduce new metabolic routes in microbes when existing metabolic pathways cannot digest contaminants. *Pseudomonas fluorescens* (strain HK44GEM) was genetically altered and introduced into a contaminated soil site with Vector pUTK21, a naphthalene catabolic plasmid [70]. This was the first GEM to be released for hydrocarbon degradation field experiments. A transposon-based lux gene coupled with a promoter for naphthalene catabolic genes resulted in bioluminescence in this strain. As a result, this strain generates bioluminescence anytime it is activated to degrade hydrocarbon contaminants, and it can be used as a reporter gene for naphthalene availability and degradation.

Application of Bioremediation

1. A natural and ecological process; when contaminants are present, the number of existing microbes grows, and the microbial population naturally reduces as the contaminants disintegrate. The residues obtained from biological treatment, such as carbon dioxide, fatty acids, and water, are normally nonhazardous, and plants can utilize the CO_2 for photosynthesis.
2. Bioremediation is the act of destroying rather than moving the target compounds from one area to another.
3. Bioremediation is more expensive than other approaches for removing hazardous material. Cleaning up the Exxon Valdez disaster, for example, was more expensive than bioremediating 120 km of beach front in a single day.
4. Bioremediation is an *in-situ* treatment that eliminates the need for a large amount of polluted waste to be transported off-site and reduces transportation risk.
5. Nutrient formulation can boost microbial effectiveness in the bioremediation process.
6. Fatty acids, CO_2, water, and other residues derived through biological treatment are generally nonhazardous. It is a less expensive approach to toxic waste remediation than other options.

Hydrocarbons have a lower polarity and solubility than polar substances like methanol due to their different solubility [36, 71]. Hydrocarbon degradation is determined by more than just solubilization. Numerous microbes, including *Bacillus subtilis*, *Pseudomonas aeruginosa*, *Pseudomonas putida*, *Bacillus licheniformis*, *Bacillus cereus*, and *Bacillus laterosporus*enhance the surface area of the substrate by excreting emulsifiers. Microorganisms increase the absorption of hydrophobic substances by changing their cell surface. Modifying the physicochemical properties of hydrocarbons can change the behavior and quality of individual hydrocarbons and combinations [42].

Advantages and Disadvantages of Bioremediation

Advantages and limitations of the biodegradation of hydrocarbons are listed in Table 2.

Table 2. Advantages and limitations of the biodegradation of hydrocarbons

Advantages	Disadvantages
Bioremediation is a natural process that is widely recognized as an efficient method for removing hydrocarbon waste. The harmless biodegraded compounds can be incorporated into the environment (carbon dioxide, water, and biomass.)	The bioremediation process is restricted to biodegradable substances. Not every hydrocarbon pollutant is biodegradable or susceptible to microbial degradation. For instance, chlorinated hydrocarbons
Transferring contaminants may result in leaching and additional contamination; however, bioremediation aids in degrading the pollutants on-site without adding to the danger.	The growth of microorganisms for bioremediation of a polluted site is frequently very specific and demanding; the factors affecting the growth of microorganisms must be optimal for microbes to degrade contaminants efficiently.
Bioremediation ensures that biodegradation waste can be reintroduced into the environment and does not need to be transported off-site for disposal.	Sites containing multiple types of contaminants in various phases (solid, liquid, gas) that require special treatment or a combination of native or genetically modified microorganisms.
It is relatively more cost-effective than other hazardous waste removal techniques.	It is a time-consuming process that may require additional pre-treatment before microorganisms can degrade them (excavation, incineration). This makes the procedure tiresome.

Future Prospectus

Over the last few decades, bioremediation of petroleum contaminated sites through the activity of microorganisms has gained widespread attention as a favorable remediation strategy [36]. Further research in this field could develop the most efficient and time-saving microbial technologies, which is vital for both developed and developing countries like Sri Lanka, which has accelerated the development of international shipping harbors and fishing harbors along its coast. All such development projects and activities enhance environmental pollution either by direct and indirect way. Hydrocarbon pollution is one of the most prominent ways of aquatic pollution in coastal areas and we are far away behind on the research bioremediation of hydrocarbon pollutant with ecofriendly techniques. Hence introduction, establishment and development of such novel ecofriendly, cost effective bio technologies for the world is timely important to help for the development. Therefore, future studies should continue to combine complementary approaches in order to advance our understanding of microbial processes and create a stronger basis for assessing biodegradation in the natural environment.

Conclusion

Because of the world's rapid development, hydrocarbon pollutants spread quickly on land, the aquatic environment, and the atmosphere. This has been an issue since the use of fossil fuels and the beginning of the industrial revolution. With unprecedented population growth, frequent oil spills, pipeline leaks, and widespread pesticide use, pollution levels have risen dramatically. Animals and natural microbiological populations on land, air, water surfaces, and underground ecosystems are at risk. As a result, the most crucial part of recovering from hydrocarbon pollution is environmental cleanup. This chapter discusses the causes of hydrocarbon pollution and the ways to resolve it. Phytoremediation, bioremediation, and chemical-physical treatments are some of the remediation strategies discussed in this chapter. With the industrial and green revolutions leads to increase the levels of hydrocarbon pollutants in the environment. That also act as a selective pressure to evolve a larger number of hydrocarbon-degrading microbial species in modern environment. The recent developments in the field of molecular microbiology in particular modern DNA sequencing approaches also immensely help to isolate and characterize hydrocarbon-degrading microorganisms, metabolites and metabolic pathways.

References

[1] Ossai, I.C., Ahmed, A., Hassan, A. and Hamid, F.S., 2020. Remediation of soil and water contaminated with petroleum hydrocarbon: A review. *Environmental Technology and Innovation*, 17, p. 100526.

[2] Villegas, L.G.C., Mashhadi, N., Chen, M., Mukherjee, D., Taylor, K.E. and Biswas, N., 2016. A short review of techniques for phenol removal from wastewater. *Current Pollution Reports,* 2(3), p. 157-167.

[3] Nahak, B.K., Preetam, S., Sharma, D., Shukla, S.K., Syväjärvi, M., Toncu, D.C. and Tiwari, A., 2022. Advancements in net-zero pertinency of lignocellulosic biomass for climate neutral energy production. *Renewable and Sustainable Energy Reviews,* 161, p. 112393.

[4] Zheng, Y., Wang, J., Wang, D. and Zheng, Z., 2022. Advanced catalytic upgrading of biomass pyrolysis vapor to bio-aromatics hydrocarbon: A review. *Applications in Energy and Combustion Science,* p. 100061.

[5] Queiroz, R.N., Prediger, P. and Vieira, M.G.A., 2022. Adsorption of polycyclic aromatic hydrocarbons from wastewater using graphene-based nanomaterials synthesized by conventional chemistry and green synthesis: A critical review. *Journal of Hazardous Materials,* 422, p. 126904.

[6] Li, N., Jiang, Q., Wang, F., Cui, P., Xie, J., Li, J., Wu, S. and Barbieri, D.M., 2021. Comparative assessment of asphalt volatile organic compounds emission from field to laboratory. *Journal of Cleaner Production,* 278, p. 123479.

[7] Thomas, A.R., 2018. *American Shale Energy and the Global Economy: Business and Geopolitical Implications of the Fracking Revolution.* Springer cham.

[8] Amaral, L.V., Santos, N.D.S.A., Roso, V.R., de Oliveira Sebastião, R.D.C. and Pujatti, F.J.P., 2021. Effects of gasoline composition on engine performance, exhaust gases and operational costs. *Renewable and Sustainable Energy Reviews,* 135, p. 110196.

[9] Kuppusamy, S., Maddela, N.R., Megharaj, M. and Venkateswarlu, K., 2020. An overview of total petroleum hydrocarbons. *Total petroleum hydrocarbons,* p. 1-27.

[10] Zhang, Y., Huang, C., Kong, F., Wang, Y., Shi, Q. and Zhang, L., 2022. Selective molecular characterization of olefins in hydrocarbon mixtures by Ag+ complexation ESI high-resolution mass spectrometry. *Fuel,* 319, p. 123760.

[11] Witte, K.A., Fiers, P., Sheets-Singer, A.L. and Collins, S.H., 2020. Improving the energy economy of human running with powered and unpowered ankle exoskeleton assistance. *Science Robotics,* 5, p. 9108.

[12] Kalair, A., Abas, N., Saleem, M.S., Kalair, A.R. and Khan, N., 2021. Role of energy storage systems in energy transition from fossil fuels to renewables. *Energy Storage,* 3, p. 135.

[13] Ede, A., (2019). *Technology and society: a world history.* Cambridge University Press.

[14] Fattouh, B., Poudineh, R. and West, R., 2019. The rise of renewables and energy transition: what adaptation strategy exists for oil companies and oil-exporting countries? *Energy transitions*, 3, p. 45-58.

[15] Pesa, I. and Ross, C., 2021. Extractive industries and the environment: Production, pollution, and protest in global history. *The Extractive Industries and Society*, 8, p. 100933.

[16] Bradu, P., Biswas, A., Nair, C., Sreevalsakumar, S., Patil, M., Kannampuzha, S., Mukherjee, A.G., Wanjari, U.R., Renu, K., Vellingiri, B. and Gopalakrishnan, A.V., 2022. Recent advances in green technology and Industrial Revolution 4.0 for a sustainable future. *Environmental Science and Pollution Research*, 1, p. 1-32.

[17] Babarinde, T.O., Akinlabi, S.A. and Madyira, D.M., 2021. The Use of Hydrocarbon Refrigerants in Combating Ozone Depletion and Global Warming: A Review. *Trends in Mechanical and Biomedical Design*, 1, p. 731-739.

[18] Sharma, P., 2022. Microbial communication during bioremediation of polyaromatic hydrocarbons. *Systems Microbiology and Biomanufacturing*, p. 1-15.

[19] Varadarajan, R. and Bathusha, A.M., 2019. Non-Point Source Pollutants from Motor Vehicles: Effects and Control. In *Global Perspectives on Air Pollution Prevention and Control System Design*, p. 227-238.

[20] Muller, A., Österlund, H., Marsalek, J. and Viklander, M., 2020. The pollution conveyed by urban runoff: A review of sources. *Science of the Total Environment*, 709, p. 136125.

[21] Cerniglia, C.E. and Sutherland, J.B., 2001. Bioremediation of polycyclic aromatic hydrocarbons by ligninolytic and non-ligninolytic fungi. In: Gadd GM (ed) *Fungi in bioremediation.* Cambridge University Press, Cambridge, 23, p. 136 187.

[22] Bhattacharjee, S. and Dutta, T., 2022. An overview of oil pollution and oil-spilling incidents. *Advances in Oil-Water Separation*, 1, p. 3-15.

[23] Varsami, C., 2019. Analysis of Building Requirements for Oil Tankers in Order to Prevent Maritime Accidents. *Constanta Maritime University Annals*, 18, p. 223.

[24] Hungerford, N.L., Fletcher, M.T., Tsai, H.H., Hnatko, D., Swann, L.J., Kelly, C.L., Anuj, S.R., Tinggi, U., Webber, D.C., Were, S.T. and Tan, B.L., 2021. Occurrence of environmental contaminants (pesticides, herbicides, PAHs) in Australian/Queensland Apis mellifera honey. *Food Additives and Contaminants: Part B*, 14, p. 193-205.

[25] Bi, C., Wang, X., Jia, J. and Chen, Z., 2018. Spatial variation and sources of polycyclic aromatic hydrocarbons influenced by intensive land use in an urbanized river network of East China. *Science of the Total Environment*, 627, p. 671-680.

[26] Yu, H., Liu, Y., Han, C., Fang, H., Weng, J., Shu, X., Pan, Y. and Ma, L., 2021. Polycyclic aromatic hydrocarbons in surface waters from the seven main river basins of China: Spatial distribution, source apportionment, and potential risk assessment. *Science of the Total Environment*,752, p. 141764.

[27] Das, N. and Chandran, P., 2011. Microbial degradation of petroleum hydrocarbon contaminants: an overview. *Biotechnology research international*, 2011, p. 1-13.

[28] Gibson, D.T. and Parales, R., 2000. Aromatic hydrocarbon dioxygenases in environmental biotechnology. *Current Opinion in Biotechnology,*11, p. 236 243.

[29] Deeduah, F. and Iwuoha, G., 2021. Genotoxicity and carcinogenicity of traditionally roasted meat using indicator polycyclic aromatic hydrocarbons (PAHs), Port Harcourt, Nigeria. DF Mene and GN Iwuoha. Genotoxicity and Carcinogenicity of Traditionally Roasted Meat Using Indicator Polycyclic Aromatic Hydrocarbons (PAHs), Port Harcourt, Nigeria. *Chemistry International*, 7, p. 217-223.

[30] Onyegeme-Okerenta, B.M., West, O.L. and Chuku, L.C., 2022. Concentration, dietary exposure and human health risk assessment of total petroleum and polycyclic aromatic hydrocarbons in seafood from coastal communities in Rivers State, Nigeria. *Scientific African*, 16, p. 01186.

[31] Da Silva Junior, F.C., Felipe, M.B.M.C., de Castro, D.E.F., da Silva Araújo, S.C., Sisenando, H.C.N. and de Medeiros, S.R.B., 2021. A look beyond the priority: A systematic review of the genotoxic, mutagenic, and carcinogenic endpoints of non-priority PAHs. *Environmental Pollution*, 278, p. 116838.

[32] Darwish, W.S., Chiba, H., El-Ghareeb, W.R., Elhelaly, A.E. and Hui, S.P., 2019. Determination of polycyclic aromatic hydrocarbon content in heat-treated meat

retailed in Egypt: Health risk assessment, benzo [a] pyrene induced mutagenicity and oxidative stress in human colon (CaCo-2) cells and protection using rosmarinic and ascorbic acids. *Food chemistry*, 290, p. 114-124.

[33] Ahmed, F. and Fakhruddin, A.N.M., 2018. A review on environmental contamination of petroleum hydrocarbons and its biodegradation. *International Journal of Environmental Sciences & Natural Resources*, 11, p. 1-7.

[34] Gerba, C.P., 2019. Environmental toxicology. In *Environmental and pollution science* Academic Press, p. 511-540.

[35] Soni, V., Singh, P., Shree, V. and Goel, V., 2018. Effects of VOCs on human health. In *Air pollution and control*. Springer, Singapore, 1, p. 119-142.

[36] Alvarez, A., Saez, J.M., Costa, J.S.D., Colin, V.L., Fuentes, M.S., Cuozzo, S.A., Benimeli, C.S., Polti, M.A. and Amoroso, M.J., 2017. Actinobacteria: current research and perspectives for bioremediation of pesticides and heavy metals. *Chemosphere*, 166, p. 41-62.

[37] Olowoyo, J.O. and Mugivhisa, L.L., 2019. Evidence of uptake of different pollutants in plants harvested from soil treated and fertilized with organic materials as source of soil nutrients from developing countries. *Chemical and Biological Technologies in Agriculture*, 6, p. 1-11.

[38] Li, C., Zhou, K., Qin, W., Tian, C., Qi, M., Yan, X. and Han, W., 2019. A review on heavy metals contamination in soil: effects, sources, and remediation techniques. *Soil and Sediment Contamination: An International Journal*, 28, p. 380-394.

[39] Geueke, B., Groh, K. and Muncke, J., 2018. Food packaging in the circular economy: Overview of chemical safety aspects for commonly used materials. *Journal of Cleaner Production*, 193, p. 491-505.

[40] Fernandez, V.I., Stocker, R. and Juarez, G., 2022. A tradeoff between physical encounters and consumption determines an optimal droplet size for microbial degradation of dispersed oil. *Scientific reports*, 12, p. 1-10.

[41] Elijah, A.A., 2022. A Review of the Petroleum Hydrocarbons Contamination of Soil, Water and Air and the Available Remediation Techniques, Taking into Consideration the Sustainable Development Goals. *Earthline Journal of Chemical Sciences*, 7, p. 97-113.

[42] Srivastava, M., Srivastava, A., Yadav, A. and Rawat, V., 2019. Source and control of hydrocarbon pollution. In *Hydrocarbon Pollution and its Effect on the Environment*. IntechOpen, p. 9. London, UK.

[43] Baciocchi, R., D'Aprile, L., Innocenti, I., Massetti, F. and Verginelli, I., 2014. Development of technical guidelines for the application of in-situ chemical oxidation to groundwater remediation. *Journal of cleaner production*, 77, p. 47-55.

[44] Al-Majed, A.A. Adebayo, A.R. and Hossain, M.E., 2012. A sustainable approach to controlling oil spills. *Journal of environmental management*, 113, p. 213-227.

[45] Dubchak, S. and Bondar, O., 2019. Bioremediation and phytoremediation: best approach for rehabilitation of soils for future use. In *Remediation measures for radioactively contaminated areas*. Springer, Cham, p. 201-221.

[46] Pandey, B. and Fulekar, M.H., 2012. Bioremediation technology: A new horizon for environmental clean-up. *Biology and Medicine*, 4, p. 51.

[47] Thapa, B., Kc, A.K. and Ghimire, A., 2012. A review on bioremediation of petroleum hydrocarbon contaminants in soil. *Kathmandu university journal of science, engineering and technology*, 8, p. 164-170.

[48] Singh, R., 2014. Microorganism as a tool of bioremediation technology for cleaning environment: a review. *Proceedings of the International Academy of Ecology and Environmental Sciences*,4, p. 1.

[49] Xenia, M.E. and Refugio, R.V., 2016. Microorganisms metabolism during bioremediation of oil contaminated soils. *Journal of Bioremediation and Biodegradation*, 7, p. 20-25.

[50] Kour, D., Kaur, T., Devi, R., Yadav, A., Singh, M., Joshi, D., Singh, J., Suyal, D.C., Kumar, A., Rajput, V.D. and Yadav, A.N., 2021. Beneficial microbiomes for bioremediation of diverse contaminated environments for environmental sustainability: present status and future challenges. *Environmental Science and Pollution Research*,28, p. 24917-24939.

[51] Ekta, P. and Modi, N.R., 2018. A review of phytoremediation. *J Pharmacogn Phytochem*, 7, p. 1485-1489.

[52] Padmavathiamma, P.K. and Li, L.Y., 2007. Phytoremediation technology: hyper-accumulation metals in plants. *Water, Air, and Soil Pollution*, 184, p. 105-126.

[53] Huang, X.D., El-Alawi, Y., Gurska, J., Glick, B.R. and Greenberg, B.M., 2005. A multi-process phytoremediation system for decontamination of persistent total petroleum hydrocarbons (TPHs) from soils. *Microchemical Journal*, 81, p. 139-147.

[54] Sharma, S., Chatterjee, S., Datta, S., Mitra, A., Vairale, M.G., Veer, V., Chaurasia, A. and Gupta, D.K., 2014. In vitro selection of plants for the removal of toxic metals from contaminated soil: Role of genetic variation in phytoremediation. *Heavy Metal Remediation Transport and Accumulation in Plants; Gupta, DK, Chattejee, S., Eds*, 1, p. 155-177.

[55] Meagher, R.B., 2000. Phytoremediation of toxic elemental and organic pollutants. *Current opinion in plant biology*, 3, p. 153-162.

[56] Gerhardt, K.E., Huang, X.D., Glick, B.R. and Greenberg, B.M., 2009. Phytoremediation and rhizoremediation of organic soil contaminants: potential and challenges. *Plant science*, 176, p. 20-30.

[57] Marmiroli, N., Marmiroli, M. and Maestri, E., 2006. Phytoremediation and phytotechnologies: a review for the present and the future. *Soil and water pollution monitoring, protection and remediation*, 1, p. 403-416.

[58] Abha, S. and Singh, C.S., 2012. Hydrocarbon pollution: effects on living organisms, remediation of contaminated environments, and effects of heavy metals co-contamination on bioremediation. *Introduction to enhanced oil recovery (EOR) processes and bioremediation of oil-contaminated sites*, 1, p.185-206.

[59] Hua, F. and Wang, H.Q., 2014. Uptake and trans-membrane transport of petroleum hydrocarbons by microorganisms. *Biotechnology & Biotechnological Equipment*, 28, p. 165-175.

[60] Rhykerd, R.L., Crews, B., McInnes, K.J. and Weaver, R.W., 1999. Impact of bulking agents, forced aeration, and tillage on remediation of oil-contaminated soil. *Bioresource Technology*, 67, p. 279-285.

[61] Omotayo, A.E., Ajayi, A. and Amund, O.O., 2017. Occurrence of hydrocarbon-degrading nitrogen-fixing bacteria in the rhizosphere of Paspalum vaginatum Sw. *Journal of BioScience and Biotechnology*, 6, p. 119-127.

[62] Latha, R., Kalaivani, R., 2012. Bacterial Degradation of Crude Oil by Gravimetric Analysis, *Advances in Applied Science Research*, 3, p. 2789-2795.

[63] Liyanage, G.Y. and Manage, P.M., 2016. Isolation and characterisation of oil degrading bacteria from coastal waters and sediments from three locations in Sri Lanka. *Journal of National Science Foundation, Sri Lanka*, p. 351-358.

[64] Cerniglia, C.E., 1997. Fungal metabolism of polycyclic aromatic hydrocarbons: past, present and future applications in bioremediation. *Journal of Industrial Microbiology and Biotechnology*, 19, p. 324-333.

[65] Kadri, T., Rouissi, T., Brar, S.K., Cledon, M., Sarma, S. and Verma, M., 2017. Biodegradation of polycyclic aromatic hydrocarbons (PAHs) by fungal enzymes: A review. *Journal of environmental sciences*, 51, p. 52-74.

[66] Sayed, K., Baloo, L. and Sharma, N.K., 2021. Bioremediation of total petroleum hydrocarbons (TPH) by bioaugmentation and biostimulation in water with floating oil spill containment booms as bioreactor basin. *International Journal of Environmental Research and Public Health*, 18, p. 2226.

[67] Shahi, A., Aydin, S., Ince, B. and Ince, O., 2016. Reconstruction of bacterial community structure and variation for enhanced petroleum hydrocarbons degradation through biostimulation of oil contaminated soil. *Chemical Engineering Journal*, 306, p. 60-66.

[68] Ramos-García, Á.A., Walecka-Hutchison, C. and Freedman, D.L., 2022. Effect of biostimulation and bioaugmentation on biodegradation of high concentrations of 1, 4-dioxane. *Biodegradation*, 33, p. 157-168.

[69] Rao, M.A., Scelza, R., Scotti, R. and Gianfreda, L., 2010. Role of enzymes in the remediation of polluted environments. *Journal of soil science and plant nutrition*, 10, p. 333-353.

[70] Sayler, G.S. and Ripp, S., 2000. Field applications of genetically engineered microorganisms for bioremediation processes. *Current opinion in biotechnology*, 11, p. 286-289.

[71] Dowaidar, A.M., El-Shahawi, M.S. and Ashour, I., 2007. Adsorption of polycyclic aromatic hydrocarbons onto activated carbon from non-aqueous media: 1. the influence of the organic solvent polarity. *Separation Science and Technology*, 42, p. 3609-3622.

Chapter 7

Phytoremediation and Phytostabilization: The Current Scenario and Future Prospects

M. S. Ekanayake[1,2]
K. R. V. Bandara[1,4]
F. S. Idroos[1,3]
and P. M. Manage[1,4,*]

[1]Centre for Water Quality and Algae Research, Department of Zoology, University of Sri Jayewardenepura, Sri Lanka
[2] Department of Aquaculture and Fisheries, Faculty of Livestock, Fisheries and Nutrition, Wayamba University of Sri Lanka, Sri Lanka
[3]Department of Zoology, Faculty of Applied Sciences, University of Sri Jayewardenepura, Sri Lanka
[4]Faculty of Graduate Studies, University of Sri Jayewardenepura, Gangodawila, Nugegoda, Sri Lanka

Abstract

The upsurge of living standards and life expectancy has given rise to a multitude of industries that alter the soil, water, and air by adding hazardous materials in a direct or indirect manner. Most of these chemicals are resistant to natural degradation processes, allowing them to persist in the environment for decades, disrupting natural biogeochemical cycles and threatening human health. Therefore, waste management containing various hazardous chemicals must remain a top priority. Physicochemical treatment methods such as activated carbon, membrane filtration, photochemical and electrochemical destruction, irradiation, and ion exchange are expensive, produce secondary

* Corresponding Author's Email: pathmalal@sjp.ac.lk.

In: Microbial Technologies and Their Applications
Editors: Santosh Kumar Karn and Anne Bhambri
ISBN: 979-8-88697-913-8

pollution, and have limited applicability. As a result, the development of cost-effective, efficient, and environmentally friendly remediation technologies is critical. Phytoremediation, the use of plants and/or their associated microflora for the treatment of contaminants, is a growing concern to remove pollutants or reduce their bioavailability in the environment. It is an aesthetically pleasing, low-cost, easily monitored technique that preserves topsoil, reduces erosion, and pollution leaching. Plants and endophytes have the ability to uptake pollutants from contaminated soil or liquid media at low concentrations and then degrade, transform, or immobilize the pollutants through various mechanisms such as phytodegradation, rhyzodegradation, phyto-stabilization, phytoextraction, and phytovolatilization. Even though phytoremediation has a long history in contaminated sites, the current concern and approach have been to find novel and developed enzyme-based phytoremediation pathways to overcome limitations such as the need for a long incubation time plant necrosis due to toxicity, and so on. Understanding the basic physiological and biochemical processes in plants that underpin different phytoremediation processes is a promising way to improve plants' practical applicability in phytoremediation. As a result, this chapter will focus on phytoremediation of pollutants, specifically its applications, remediation mechanisms, plant associated endophytes, synergistic strategies, enzymes, benefits, limitations, and future directions.

Keywords: phytoremediation, bioremediation, phytostabilization, pollution management, enzymatic degradation, endophytes

Introduction

The Earth is currently facing numerous negative consequences and challenges due to diverse anthropogenic activities, which have led to harmful environmental consequences. Among many human-caused factors, the industrial revolution was the most significant action that impacted the environment. Various anthropogenic activities including endless combustion of fossil fuels, toxic wastewater generated from manufacturing or chemical processing industries (leather, textile, and distilleries), bombardment of the stable nucleus, radiological events, mining, deforestation, and plastic waste littering, are major polluters of the water, causing irreversible damage to the Earth [1-3]. Furthermore, wastewater generated from residential activities is also responsible for raising serious eco toxicological problems [4].

Due to the abundance, non-biodegradable and persistent nature in the environment, most of these contaminants are considered as highly notorious pollutants. Hence, these contaminants adulterate the air, water and soil and lead to severely toxic, teratogenic, mutagenic, and genotoxic effects on living organisms [5, 6]. Most of the xenobiotic contaminants in the environment cause endocrine disruptions and neurological disorders, which result in different physical deformities, low embryonic development, delayed hatching, oxidative stress in aquatic organisms and ultimately lead to the death [5, 7].

The heavy metal contamination in the environment results in the reduction of hematological indices and glycogen reserves in fish, making the fish feeble, anemic, and susceptible to sickness [8, 9]. Further, some organic and inorganic contaminants directly affect the vast array of behavioral changes including sociability, exploration, avoidance, sexual, aggressiveness, and feeding habits in most aquatic organisms, leading to the collapse of the ecosystems [9]. These pollutants exert harmful effects both on fauna and flora. Contamination of colored contaminants, including textile dyes, reduces sunlight penetration in the aquatic environment. It reduces photosynthesis activities which may mainly be associated with the loss of the productivity of receiving water bodies [10]. Few studies have recorded the changes in nitrogenase, alcohol dehydrogenase, nitrate reductase, hydrolytic (ribonuclease and phosphatase) and carboxylation (ribulose-1,5-bisphosphate carboxylase, phospho-enolpyruvate carboxylase) enzymes in terrestrial plants subsequently the exposure to various inorganic and organic pollutants [11, 12]. Further, many developing countries employ irrigation of food crops in farm fields with water and soil contaminated with harmful industrial effluents, are at risk of ensuring the safety of their food crop. It may allow potentially dangerous metals and other contaminants to transport from polluted soil to edible crops. They eventually enter into the animal or human body via intake and then create severe toxic effects. Contamination of receiving water bodies, pollution of groundwater and surrounded soil layers, accumulation of toxic substances in biota, and flow through higher levels via food chains are all associated with the risk of such pollutants [13, 14].

Environmental contaminants may percolate through the soil and soil-forming rocks and create groundwater pollution [15]. The organisms that live in the soil are affected by the regular infiltration of contaminants into the soil. Despite all of these negative consequences of pollution, little to no effort has been made to address its environmental repercussions and human health hazards consciously and pragmatically [3, 15]. Changes in the environment's makeup ultimately affect the biosphere's sensitivity [16, 17]. Therefore, environmental pollution is a primary ecological concern worldwide, with a growing public demand for the safest and healthiest environment possible. Hence, pollution treatment and management are critical challenges for pollution control authorities.

Pollution reduction is critical for removing pollutants, and attempts have been made to develop various physical, chemical, and biological treatment techniques [18]. For decades, researchers have been focused on developing strategies to remediate air, soil, water and groundwater contaminated with various hazardous chemicals and other organic wastes. Unfortunately, traditional physicochemical waste treatment methods such as incineration, advanced electrochemical oxidation, chemical stabilization of contaminants, cultivation-based processes, and even nanotechnology are highly expensive, technically more fitted to small areas, form hazardous by-products which lead to secondary environmental pollution, demand intensive energy requirements, and thus not applicable worldwide, especially for developing countries where the most industries are located [19].

The microbial remediation of various pollutants has been recorded for multiple chemical contaminants such as textile dyes [10], oil [20, 21], heavy-metal [22], etc. However, some of these microbe-based treatment methods cannot be successfully applied under field conditions. Extreme weather fluctuations would prevent microorganisms from completely removing contaminants from the environment over time, preventing natural soil amendments from being reclaimed. Hence, phytoremediation is gaining popularity, as a non-intrusive, environmentally benign, and attractive treatment tool for removing toxic pollutants from contaminated sites [3, 23].

Phytoremediation

Phytoremediation is the process that uses the plant and/or its associated microflora for the treatment of polluted soil, sediments or water [24]. Plants work with the soil to transform organic, nutrient or metal pollutants into harmless forms. Some plants, known as "green livers" have a strong ability to break down many resistant xenobiotics, acting as a significant sink for environmentally harmful substances [25]. Plants are one-of-a-kind organisms with superior metabolic activity, absorption capacity, and well-developed transport mechanisms that can selectively extract nutrients or pollutants presence in the growth matrix, water or soil. The mechanism of pollution degradation by plants originated from the defense system against various allelo-chemicals released by competing organisms or due to the toxicity of the harmful organic molecules [24].

The remediation technique may vary depending on the type of the plant and the type of pollutants used in the phytoremediation process, except few previously recorded studies where phytoremediation has effective regardless of the source or characteristics of the contaminant [26]. Therefore, proposing a single technique for the phytoremediation of the same pollutants is not reasonable for all plants. By acting as filters or traps, even a single plant could use a few different mechanisms to degrade organic compounds or extract and then stabilize different inorganic contaminants. The best way to improve the applicability of plants in the phytoremediation processes is to have a broad knowledge of the basic physiological and biochemical processes in plants that underpin different phytoremediation techniques [27].

Phytoremediation Techniques and Related Mechanisms

Plants could be used in various ways to clean up or remediate contaminated environments. Contaminants enter plants mainly through the root system, which also houses the primary toxicity prevention mechanisms [27, 28]. Water, nutrients, and other non-essential pollutants are absorbed and stored within the root system. The fate of these contaminants within the plant body may differ as it depends on the type of plant employed and the type of the target contaminant. The basic techniques that plants can use for concentration, transformation, or elimination of toxic pollutants from polluted soil, water, and air are depicted in Figure 1.

When planning phytoremediation-based experiments, knowledge of each phytoremediation technique is critical. Therefore, the basic characteristics, potential applications and possible mechanisms will be discussed.

Phytoextraction

Organic compounds, heavy metals, and radionuclides like contaminants highly resistant to plant metabolism are usually treated by phytoextraction, also known as phytoaccumulation. Plant roots uptake and translocate pollutants from the soil into the above-ground parts of the plant, and some of these contaminants are recoverable. Phytoextraction would be most commonly used to clean up contaminated soils. In this technology, hyperaccumulator plant species first absorb contaminants into their body, concentrate, and then precipitate in the above-ground parts such as shoots and leaves [29]. A hyperaccumulator is a plant that can accumulate a hundred times the amount of metal as a non-accumulating plant. Compared to conventional methods, phytoextraction is relatively inexpensive and permanently removes the contaminants from the site. The type of contaminant uptake by roots, a proportion of contaminants are fixed within the roots, the rate of xylem translocation to shoots, and cellular tolerance to toxic pollutants are the main factors that limit bioavailability of contaminants within the rhizosphere for phytoextraction process [29]. Though metal hyper-accumulating plants have been potentially helpful for removing metal from contaminated sites, the low annual biomass production leans towards limiting the recovery of metals from plants after the remediation process [30].

According to previous study, hyperaccumulation and metal recovery are only possible when plants are healthy and produce at least 3 t dry matter per hectare [23]. In the case of organic pollutants, the plants are harvested and disposed of after a specified period, burned or composted for recycling. Following burning, the plants need to be disposed of in a hazardous waste landfill, considering the bioavailability of the byproducts. Recent studies recorded the ability of the recovery of Cu and Cd using *Broussonetia papyrifera* plant through adsorption followed by pyrolysis [31].

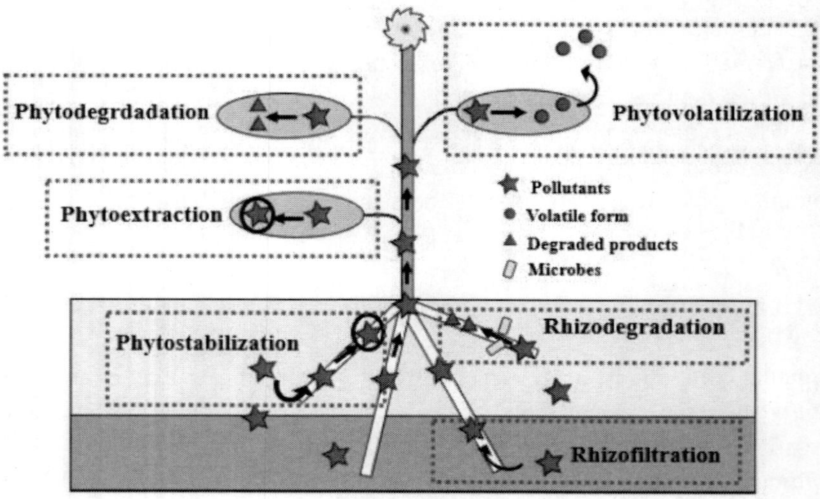

Figure 1. The key techniques involved in phytoremediation of pollutants.

Plants have developed methods to tolerate and survive in large concentrations of potentially hazardous substances due to an evolutionary process. Metal detoxification by plants has been proposed as a four-step mechanism, as follows [32].

1. Metal bind to the cell wall.
2. Reduce uptake or efflux pumping of metals at the plasma membrane.
3. Chelation of the metal in the cytosol by metallothioneins, phytochelatins, and metal-binding proteins.
4. Tonoplast-located transporters compartmentalize metals in the vacuole.

Further, plant's toxic tolerance and regulating mechanisms may be related to bioaccumulation of metal via chelating or sequestration through the role of phytochelatins and metallothioneins, antioxidant enzymatic activity through peroxidase, catalase, ascorbate, guaiacol, like enzymes, non-enzymatic antioxidant via carotenoids, ascorbic acid, glutathione, and homeostasis in the diffusion of cations, as well as by the bacteria live in the rhizosphere [32]. When considering the detoxification of organic matter, the mechanism may involve the reduction, enzymatic oxidation, hydrolysis or conjugation of the organic pollutants and these biotransformation products are put in vacuole storage where they are dispersed within the plant or retained in lignin-like insoluble cellular structures.

Rhizofiltration

Rhizofiltration, the use of plant roots which grown in aerated water to precipitate and concentrate toxic contaminants, can be applied for remediation of polluted surface water bodies, wetlands, wastewater, groundwater, and landfill sites. Various inorganic and organic pollutants, including heavy metals and dyes, can be removed via this technique. Rhizofiltration is more similar to phytoextraction process, except that the plants are employed to treat contaminated groundwater instead of soil. Plants for rhizofiltration are grown in green-houses with their roots submerged in water rather than soil, acclimatized, introduced into the contaminated sites, and harvested after a certain period when the roots are saturated with the targeted contaminants. By this method, both aquatic and terrestrial plants can be used either as *in-situ* or *ex-situ* methods. Further, some plants other than hyperaccumulators, are also successful in this process, as translocation of the contaminants into the shoots is not essential [29]. The constant need to modify pH, the necessity of nursery or greenhouse grown plants, continuous harvesting and disposal of plants, requirement of properly constructed tank system, and firm grasp of chemicals are all disadvantages and limits rhizofiltration.

Phytostabilization

In Phytostabilization, plants immobilize pollutants presence in the soil or groundwater by absorption or adsorption onto the surface of the roots, or the production of insoluble compounds and precipitation within the rhizosphere [29]. Therefore, it is an in-place inactivation of toxic contaminants that are widely used to remediation of contaminated soil, sludge, or sediments. Sorption, precipitation, complexation, or metal valence reduction can be identified as the main mechanisms involved in the phytostabilization technique [29]. This technique decreases the mobility of pollutants, preventing their migration via wind, surface soil movements, soil leaching, and contamination of groundwater or air [28]. It is dependent on the roots' ability to control pollutant movement in the soil [28]. In addition, soil additives promote the formation of insoluble metal complexes, limiting biological availability and plant absorption, and preventing metals from entering the food chain [23]. Changing the physicochemical characteristics of the metal–soil complex by introducing a multifunctional anion, such as phosphate increases metal adsorption via an anion-induced negative charge. Thus, metal precipitation is one strategy for assisting such immobilization [23].

Addition humified organic matter and lime to elevate soil pH is typical for facilitating revegetation of contaminated soils and for the improvement of soil conditions [19, 28]. It reduces water percolation through the soil matrix and produces toxic leachate and avoids soil erosion and the spread of harmful metals to neighbouring areas. This can reestablish a vegetative cover at sites lacking natural vegetation due to high pollutant concentrations. Grasses, reeds, and sedges like perennial plants with fast-growing ability, dense or deep root system, high transpiration rate and ability to form tight cover on the surface soil layer are more applicable for phytostabilization technique.

It is more about ensuring environmental and human health stability than removing metal pollutants from specific locations. Through phytostabilization, the exposure to hazardous substances can be reduced to safe levels. This is one of the most intrusive soil remediation technology which results less disruption to the site activities. Phytostabilization is most effective in fine-textured soils with high organic-matter content, although it can be used to treat a variety of locations with extensive areas of pollution. Phytostabilization is less expensive, easily applicable, and more visually pleasing than other soil remediation techniques. Phytostabilization is beneficial when decontamination procedures are impracticable due to the size of the contaminated region or a lack of suitable funds. It could also be used as a stopgap measure to reduce risk in areas where problems prevent using the best approach. Plant growth and survival are difficult in some highly contaminated environments. Therefore, phytostabilization is not an option for such occasions.

Tolerance to high levels of the target contaminant(s), increased root biomass which capable of immobilizing these contaminants through uptake, precipitation, or reduction, retention of relevant contaminants in roots rather than transfer to shoots to avoid special handling and disposal of shoots are all characteristics of plants suitable for phytostabilization at a given location. Some plants including *Verbascum speciosum, Chenopodium botrys, Cousinia bijarensis, Scariola orientalis, Stipa barbata*, and *Chondrila juncea* have been successfully employed for phytoextraction of Mn. However, considering the bioavailability of the adsorbate, plants are harvested and disposed of or subjected to burning. In the case of organic pollutants, plants are composted for recycling at the end of the treatment process [23].

Phytovolatilization

Another type of phytotransformation is phytovolatilization, which involves using plants to absorb contaminants from the soil, convert into volatile forms, and release them into the atmosphere [29]. By this method, contaminants are taken up into the plant's body. Leaves exhale only the contaminant's volatile or degraded volatile form as water vapor [23]. Phytovolatilization also refers to the transfer of pollutants from stems or other plant parts to leaves. Phytovolatilization can be used for remediation of soil, sediment, or water contaminants. The applicability of this technique has been investigated for volatile organic compounds such as trichloroethylene and inorganic substances with volatile forms such as Se and As [29]. Mercury has been the most commonly used metal pollutant in this technique. However, Hg released into the atmosphere is likely to be recycled by precipitation and re-deposited in lakes and oceans, effectively repeating the anaerobic bacteria's methylmercury production [29]. Although transferring toxins from the soil or groundwater to the atmosphere is not as desirable as *in-situ* degradation, phytovolatilization may be preferable to prolonged soil exposure and the risk of contamination of groundwater [23].

Rhizodegradation

Rhizodegradation, also known as phytostimulation, is a pollutant remediation method in which rhizosphere microbes convert the contaminant into less or non-toxic forms. Plant roots manipulate soil redox conditions, organic content, moisture, and other soil variables through their activity in the rhizosphere. Furthermore, the roots provide more surface area for bacteria to thrive on and a channel for oxygen transfer from the environment. Rhizosphere is rich of different microorganisms as plant releases sugars, enzymes, amino acids and other substances that may drive bacterial development. Deep-rooted trees and other annual species use rhizodegradation to improve the removal of mostly organic chemicals, Polycyclic Aromatic Hydrocarbons (PAHs), pesticides, chlorinated solvents, benzene, ethylbenzene, Polychlorinated Biphenyls (PCBs), xylenes, and petroleum hydrocarbons from soil [23]. Exudates/enzymes are released into the root zone (rhizosphere) to promote the breakdown process, also known as plant-assisted bioremediation [29].

Phytodegradation/Phytotransformation

Phytotransformation, also known as phytodegradation, is a process that goes beyond the uptake and storage of harmful pollutants. It entails breaking down complex organic molecules into more straightforward, less toxic, or non-toxic forms and incorporating these degraded contaminants into plant tissues [27, 28]. Pollutants that the plant has taken up are broken down when the phytodegradation system is active. Like phytovolatilization and phytoextraction techniques, plant uptake occurs only when the pollutants' solubility and hydrophobicity fall within a specified permissible range. Phytodegradation, which can address soil, sediment, or groundwater, to remediate herbicides, chlorinated solvents, and munitions, efficiently [29].

During the phytoremediation process, the roots absorb contaminants present in the water. The cation exchange through the cell membrane and induced by the carboxyl groups' presence in the roots system, may involve the movement of contaminants into the plants. This absorption may happen either via direct absorption through the root system or foliar absorption, or by the passive absorption through stoma cells and cracks in the cuticle.

The foreign compound metabolism in plant systems is widely thought to be a 'detoxification' process comparable to human xenobiotic compound metabolism. According to the "Green liver model," xenobiotic detoxification occurs in transformation, conjugation, and sequestration. Initial processes; including reduction, oxidation, and hydrolysis, produce favorable molecules for conjugation reactions later [23]. The oxidation processes are the most common transformation reactions in pesticide metabolism; however, several nitro-aromatic chemicals have also been discovered to undergo these reactions. The xenobiotic compounds conjugate with the organic molecules present in the plants, which finally results from the detoxification of such xenobiotics. Then these resulted compounds subjected to different sequestration processes within the plants. The frequent strategy is to store these conjugates in plant organelles like vacuoles, at which point the compound would not interfere with the plant's cell functions. Further, these xenobiotic conjugates can be incorporated with the lignin like biopolymers and become non-extractable.

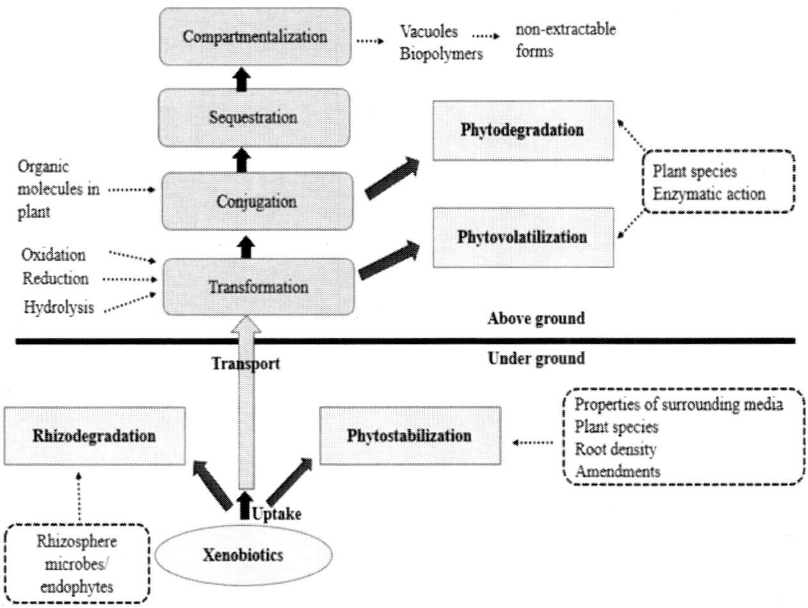

Figure 2. Possible mechanisms involve in phytoremediation of xenobiotics.

The metal detoxification mechanism involving plants is followed by four significant steps [32]. Metal chelation in the cytosol by various ligands (metallothioneins, phytochelatins) and metal-binding proteins, followed by compartmentation of metals in the vacuole by tonoplast-located transporters.

A single contaminant may show different remediation pathways. For instance, the organic contaminants can be translocated to different plant tissues and then volatilized, or contaminants can be complete or partially degraded or converted into less toxic forms and retained in the plant tissues (Figure 2). Contingent upon the specific interactions of the selected contaminant with the environment where it is present; soil, water or air, and physiological properties of the plants will determine the remediation process whether the contaminant undergoes; phytodegradation, phytovolatilization, phytoextraction, or rhizodegradation.

Applications of Phytoremediation and Phytostabilization

Domestic Waste Management

Domestic wastewater with high organics, suspended particles, oil and grease is being produced in large quantities practically in urban areas. Hence, municipalities are finding it difficult to manage proper disposal and treatment issues. Moreover, the economic crisis faced by growing cities has made them inaccessible to high-cost disposal systems and treatment facilities [33]. Therefore, domestic wastewater is channeled into low-lying areas and aquatic bodies such as ponds and lakes, posing a significant danger to water quality. Disposal of domestic wastewater into freshwater bodies continuously adds nutrients to the water, primarily responsible for increases in nitrogen and phosphorus concentrations that are undesirable and form toxic effects on many aquatic organisms, including fish.

On the other hand, the depletion of dissolved oxygen in a water body is problematic to aquatic ecosystems. Maintaining the desired level of dissolved oxygen is vital for the survival of almost all living forms other than anaerobic microorganisms. Due to the heavy load of nitrogen and phosphorus, eutrophication is one serious problem that most domestic wastewater receiving ponds face today [4]. These pollutants may also contact groundwater and deteriorating their quality [33]. Proper treatment before discharge to the environment will lower these consequences. An aquatic macrophyte based wastewater treatment system containing either floating or submerged plants is one of the low-cost and environmentally friendly approaches for domestic wastewater treatment [27].

The treatment of residential wastewater by *Lemna minor* recorded that after seven days of treatment by *L. minor*, the pH, dissolved oxygen, and oxygen saturation values have increased. In contrast, the values for the other examined Physico-chemical parameters (dissolved oxygen, alkalinity, chloride, and COD values) declined dramatically [34]. Treatment of grey water using a mini wetland comprised of *Lepironia articulata* was recorded by another study showing high removal performance of COD (85%), BOD (81%), suspended solids (54%), and turbidity (45%) suggesting the effectiveness of such application rather than high-cost physic-chemical treatment methods [35]. The efficacy of a vertical subsurface flow constructed wetland (VSSFCW) for sewage effluent treatment over eight months have been recorded utilizing *Phragmites australis* [36]. Around 75% of COD, 84% of BOD, 75% of TSS, 32% NH4, and 22% of total phosphorous removal have been recorded throughout the study period, suggesting the proposed method is applicable for repeated wastewater treatment cycles. Further, the potential applicability of *Canna x generalis* reeds in a constructed wetland for domestic water treatment under different pH conditions has been evaluated [4]. The aquatic macrophyte; *Monochoria korsakowii* is another potential biological agent for removing total nitrogen in domestic wastewater [37].

Industrial Waste Management

Significant amounts of waste, including high levels of harmful contaminants, have resulted from rapid industrialization and urbanization [1, 38]. Organic and inorganic pollutants have well-documented adverse effects on ecosystems and human health, and industrial treatment technologies are expensive to avoid or minimize discharges. Physicochemical treatment technologies are costly, so phytoremediation is one of the best alternatives for removing contaminants from industrial wastewaters prior release to the natural environment [39].

The radioactive wastes produced by nuclear reactors, fuel fabrication and reprocessing plants, radionuclide production and applications based research laboratories, coal-fired power plants, and accidents and disasters, etc. produce vast amounts of fly ash containing radionuclides with long half-lives of over a thousand years. Due to their high mobility and solubility in the environment, contaminants become more dangerous and can enter the food chain, resulting in biomagnification at higher levels of the food chain [6]. One of the most effective and low-cost treatment methods for reducing the risks associated with radioactive wastes is phytoremediation. When selecting a plant for such an application, the following factors should be considered:

1. Study the basic features of radioactive waste that need to be remediated
2. Study the plant class and its composition
3. Determination of the concentration of the desired radionuclide in the plant
4. Restrain the concentration of the desired radionuclide in waste [40].

P. australis, Macleaya cordata, *Azolla imbricate,* and *Eichhornia crassipes* plants have been identified as better candidates for remediation of U, Cs, Co and Ra contaminated soil [41]. Further, some enzymes in plants are involved in converting the inorganic Se like heavy metals to the volatile form (dimethyl selenide) by various biochemical processes.

Heavy metal pollution that can happen via natural and anthropogenic activities can be effectively treated via phytoremediation (Figure 3). Natural sources of heavy metal contamination include volcanic eruptions, forest fires, and rock seepage. Heavy metal contamination is caused mainly by human activities such as mining, smelting operations, and various other sectors such as thermal power plants and battery manufacturing. Heavy metals are also polluted through the disposal of coal combustion byproducts in terrestrial systems [42]. Heavy metal pollution in soil and crop plants is caused by atmospheric deposition, municipal wastes, sewage water as fertilizer, and phosphate fertilizers used in agriculture. When some Hg and Pb like heavy metals get introduced into the environment, it causes very dangerous effects due to bioaccumulation and biomagnification.

Some strategies have developed in higher plants to help them survive and reproduce on soil heavily contaminated with different heavy metals [42]. Willow plants, for example, have identified as a potential plant for heavy metal phytoremediation of soil and water [43]. Toxic metals accumulate in significant amounts in many areas of these plants. Metal detoxification is enhanced by numerous antioxidant enzymes and other cellular antioxidants such as cysteine and thiols, resulting in high metal tolerance in plants. Antioxidants are thought to be a critical line of defense against metal poisoning. The hyperactivity of enzymes helps plants overcome heavy metal toxicity through detoxifying and hyperaccumulation [6].

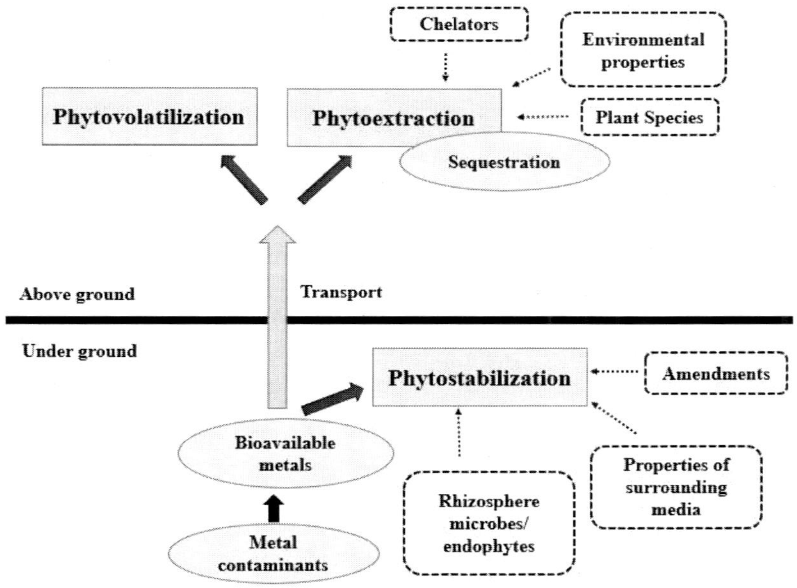

Figure 3. Possible mechanisms involve in phytoremediation of heavy metals.

Metal accumulator plants' extraction potential increases up to a certain amount. The bioaccumulation coefficient (phytoextraction rate of metal) decreases as metal concentrations increase [44]. Hg is also remediated by phytovolatilization, which is known to cause many neurodegenerative diseases. The methylated form of Hg (MeHg) is of particular concern because of its biomagnification in the food chain [19]. Plants can absorb metal through their roots, transport it to the upper plant parts via the vascular system, and exhale it. The metal is converted to a volatile form by enzymes within the plant (HgO). Reports on Hg volatilization through the root system of transgenic plants are also available. The pearl grass (*Micranthemum umbrosum*) like some ornamental plants are proved to be very effective in accumulating heavy metals such as As and Cd from polluted water streams [6].

In addition, terrestrial plants are important in the phytoremediation of heavy metals and radionuclides in contaminated soil and water. Several plant species have been successfully used to remediate radionuclide-contaminated terrestrial and aquatic systems. Plants like reeds (*P. australis*), grass

(*Polygonum* sp.), and bulrush (*Typha latifolia*) have been identified as radioactivity hyperaccumulators in U-polluted water [6]. Water quality in streams receiving discharge from U mines and heavy metals has been improved using wetlands designed with these plant species.

Excessive metal toxicity causes plants to either stabilize the root zone, prevent the metal transfer to the plant's above-ground sections, or collect large amounts of metals in stems and leaves [44]. The plant (hyperaccumulator or excluder) used determines the route or type of phytoremediation. Excluder plant species confine pollutants to the root system rather than the shoot system whereas hyperaccumulator plant species accumulate metals primarily in the shoot rather than the root [45]. Hyperaccumulators can collect a large amount of metals from contaminants. Some bryophytes include *Scopelophilaca taractae*, *Bryummuehlem beckii*, *Sphagnum perichaetiale*, and *Ricciocar posnatans*, showed higher potential for removing over 95% of Fe, As, Cu, and Cr from aquatic environments [44, 46].

Even though textile dyes are one of the most common water pollutants globally, the removal of textile dyes via plants has been overlooked compared to other phytoremediation processes such as organic contaminant and metal removal [39]. Hairy roots have been used as a successful tool for the removal of various pollutants such as textile dyes, phenolics, heavy metals, and radionucleotides in recent years, with the involvement of various pathways such as Rhizodegradation, Rhizofiltration, Phytoaccumulation, Phytotransformation, Phytovolatilization, and Evapotranspiration [24, 47-49]. Furthermore, the effects of plant-derived enzymes like laccase and peroxidase [27, 48] as well as some endophytes on phytoremediation processes have been studied [50, 51]. However, because phytoremediation processes involve multiple pathways, understanding the dye decolorization process is more complicated. Most hairy roots have a high potential for sorption or binding pollutants within their hairy roots [52]. Further, few studies recorded that removing textile azo dyes by hairy root aquatic plants such as *E. crassipes, B. juncea* and *Pistia* sp., involved both bio-sorption and enzymatic degradation [27, 53].

Factors Influencing Phytoremediation

Both biotic and abiotic factors in the environment greatly influence the phytoremediation process [6]. Several factors that can affect the uptake mechanisms of different pollutants are worth discussing.

Plant Species

Plant tissues have a high potential for uptake and accumulation of various essential and non-essential metals. They are frequently unable to differentiate between metals with similar physicochemical properties of metal isotopes [6]. Because of this property, various plants are commonly employed in the removal of contaminants from terrestrial and aquatic systems [6]. The selection of a plant is one of the primary vital points which determine the success of the phytoremediation technique. The plant used for the phytoremediation should have the ability to grow fast, have a dense root and shoot system, high tolerance towards toxic pollutants, resistance to diseases and pests, tolerance to stressed conditions, etc.

Furthermore, such plants should be unattractive to animals, resulting from reduced pollutant transfer to higher trophic levels via food chains [27, 28, 40, 50]. The selection of plants that are not specific to a single contaminant may advantageous as it can remove mixture of different pollutants that can be remediated simultaneously, especially during the treatment of natural pollutants [6]. Researchers have recognized various plant groups, from mosses to flowering plants, including grasses, shrubs, and trees, as having the potential to remediate organic and inorganic contaminants in different polluted sites in both soil and water resources.

Despite a lack of roots and other true tissues, some bryophytes, most of which are found in misty environments or on rocks along streams and rivers, can withstand harsh environmental conditions. As a result, this species could be used as a bioindicator and phytoremediator for environmental contaminants. Several studies have documented the potential use of aquatic and wetlands bryophytes such as *Bryum muehlembeckii*, *Ricciocarpos natans*, *Scopelophila cataractae*, *Brachythecium plumosum*, and *Rhynchostegium riparioides* for heavy metal removal [44, 46]. Some Pteridophyte plants, the non-flowering plants which use spores for reproduction, have been recorded as potential hyperaccumulators of pollutants, especially heavy metals. Removal of Cu, Pb, Zn and Ni by *Nephrolepis cordifolia* and *Hypolepis muelleri*, effectively [54]. Further, *Pteris vittata* has been reported as a potential hyperaccumulator of As and Pb [55].

Several spermatophytes have been identified as possible contaminants of phytoremediation agents. The most common angiosperm families that absorb and concentrate various organic and inorganic toxins in their roots are Asteraceae, Caryophyllaceae, Brassicaceae, Lamiaceae, Violaceae, Cyperaceae, Fabaceae, Flacourtiaceae, Cunoniaceae, Poaceae, and

Euphorbiaceae. The nitrogen-fixing legume *Lupinus albus* has been discovered to absorb arsenic, particularly its root structure. This legume thrives in acidic soils with low nutrient availability and soils with heavy metal remediation [56]. *Arrhenatherum elatius* and *Sonchus transcaspicus* are two perennial types of grass recommended for phytostabilization of metal-polluted soils [57]. Metals such as Ni, Cu, Cd, Co, Mn, Pb, Cr, and Zn were accumulated in roots in *A. elatius*, whereas metals were accumulated preferentially in shoots in *S. transcaspicus*.

Not only the metals, some spermatophytes: *Alfalfa cultivarriley*, *Medicago falcate* and *Medicago sativa* have been recorded in effective remediation of petroleum-like hydrocarbon contaminated soil [58]. Phytoremediation applications of some grasses like Vetiver (*Vetiveria zizanioides*), Carabao grass (*Paspalum conjugatum*), Cogon grass, (*Imperata cylindrica*) have been recorded for a wide range of metals such as Cr, Pb, Ni, Cu, As, Hg, Zn, and Se [59, 60]. Among them, Vetiver; perennial grass with a strong dense and vertical root system, shows the high capability of tolerance and phytoremediation of a wide range of acidic, alkaline, saline levels in the soil as well as for high concentrations of Al, Mn, As, Cr, Ni, Pb, Se, and Hg contaminants in the soil. Further, it is considered a high potential phytoremediator as it can grow in hydrophilic and xerophytic conditions [60]. Moreover, some floating aquatic plants: *E. crassipes*, *Pistia stratiotes*, *Hydrocotyle umbellate*, and *Lemna minor*, have been recorded as potential phytoremediation agents for remediation of both inorganic and organic pollutants via different phytoremediation mechanisms [6, 27, 61-63].

When choosing a plant for phytoremediation, the root zone is important because it is where most toxins enter the plant tissue and, in certain situations, store or degrade. For phytoremediation, the necessities of a well-established root system, particularly for rhizofiltration have been identified [45]. Plants with a well-established deep root system may withstand extreme climate fluctuations like flooding, cold, fire, heat waves, and protracted drought. Furthermore, genes play an important role in making plants effective at remediation. For example, metal tolerance in plants is assisted by glutathione and organic acid metabolism pathways [45].

Properties of the Medium

Factors such as temperature, moisture content, pH, soil structure, texture, weathering process, availability of water and oxygen, nutrients, solar radiation, as well as the concentration of the contaminants in the medium can directly affect the efficiency of the phytoremediation process by plants [27, 45]. Usually, the influence of most of these factors on the phytoremediation process may vary according to geographical locations.

When considering the effect of pH, it has recorded that dye decolorization by *E. crassipes* and *Pistia* sp., reduced at pH 3 and 13, as plants cannot tolerate and withstand such extreme pH conditions [61]. In the same way, dye uptake by plants and algae has increased at pH 3-7 and then started to decrease [64]. Metals and metalloids' chemistry is generally pH-dependent—metals which strongly bind to inorganic phases at higher pH and become more accessible at lower pH. As metal salts are soluble under acidic circumstances, a low pH can improve metal bioavailability [65].

As a result of the higher initial dye concentrations, the time required to reach dye decolorization increased [49]. When the concentration of Cd is increased to a certain level (10 mg Cd/L), toxic symptoms in *Pistia stratioltes* have been evidenced [66], implying that increasing contaminants concentrations may be toxic to the plants.

The structure, texture, and organic matter content, all play a role in determining soil type. Plant development and survival are heavily influenced by soil structure [28]. The carabao grass: *Paspalum conjugatum*, one of the great examples that can be used to remediate wide range of heavy metals (Pb, Cr, Ni, As) and some organic matter since these plants can adapt to a wide range of soils [60]. Cogon grass (*Imperata cylindrica*) continues to thrive in light-textured acidic soils with a clay subsoil. It can withstand soil pH levels ranging from strongly acidic to slightly alkaline. It's a hardy plant that can withstand both shade and high salinity [59, 60].

The bioavailability of pollutants is also influenced by soil texture. There is an order of pollutant bioavailability in different soil types: loam or sand > silt > clay loam > clay soils [67]. Fine-textured soils, such as clay loam and clay, typically have a higher concentration of contaminants than coarse-textured soils (e.g., sand and silt). These fine-textured soils have a significantly higher electrostatic attraction of ions to soil particles than coarse-textured soils due to their large specific surface area and big charge [67, 68]. For instance, clay can bind molecules more effectively than silt or sand. Because soil structure and texture play a role in pollutant bioavailability, choosing the right soil type is critical to a phytoremediation mechanism's performance. It has been discovered that soil with a high organic carbon content (>5%) has substantial adsorption, which reduces availability of contaminants. In comparison, soil with a moderate organic carbon concentration (1–5%) has limited availability [28]. Hence, soil type may directly impact phytoremediation efforts.

At a specific temperature, phytoremediation systems work their fastest. Pre-rooted weeping willows (*Salix babylonica*) were used in an experiment to evaluate cyanide uptake and metabolism in response to temperature changes [69]. These weeping willows showed 2.78 mg CN/(kgd) at 32^0C of cyanide metabolism rate, while the lowest value was 1.20 mg CN/(kgd) at low temperature around 11^0C. Finally, temperature variations have significantly impacted plant cyanide uptake and metabolism.

Solar radiation's impact on the phytoremediation process is also of particular concern. Before uptake by the Plant, PAHs can be photo modified by ultraviolet light in polluted water or on the soil's surface, increasing the polarity, water solubility, and toxicity of the pollutants [28].

Evapotranspiration, leaching, hydrolysis, volatilization, photo modification and contaminant biotransformation are some of the weathering processes included in phytoremediation technology. These methods can diminish the concentration of easily degradable pollutants while leaving more resistant substances in the soil. Nonvolatile or semi-volatile chemicals preferentially partition to soil organic matter or clay particles, limiting their bioavailability and decomposition [28].

Amendments

The contaminated sites that need to be remediated with plants often have low organic matter content and nutrient levels, resulting in low soil fertility and slow growth of plants [68]. When plants and their associated microbes grow under significant stress from harmful pollutants, nutrients are considered a vital component of their growth. The impact of fertilizers on biostimulation and phytoremediation on habitat restoration and oil degradation in petroleum-contaminated wetlands has proven substantial [28].

Organic additives have been explored further in phytoremediation because they can address both challenges simultaneously. Organic amendments can achieve five key effects: increased soil organic matter, increased soil nutrients (N, P, K), decreased metal bioavailability and leaching potential, increased soil cation exchange capacity, and increased soil water holding capacity. These direct impacts result in improved plant growth and biomass yield and lower metal concentrations in harvestable plant biomass. Sewage sludge from the food industry, biochar, and compost manure from poultry provide plant nutrients, increase water storage capacity, lower Zn, Pb and Cd levels in plant biomass increase biomass and improve soil fertility [68].

Because of the high annual output volume, some industrial waste including red mud, fly ash, and coal gangue, is appealing for use as a soil supplement. Using these materials in phytoremediation procedures could be a promising strategy to manage them appropriately. These wastes serve various purposes in this process, including giving nutrients for plant growth and minimizing metal uptake by raising soil alkalinity. Coal fly ash is an amorphous mixture of ferroalumino-silicate minerals that contains vital nutrients such as Ca, K, Mg and B which are important on plant growth. It is a byproduct of coal combustion in power plants [68].

The majority of the contaminants handled during the phytoremediation are toxic and create toxic effects on plants. The plant response towards such toxic contaminants present in the environment may vary according to the properties of each plant, as well as the toxic concentration in the site. Recent study recorded the reduction of photosynthetic pigments content of *E. crassipes* and *Pistia* sp. due to the toxicity of CI Direct Blue 201 azo dye (50 ppm initial concentration) [27]. *Ceratophyllum demersum* and *Hydrilla verticillata* both had considerable cadmium accumulation (7,381 and 7,942 mg Cd/kg, respectively), as well as a high BCF (> 2,500) at 1 mg/L with increasing metal concentrations, both species' development was significantly reduced. However, cadmium was less harmful to *H. verticillata*, and it had a more significant cadmium accumulation capability [66].

Chelating Agents

Increased bioavailability of contaminants can be influenced by adding biodegradable physiochemical factors like chelating agents and micronutrients to increase crop uptake [40]. Chelating agents' effectiveness is determined by the metals to be removed and the plant species used [65]. When biodegradable amendments like citric acid, oxalic acid, NH_4-citrate/citric acid, or nitrilotriacetic acid were applied to Indian mustard and ryegrass, the uptake of Cd, Cr, Cu, U, Zn, and Pb was increased [70]. The degradation rate becomes particularly important while working with increased metal mobility as it directly influences leaching probability [65]. This has been identified as one of the inherent risks of this technology.

Endophytes' Role in Phytoremediation

Even though plants suited for phytoremediation must be adapted to the polluted environment, the presence of toxic contaminants in soil lowers the growth of plants and as a result, it will reduce the efficiency of the phytoremediation process [27]. Plant-microbe partnerships, in which plants are utilized with pollutant-degrading, plant growth-promoting microbes, or both, to clean up polluted soil and water, may overcome these restrictions [71]. Endophytes are the microbes that live within plant tissues instead of rhizospheric microbes that live on or around plant roots [50]. Endophytes have some benefits over rhizospheric bacteria. Endophytic bacteria invade the plant interior without causing pathogenicity to their host plant, whereas rhizobacteria populate the proximity of roots. On the other hand, utilizing endophytic bacteria to boost phytoremediation efficacy is a relatively new idea of utilizing endophytic bacteria. Controlling a rhizospheric population is difficult, and competition among microorganisms frequently diminishes the number of desired strains unless the pollutant's metabolism is selective for the other bacteria.

On the other hand, endophytes reside in the plant's internal tissues and are selected or regulated by the plant. Furthermore, endophytic bacteria with plant growth-promoting properties may help plants adapt to and grow in polluted soil. Endophytic bacteria can interact more closely with their host plant than rhizobacteria since they colonize the plant interior [71]. One example is *Exiguobacterium profundum* bacteria strain isolated from *Amaranthus spinosus* plant to decolourize Reactive Black-5 dye [50].

Table 1. Overview of endophytes involved in phytoremediation of different contaminants

Host plant	Plant tissue	Endophyte/s	Target contaminant	Reference
Centella asiatica	Root	*Bacillus fermus*	Direct Blue dye	[73]
Allium macrostemon	Stem	*Enterobacter* sp.	Pyrene	[74]
Lolium multiflorum	Root	*Pseudomonas* sp.	Alkanes	[75]
Phragmites australis, Ipomoea aquatica, Vetiveria zizanioides	NM	*Achromobacter xylosoxidans*	Catechol and phenol	[76]
Populus sp.	Root, Stem	*Pseudomonas tolaasii,* P. *Jessenii,* P. *Rhodesiae,*	Phenol, toluene	[68]
Poplar	NM	*Pseudomonas putida*	TCE	[77]

NM: Not Mentioned.

Further, *Marasmius cladophyllus*; a bascidiomycetes fungi isolated from *Melastoma malabathricum* plant, has effective decolorization of 97% Remazol Brilliant Blue R dye and 56% of Methyl Red dye [51]. In addition, an endophyte fungi isolated from leaves of *Eleusine indica* were involved in the decolorization of Malachite Green dye in a preliminary study [72]. Several studies have recently revealed the expression of toxic pollutant degrading genes from endophytic microbes involved in the phytoremediation of various inorganic and organic pollutants. Table 1 summarizes some recent studies focused on endophytes-based phytoremediation of other toxins.

Phytoremediation Technology Models: Land Vegetation, Hydroponics, and Constructed Wetlands

Poor fertility, low organic matter content and toxicity are the main issues that bind to contaminated environments while serving as a substrate for numerous biological activities. Soil washing, physical or chemical-based stabilization, chemical oxidation or reduction, electro-kinetic treatment, excavation and off-site treatments, incineration and recovery are the most common physico-chemical treatment methods for remediation of contaminated soil [78]. Compared to these methods, phytoremediation based land vegetation is one of the best methods that can be used for remediation of such contaminated soil with the aims of phytoextraction of metals, phytostabilization of metals to reduce the risk of bioaccumulation and biomagnification, enhance the soil fertility and maintain its ecosystem services. A successful story has been recorded for application of poplar trees for the remediation of timber waste piles in New Zealand [79].

Phytocapping, the use of plant cover over the final coverage of landfills as a post-closer technique, have identified as the best alternative for existing physicochemical based capping. The soil layer associated with established plant roots can operate like a sponge to absorb infiltrating rainwater, which can then be pumped out of the soil layer by evapotranspiration of above ground plant foliage, regenerating the cap's retention capacity [80]. Rapid waste stabilization, reduced gas generation after 20 years, quicker access to the site for alternate purposes (municipal building construction, parkland, recreational activities), and economic effects were all advantages of vegetative caps [79].

Not only for the photocopying, but phytoremediation can also be used for treatment of landfills and its leachate. It may depend on the type of application, the plant employed, and the toxicity of the target contaminant [79]. This may either be via phytodegradation or phytostabilization. Phytoremediation of landfill-derived leachate has several potential benefits, including providing a water and nutrient source for improved plant performance, reducing leachate volume and thus mitigating the migration of groundwater containing elevated levels of contaminants and removing contaminants in the leachate and thus improving the leachate quality [81]. Further, applying energy crops and some short rotation crops plant species for phytoremediation of contaminated soil have growing concerns as such biomass may contribute to the energy supply [78]. These energy crops may either annuals (*Sorghum bicolor, Brassica carinata, Hibiscus cannabinus, Brassica napus*) or perennials (*Populus* sp, *Phalaris arundinacea, Eucalyptus camalduensis, Robinia pseudoacacia*).

However, in the case of land phytoremediation, most hyperaccumulator plants grow slowly with a lesser amount of biomass. To address the issue, hydroponic systems come into concern. The potential applicability of *Ricinus cmmunis*, a perennial terrestrial plant, have been studied on the removal of Cd by a hydroponic system, and the highest accumulation has been recorded in plant roots, followed by leaves and stem [82, 83]. The application of Vetiver (*Chrysopogon zizanioides*) in a floating hydrophonic system has been recorded for the remediation of paper board mill effluent wastewater [84]. A pilot-scale investigation showed the efficient reduction of EC (56%), total soluble salts (81%), total dissolved solids (56%), COD (58%), total nitrogen (59%), phosphorous (48%) and some heavy metals (65% of Pb and 64% of Cd) as well.

Constructed wetlands are one of the best plant-based approaches for the removal of a wide range of waterborne contaminants, including domestic, agricultural and industrial oriented. The multi-pollutant treatment capability, easy operational and maintenance process and the comparatively low cost make them more attractive in wastewater treatment [28]. Like natural wetlands, constructed wetlands also comprise a complex ecosystem including plants, sediment, and associated microbes that mutually work to transform toxic contaminants into more minor toxic forms or immobilize them within the system [85]. The ecological function of constructed wetland is determined by hydrological elements such as nutrient availability and physicochemical properties such as soil, water, pH, and soil anaerobiosis [86]. These parameters are managed in constructed wetlands to ensure the effectiveness of the biotic processes that occur there. Wetland ecosystems rely on water, and the water budget of these systems directly impacts biological processes, resulting in significant seasonal changes. The hydrology of specific wetlands determines species variety, productivity, and nutrient cycling, which is critical for flora and fauna diversification and pollution control [28].

The hydroperiod of a constructed wetland is directly related to its stability. A wetland's hydroperiod is seasonally changed between surface and subsurface water levels. The hydrologic cycle is the most crucial hydrologic process in any wetland system. Precipitation, surface water flow, groundwater flow, and evapotranspiration are significant components of the hydrologic cycle, which must be studied before construction. The inflows and outflows of water within a built wetland make up the water budget.

Another benefit of such gravel-bed wetland is extensive partitioned micro-environments near the root surfaces, favoring metal-binding like contaminants. Further, there is a high possibility of the formation of biofilms on the root surfaces and on the gravel to enhance the binding capacity and microbes assisted degradation [85]. Although artificial wetlands are a less expensive option than other water treatment systems, they must be tuned to improve efficiency, and reproducibility and limit the risk of ecotoxic contamination. The silt immobilizes the majority of the pollutants removed from the waste stream. To build a sustainable phytoremediation system, post-harvest tactics must be combined with pre-harvest procedures. To forecast and validate that wetland plants have almost detoxified pollutants with minimum remaining dangers to humans and the environment, knowledge, experience, and field testing are required. Residual management is essential to remove various public constraints. Various submerged plants; *Hydrilla verticillata, Ceratophyllum demersum, Heteranthera dubia, Myriophyllum spicatum,* and floating plants; *Eichhornia crassipes Salviniaherzogii, Wolffia columbiana, Lemna valdiviana, Juncus effuses, Nymphaea* spp., *Nupharadvena, P. stratiotes, Lemna* sp. have been recorded as potential phytoremediation agents against different inorganic and organic contaminants including textile dyes, VOCs, PAHs, organochlorines, and some pharmaceuticals [27, 28, 87]. The phytoremediation potential of *Penisetum purpureum, Brachiaria decumbens* and *P. australis* plants in hydroponically grown experimental gravel beds have been recorded for removal of 10 and 20 mg Cr with 97- 99% efficiency within 24h [85]. The structural characteristics of floating plants prevent them from sinking in water. They can absorb the maximum sunlight and easily exchange gases with the atmosphere by staying afloat. Floating plants can reduce the number of submerged species that compete for nutrients by shading the water below [28].

Constructed wetlands can be classified by type of plant employed (free-floating, submerged, emerged), the direction of flow (horizontal, vertical), or wetland hydrology (free water surface flow, subsurface flow) [87]. It is feasible to enhance the unique advantages of each system by combining several types of constructed wetlands as it may be more resilient treatment systems. Compared to traditionally constructed wetlands, this is due to the greater biological complexity, generally more considerable land area utilization, and related longer hydraulic residence time of integrated engineered wetlands [87]. The selection of aquatic plants is one of the critical points in constructed wetland design. Usually, the selected plant should be disease-free with high tolerance towards local environmental conditions, climate, diseases, pests, and water logging. Further, as with other phytoremediation agents, these plants should exhibit a rapid growth rate, propagation rate, and high tolerance towards the toxic pollutants.

Advantages and Limitations of Phytoremediation Techniques

Phytoremediation has a lot of benefits and limitations [27, 28, 68, 69]. The following are the most positive aspects, including performance benefits:

- More cost-effective due to low capital and operating cost
- Environmental friendliness
- Widely applicable for vast range of toxins
- Aesthetic pleasing
- Permanent treatment solution
- Capable of mineralizing organics
- Highly specialized and personal not required
- Relatively easy to implement
- *In-situ*, application avoids excavation and transport of polluted media
- Eliminate secondary pollutants.

On the other hand, phytoremediation presents some limitations as follows;

- Time consuming
- Mostly limited for surface soils
- Not possible to reduce contaminations completely
- Requirement of technical assistant for cultivars for specific metals and regions
- Depend on environmental conditions.

Phytostabilization that helpful at sites with external contamination and relatively low contamination. The following are some of the advantages of phytostabilization:

- Phytostabilization does not produce secondary wastes
- Handling is easy and cost-effective
- This technology enhances the soil fertility
- Aesthetic quality is improved.

Limitations of phytostabilization can be listed as follows:

- The contaminated area has to be continuously monitored to maintain the stabilization
- Phytostabilization does not apply to highly contaminated areas because high concentrations of contaminants may be toxic to plant growth, while phytostabilization is slow
- If soil additives are used, they may need to be reapplied repeatedly to maintain their effectiveness in immobilizing contaminants.

Future Perspectives

The limitations of phytoremediation and phytostabilization compel researchers to investigate new directions in the field, and modifying the existing techniques into more advanced methods. The chelate-assisted or induced assisted phytoextraction methods can be considered as one of the key areas that has to be developed, specially for non-hyperaccumulated plants. More research on synthetic chelating agents such as Ethylene Diamine Tetra–acetic Acid (EDTA), can be used for enhancement of the bioavailability of heavy metals and in turn the metal uptake capacity by the plant[5]. However, the toxicity effects related to higher concentrations of metals in plant's body should be investigated and addressed carefully. The microbial-assisted phytoremediation; use of rhizosphere and endophytic microbes, has growing concern. Use of whole microbes or their enzymes is an area that needs to be developed. When considering the role of microbes on phytoremediation, use of Growth Promoting Microbes (GPM) is another interesting approach. GPM help plant to grow under harsh conditions, supporting the plants through various mechanisms such as specific enzyme activities, reducing ethylene production under stressed conditions, nitrogen fixation mechanisms, etc. [5]. Therefore, more research on development of such microbes based techniques help the plants to remediate the contaminants present in the surrounding media efficiently, than the conventional phytoremediation applications. Use of transgenic plants can be used to enhance both phytodegradation and phytostabilization processes. The studies on the gene expressions which are related to tolerance, uptake, translocation, and sequestration of toxic contaminants can be play a vital role for the improvement of phytoremediation efficiency. Therefore, as similar to many other fields, genetic engineering can be considered as a potential aspect on phytoremediation in future. Some recent studies emphasis the involvement of enzymes on phytoremediation process [27]. Therefore, further studies on enzyme assisted phytoremediation is important as it can be developed in more smart ways, without using the whole plant.

Conclusion

Contamination of the environment by various anthropogenic pollutants is a primary environmental and public health concern today because of the potential for food chain contamination and other health risks. Application of different physical and chemical methods for decontamination of pollutants appear to be economically impractical, as well as in most cases secondary pollution can be happened. Therefore, phytoremediation techniques have growing concern as an ecofriendly and low-cost treatment method which provide aesthetic value to the environment. Toxic metals and other inorganic and organic soil contaminants can be effectively extracted and harvested by hyperaccumulator plants. However, traditional phytoremediation methods are less cost-effective on a large scale as most of the naturally available hyperaccumulator plants show slow-growing and produce less harvestable aboveground biomass as a response to the high toxicity in their growing media. As a result, a genetic engineering approach to developing transgenic plants with high biomass production, higher metal accumulation, metal toxicity tolerance, and excellent adaptability to a wide range of climatic conditions may be more advantageous. Other recent phytoremediation techniques, such as chemically assisted phytoextraction, microbially assisted phytoremediation, and enzyme-assisted phytoremediation may also be used to decontaminate the environment. The use of genetic engineering to improve the phytoremediation capabilities of transgenic plants is still at its primary stage. Therefore, deep rooted researchers on such fields are essential to understand the mechanisms and efficacy of phytoremediation techniques to make these technologies more efficient, time-saving, and economically viable.

References

[1] Ekanayake, E. M. M. S. and Manage, P. M., 2020. Green approach for decolorization and detoxification of textile dye-CI direct blue 201 using native bacterial strains. *Environment and Natural Resources Journal*, 18(1), pp. 1-8.

[2] Mahagamage M. G. Y. L., Chinthaka S. D. M. and Manage P. M., 2014. Multivariate analysis of physico-chemical and microbial parameters of surface water in Kelani river basin. *International Journal in Multidisciplinary Studies* (IJMS), 1, pp. 55-61.

[3] Saxena, G., Purchase, D., Mulla, S. I., Saratale, G. D. and Bharagava, R. N., 2019. Phytoremediation of heavy metal-contaminated sites: eco-environmental concerns, field studies, sustainability issues, and future prospects. *Reviews of Environmental Contamination and Toxicology*, 249, pp. 71-131.

[4] Ojoawo, S. O., Udayakumar, G. and Naik, P., 2015. Phytoremediation of phosphorus and nitrogen with Canna x generalis reeds in domestic wastewater through NMAMIT constructed wetland. *Aquatic Procedia*, 4, pp. 349-356.

[5] Sarwar, N., Imran, M., Shaheen, M. R., Ishaque, W., Kamran, M. A., Matloob, A., Rehim, A. and Hussain, S., 2017. Phytoremediation strategies for soils contaminated with heavy metals: modifications and future perspectives, *Chemosphere,* 171, pp. 710-721.

[6] Sharma, B., Singh, S. and Siddiqi, N. J., 2014. Biomedical implications of heavy metals induced imbalances in redox systems, *BioMed research international*, 2014.

[7] Fatima, M., Usmani, N., Mobarak Hossain, M., Siddiqui, M. F., Zafeer, M. F., Firdaus, F. and Ahmad, S., 2014. Assessment of genotoxic induction and deterioration of fish quality in commercial species due to heavy-metal exposure in an urban reservoir. *Archives of environmental contamination and toxicology*, 67(2), pp. 203-213.

[8] Wijesinghe, H., Idroos, F. S., & Manage, P. M., 2018. Heavy metal contamination status in seven fish species from reservoirs of Polonnaruwa district, Sri Lanka. *Sri Lanka J. Aquat. Sci*. 23(1), pp. 95-104.

[9] Jacquin, L., Petitjean, Q., Côte, J., Laffaille, P. and Jean, S., 2020. Effects of pollution on fish behavior, personality, and cognition: some research perspectives. *Frontiers in Ecology and Evolution*, 8, pp. 86.

[10] Ekanayake, M. S. and Manage, P., 2022. Mycoremediation Potential of Synthetic Textile Dyes by *Aspergillus niger* via Biosorption and Enzymatic Degradation: 10.32526/ennrj/20/202100171. *Environment and Natural Resources Journal*, 20(3), pp.234-245.

[11] Cotrozzi, L., 2020. Leaf demography and growth analysis to assess the impact of air pollution on plants: A case study on alfalfa exposed to a gradient of sulphur dioxide concentrations. *Atmospheric Pollution Research*, 11(1), pp. 186-192.

[12] Nagajyoti P. C., Lee K. D., Sreekanth T. V. M., 2010. Heavy metals, occurrence and toxicity for plants. *Environ Chem Lett*, 8, pp. 199–216.

[13] Garg, V. K. and Kaushik, P., 2008. Influence of textile mill wastewater irrigation on the growth of sorghum cultivars. *Applied ecology and environmental research,* 6(1), pp. 1-12.

[14] Khan, S. and Malik, A., 2018. Toxicity evaluation of textile effluents and role of native soil bacterium in biodegradation of a textile dye. *Environmental Science and Pollution Research*, 25(5), pp. 4446-4458.

[15] Ajibade, F. O., Wang, H. C., Guadie, A., Ajibade, T. F., Fang, Y. K., Sharif, H. M. A., Liu, W. Z. and Wang, A. J., 2021. Total nitrogen removal in biochar amended non-aerated vertical flow constructed wetlands for secondary wastewater effluent with low C/N ratio: Microbial community structure and dissolved organic carbon release conditions. *Bioresource Technology*, 322, pp. 124430.

[16] Abdelgalil, S. A., Attia, A. M., Reyed, R. M., Soliman, N. A. and El Enshasy, H. A., 2018. Application of Experimental Designs for Optimization the Production of *Alcaligenes Faecalis* Nyso Laccase. *Journal of Scientific and Industrial Research*, 77(1), pp. 713-727

[17] Dubey, J. P. and Lindsay, D. S., 2003. Development and ultrastructure of Besnoitiaoryctofelisi tachyzoites, tissue cysts, bradyzoites, schizonts and merozoites. *International journal for parasitology*, 33(8), pp. 807-819.

[18] Khouni, I., Marrot, B., Moulin, P. and Amar, R. B., 2011. Decolourization of the reconstituted textile effluent by different process treatments: Enzymatic catalysis, coagulation/flocculation and nanofiltration processes. *Desalination*, 268(1-3), pp. 27-37.

[19] Muthusaravanan, S., Sivarajasekar, N., Vivek, J. S., Paramasivan, T., Naushad, M., Prakashmaran, J., Gayathri, V. and Al-Duaij, O. K., 2018. Phytoremediation of heavy metals: mechanisms, methods and enhancements. *Environmental chemistry letters,* 16(4), pp. 1339-1359.

[20] Prathiba, A. A., Idroos, F. S. and Manage, P. M., 2020. *Isolation and molecular characterization of polycyclic aromatic hydrocarbon degrading bacteria from effluent water from Weras River Park*, Sri Lanka.

[21] Liyanage, G. Y. and Manage, P. M., 2016. *Isolation and characterisation of oil degrading bacteria from coastal waters and sediments from three locations in Sri Lanka*, 44(4) 351-358.

[22] Iravani, S. and Varma, R. S., 2020. Bacteria in heavy metal remediation and nanoparticle biosynthesis. *ACS Sustainable Chemistry & Engineering*, 8(14), pp. 5395-5409.

[23] Singh, O. V. and Jain, R. K., 2003. Phytoremediation of toxic aromatic pollutants from soil. *Applied microbiology and biotechnology*, 63(2), pp. 128-135.

[24] Kimmatkar, K. S., Purohit, A. V. and Sanyal, A. J., 2017. Phytoremediation Techniques and Species for Combating Contaminants of Textile Effluents–An Overview. *International Journal of Science and Research*, 6(3), pp. 2174-2181.

[25] Erakhrumen and Agbontalor, A., 2011. Phytoremediation: an environmentally sound technology for pollution prevention, control and remediation in developing countries. *Educational Research and Review*, 2 (7), pp. 151-156.

[26] Pivetz, B. E., 2001. *Phytoremediation of contaminated soil and ground water at hazardous waste sites*. US Environmental Protection Agency, Office of Research and Development, Office of Solid Waste and Emergency Response, 12-18.

[27] Ekanayake, M. S., Udayanga, D., Wijesekara, I. and Manage, P., 2021. Phytoremediation of synthetic textile dyes: biosorption and enzymatic degradation involved in efficient dye decolorization by *Eichhornia crassipes* (Mart.) Solms and Pistia stratiotes L. *Environmental Science and Pollution Research*, 28(16), pp. 20476-20486.

[28] Herath, I. and Vithanage, M., 2015. Phytoremediation in constructed wetlands. In *Phytoremediation* (pp. 243-263). Springer, Cham.

[29] Etim, E. E., 2012. Phytoremediation and its mechanisms: a review. *Int J Environ Bioenergy,* 2(3), pp. 120-136.

[30] Zhuang, P., Yang, Q. W., Wang, H. B. and Shu, W. S., 2007. Phytoextraction of heavy metals by eight plant species in the field. *Water, Air, and Soil Pollution,* 184(1), pp. 235-242.

[31] Han, Z., Guo, Z., Zhang, Y., Xiao, X., Xu, Z. and Sun, Y., 2018. Adsorption-pyrolysis technology for recovering heavy metals in solution using contaminated biomass phytoremediation. *Resources, Conservation and Recycling*, 129, pp. 20-26.

[32] Hall, J. Á., 2002. Cellular mechanisms for heavy metal detoxification and tolerance. *Journal of experimental botany*, 53(366), pp. 1-11.

[33] Chandekar, N. and Godboley, B. J., 2017. A review on phytoremediation a sustainable solution for treatment of kitchen wastewater. *Int J Sci Res*, 6(2), pp. 1850-1855.

[34] Patel, D. K. and Kanungo, V. K., 2010. Phytoremediation potential of duckweed (lemnaminor l: a tiny aquatic plant) in the removal of pollutants from domestic wastewater with special reference to nutrients. *Bioscan*, 5(3), pp. 355-358.

[35] Wurochekke, A. A., Harun, N. A., Mohamed, R. M. S. R. and Kassim, A. H. B. M., 2014. Constructed wetland of *Lepironia articulata* for household greywater treatment. *APCBEE procedia*, 10, pp. 103-109.

[36] Abdelhakeem, S. G., Aboulroos, S. A. and Kamel, M. M., 2016. Performance of a vertical subsurface flow constructed wetland under different operational conditions. *Journal of Advanced Research*, 7(5), pp. 803-814.

[37] Fu, X. Y. and He, X. Y., 2013. Nitrogen removal from contaminated water by two aquatic plants. In *Advanced Materials Research* (Vol. 610, pp. 1829-1832). Trans Tech Publications Ltd.

[38] Nouri, J., Lorestani, B., Yousefi, N., Khorasani, N., Hasani, A. H., Seif, F. and Cheraghi, M., 2011. Phytoremediation potential of native plants grown in the vicinity of Ahangaran lead–zinc mine (Hamedan, Iran). *Environmental Earth Sciences,* 62(3), pp. 639-644.

[39] Govindwar, S. P. and Kagalkar, A. N., 2010. The use of hydroponics and plant tissue culture technologies for dye degradation. In: *Phytoremediation technologies for the removal of textile dyes: an overview and future prospects*. Nova Science: New York, pp. 29–39.

[40] Yadav, D. and Kumar, P., 2019. Phytoremediation of hazardous radioactive wastes. In *Assessment and Management of Radioactive and Electronic Wastes* (p. 29). IntechOpen.

[41] Hu, N., Ding, D. and Li, G., 2014. Natural plant selection for radioactive waste remediation. In *Radionuclide contamination and remediation through plants* (pp. 33-53). Springer, Cham.

[42] Dahmani-Muller, H., Van Oort, F. and Balabane, M., 2000. Strategies of heavy metal uptake by three plant species growing near a metal smelter. *Environmental Pollution,* 109:231–238.

[43] Mleczek, M., Mocek, A., Magdziak, Z., Gąsecka, M. and Mocek-Płóciniak, A., 2013. Impact of metal/metalloid-contaminated areas on plant growth. In *Plant-based remediation processes* (pp. 79-100). Springer, Berlin, Heidelberg.

[44] Suzuki, Y., Takenaka, C., Tomioka, R., Tsubota, H., Takasaki, Y. and Umemura, T., 2016. Accumulation of arsenic and copper by bryophytes growing in an aquatic environment near copper mine tailings. *Mine Water and the Environment*, 35(3), pp. 265-272.

[45] Singh, N. P. and Santal, A. R., 2015. Phytoremediation of heavy metals: the use of green approaches to clean the environment. In *Phytoremediation* (pp. 115-129). Springer, Cham.

[46] Tesser, T. T., Bordin, J., Da Rocha, C. M. and Da Silva, A., 2021. Application of the dry and wet biomass of bryophytes for phytoremediation of metals: Batch experiments. *Environmental Challenges*, 5, pp. 100382.

[47] Adki, V. S., Jadhav, J. P. and Bapat, V. A., 2012. Exploring the phytoremediation potential of cactus (Nopaleacochenillifera Salm. Dyck.) cell cultures for textile dye degradation. *International journal of phytoremediation*, 14(6), pp. 554-569.

[48] Agostini, E., Coniglio, M. S., Milrad, S. R., Tigier, H. A. and Giulietti, A. M., 2003. Phytoremediation of 2, 4-dichlorophenol by *Brassica napus* hairy root cultures. *Biotechnology and applied biochemistry*, 37(2), pp. 139-144.

[49] Kagalkar, A. N., Jagtap, U. B., Jadhav, J. P., Bapat, V. A. andGovindwar, S. P., 2009. Biotechnological strategies for phytoremediation of the sulfonated azo dye Direct Red 5B using *Blumeamalcolmii* Hook. *Bioresource technology,* 100(18), pp. 4104-4110.

[50] Sharma, S. and Roy, S., 2015. Isolation and Identification of a novel Endophyte from a plant Amaranthus spinosus. *International Journal of Current Microbiology App. Science* 4(2), pp. 785-798.

[51] Ngieng, N. S., Zulkharnain, A., Roslan, H. A. and Husaini, A., 2013. Decolourisation of synthetic dyes by endophytic fungal flora isolated from senduduk plant (*Melastomamalabathricum*). *ISRN biotechnology*, pp. 7.

[52] Alderete, L. G. S., Talano, M. A., Ibáñez, S. G., Purro, S., Agostini, E., Milrad, S. R. and Medina, M. I., 2009. Establishment of transgenic tobacco hairy roots expressing basic peroxidases and its application for phenol removal. *Journal of biotechnology*, 139(4), pp. 273-279.

[53] Telke, A. A., Kadam, A. A., Jagtap, S. S., Jadhav, J. P. and Govindwar, S. P., 2010. Biochemical characterization and potential for textile dye degradation of blue

laccase from *Aspergillus ochraceus* NCIM-1146. *Biotechnology and Bioprocess Engineering*, 15(4), pp. 696-703.

[54] Praveen, A. and Pandey, V. C., 2020. Pteridophytes in phytoremediation. *Environmental geochemistry and health*, 42(8), pp. 2399-2411.

[55] Rathinasabapathi, B., Rangasamy, M., Froeba, J., Cherry, R. H., McAuslane, H. J., Capinera, J. L., Srivastava, M. and Ma, L. Q., 2007. Arsenic hyperaccumulation in the Chinese brake fern *(Pteris vittata)* deters grasshopper (*Schistocerca americana*) herbivory. *New Phytologist*, 175(2), pp. 363-369.

[56] Fumagalli, P., Comolli, R., Ferre, C., Ghiani, A., Gentili, R. and Citterio, S., 2014. The rotation of white lupin (*Lupinus albus* L.) with metal-accumulating plant crops: A strategy to increase the benefits of soil phytoremediation. *Journal of Environmental Management*, 145, pp. 35-42.

[57] Lu, Y., Li, X., He, M. and Zeng, F., 2013. Behavior of native species *Arrhenatherumelatius* (Poaceae) and *Sonchus transcaspicus* (Asteraceae) exposed to a heavy metal-polluted field: plant metal concentration, phytotoxicity, and detoxification responses. *International journal of phytoremediation*, 15(10), pp. 924-937.

[58] Panchenko, L., Muratova, A. and Turkovskaya, O., 2017. Comparison of the phytoremediation potentials of *Medicago falcata* L. and *Medicago sativa* L. in aged oil-sludge-contaminated soil. *Environmental Science and Pollution Research*, 24(3), pp. 3117-3130.

[59] Paz-Alberto, A. M., Sigua, G. C., Baui, B. G. and Prudente, J. A., 2007. Phytoextraction of lead-contaminated soil using vetivergrass (*Vetiveriazizanioides* L.), cogongrass (*Imperata cylindrica* L.) and carabaograss (*Paspalum conjugatum* L.). *Environmental Science and Pollution Research-International*, 14(7), pp. 498-504.

[60] Paz-Alberto, A. M. and Sigua, G. C., 2013. Phytoremediation: a green technology to remove environmental pollutants. *American Journal of Climate Change* Vol.2 No. 1(2013), Article ID:29110,16 pages doi:10.4236/ajcc.2013.21008.

[61] Anjana, S. and Thanga, V., 2011. Phytoremediation of synthetic textile dyes. *Asian Journal of Microbiology, Biotechnology & Environmental Sciences*, 13, pp. 30-39.

[62] Ekanayake, M. S. and Manage, P., 2016. Decoloriztion of textile dye (CI Direct Blue 201) by selected aquatic plants. *The 2nd Environment and Natural Resources International Conference Proceeding*, pp. 25.

[63] Rizwana, M., Darshan, M. and Nilesh, D., 2014. Phytoremediation of textile waste water using potential wetland plant: Eco sustainable approach. *International journal of interdisciplinary and Multidisciplinary Studies* (IJIMS), 1(4), pp. 130-138.

[64] Sharma, S., Singh, B. and Manchanda, V. K., 2015. Phytoremediation: role of terrestrial plants and aquatic macrophytes in the remediation of radionuclides and heavy metal contaminated soil and water. *Environmental Science and Pollution Research*, 22(2), pp. 946-962.

[65] Couto, N., Guedes, P., Ribeiro, A. B. and Zhou, D. M., 2015. Phytoremediation and the electrokinetic process: Potential use for the phytoremediation of Antimony and Arsenic. In *Phytoremediation* (pp. 199-209). Springer, Cham.

[66] Bunluesin, S., Kruatrachue, M., Pokethitiyook, P. and Lanza, G. R., 2004. Plant Screening and Comparison of *Ceratophyllumdemersum* and *Hydrilla verticillata* for Cadmium. *Bull. Environ. Contam. Toxicol*,73, pp. 591–598.

[67] Sheoran, V., Sheoran, A. S. and Poonia, P., 2016. Factors affecting phytoextraction: a review. *Pedosphere*, 26(2), pp. 148-166.

[68] Wang, L., Hou, D., Shen, Z., Zhu, J., Jia, X., Ok, Y. S., Tack, F. M. and Rinklebe, J., 2020. Field trials of phytomining and phytoremediation: A critical review of influencing factors and effects of additives. *Critical Reviews in Environmental Science and Technology*, 50(24), pp. 2724-2774.

[69] Tong-Bin, C., Xiao-Yong, L., Ze-Chun, H., Mei, L., Wen-Xue, L., Liang-Yu, M., Zhi-Zhuang, A., Chao-Yang, W., Xi-Yuan, X. and Hua, X., 2007. Phytoremediation of arsenic-contaminated soil in China. In *Phytoremediation* (pp. 393-404). Humana Press.

[70] Duquène, L., Vandenhove, H., Tack, F., Meers, E., Baeten, J. and Wannijn, J., 2009. Enhanced phytoextraction of uranium and selected heavy metals by Indian mustard and ryegrass using biodegradable soil amendments. *Science of the Total Environment,* 407(5), pp. 1496-1505.

[71] Naveed, M., Mitter, B., Yousaf, S., Pastar, M., Afzal, M. and Sessitsch, A., 2014. The endophyte Enterobacter sp. FD17: a maize growth enhancer selected based on rigorous testing of plant beneficial traits and colonization characteristics. *Biology and fertility of soils*, 50(2), pp. 249-262.

[72] Henagamage, A. P., 2019. Bioremediation of Textile dyes by Fungal-Bacterial Biofilms. *International Journal of Environment, Agriculture and Biotechnology*, 4(3).

[73] Sandesh, K., Kumar, G., Chidananda, B. and Ujwal, P., 2019. Optimization of Direct Blue-14 dye degradation by *Bacillus fermus* (Kx898362) an alkaliphilic plant endophyte and assessment of degraded metabolite toxicity. *Journal of Hazardous Materials,* 364, pp. 742-751.

[74] Sheng, X. F., Xia, J. J., Jiang, C. Y., He, L. Y. and Qian, M., 2008. Characterization of heavy metal-resistant endophytic bacteria from rape (*Brassica napus*) roots and their potential in promoting the growth and lead accumulation of rape. *Environmental pollution*, 156(3), pp. 1164-1170.

[75] Andria, V., Reichenauer, T. G. and Sessitsch, A., 2009. Expression of alkane monooxygenase (alkB) genes by plant-associated bacteria in the rhizosphere and endosphere of Italian ryegrass (*Lolium multiflorum* L.) grown in diesel contaminated soil. *Environmental Pollution*, 157(12), pp. 3347-3350.

[76] Ho, Y. N., Mathew, D. C., Hsiao, S. C., Shih, C. H., Chien, M. F., Chiang, H. M. and Huang, C. C., 2012. Selection and application of endophytic bacterium *Achromobacterxylosoxidans* strain F3B for improving phytoremediation of phenolic pollutants. *Journal of Hazardous Materials*, 219, pp. 43-49.

[77] Weyens, N., Truyens, S., Dupae, J., Newman, L., Taghavi, S., van der Lelie, D., Carleer, R. and Vangronsveld, J., 2010. Potential of the TCE-degrading endophyte *Pseudomonas putida* W619-TCE to improve plant growth and reduce TCE phytotoxicity and evapotranspiration in poplar cuttings. *Environmental Pollution,* 158(9), pp. 2915-2919.

[78] Gomes, H. I., Dias-Ferreira, C. and Ribeiro, A. B., 2012. Electrokinetic remediation of organochlorines in soil: enhancement techniques and integration with other remediation technologies. *Chemosphere,* 87(10), pp. 1077-1090.

[79] Robinson, B. and Anderson, C., 2007. Phytoremediation in New Zealand and Australia. In *Phytoremediation* (pp. 455-468). Humana Press.

[80] Kim, K. R. and Owens, G., 2010. Potential for enhanced phytoremediation of landfills using biosolids–a review. *Journal of environmental management*, 91(4), pp. 791-797.

[81] Nagendran, R., Selvam, A., Joseph, K. and Chiemchaisri, C., 2006. Phytoremediation and rehabilitation of municipal solid waste landfills and dumpsites: A brief review. *Waste Management*, 26(12), pp. 1357-1369.

[82] Burken, J. G., 2003. Uptake and metabolism of organic compounds: green-liver model. *Phytoremediation: transformation and control of contaminants*, 59, pp. 59-84.

[83] Hadi, F., Arifeen, M. Z. U., Aziz, T., Nawab, S. and Nabi, G., 2015. Phytoremediation of cadmium by *Ricinus communis* L. in hydrophonic condition. *Cell,* 92(345), p. 8112741.

[84] Davamani, V., Parameshwari, C. I., Arulmani, S., John, J. E. and Poornima, R., 2021. Hydroponic phytoremediation of paperboard mill wastewater by using vetiver (Chrysopogonzizanioides). *Journal of Environmental Chemical Engineering,* 9(4), p. 105528.

[85] Mant, C., Costa, S., Williams, J. and Tambourgi, E., 2006. Phytoremediation of chromium by model constructed wetland. *Bioresource Technology,* 97(15), pp. 1767-1772.

[86] Tang, X., Eke, P. E., Scholz, M. and Huang, S., 2009. Processes impacting on benzene removal in vertical-flow constructed wetlands. *Bioresource technology*, 100(1), pp. 227-234.

[87] Compaore, W. F., Dumoulin, A. and Rousseau, D. P., 2020. Metal uptake by spontaneously grown *Typha domingensis* and introduced Chrysopogon zizanioides in a constructed wetland treating gold mine tailing storage facility seepage. *Ecological Engineering,* 158, p. 106037.

Chapter 8

The Role of Nano-Encapsulation Technology in Improving Plant Growth Promoting Rhizobacteria (PGPR) as Potential Biofertilizers

Anandita Govil
Pardeep Kumar
and Machiavelli Singh*
Amity Institute of Biotechnology, Amity University Haryana, Manesar, Gurugram, India

Abstract

Plant growth promoting rhizobacteria (PGPR) are soil bacteria that have the ability to influence plant growth both directly and indirectly. These organisms may be able to reduce or eliminate the use of chemical fertilisers and toxic chemical inputs. Exploring PGPR in agriculture is thus one of the more promising techniques for increasing agricultural production while minimising environmental damage. Simultaneously, nano-technological applications are having a significant impact on agriculture. Plant growth promoting rhizobacteria (PGPR) play an important role in the sustainable agriculture industry. The increasing demand for crop production while reducing the use of synthetic chemical fertilisers and pesticides is a major challenge today. The use of PGPR has been shown to be an environmentally friendly method of increasing crop yields by facilitating plant growth via either a direct or indirect mechanism. PGPR mechanisms include regulating hormonal and nutritional balance, inducing resistance against plant pathogens, and

* Corresponding Author's Email: machiavellisingh@gmail.com.

In: Microbial Technologies and Their Applications
Editors: Santosh Kumar Karn and Anne Bhambri
ISBN: 979-8-88697-913-8

solubilizing nutrients for easy uptake by plants. Furthermore, PGPR exhibits both synergistic and antagonistic interactions with microorganisms in the rhizosphere and beyond in bulk soil, which indirectly increases plant growth rate. Many bacteria species have been described in the literature as being effective at improving plant growth when acting as PGPR. However, there is a gap between the PGPR's mode of action (mechanism) for plant growth and its role as a biofertilizer—hence the importance of nano-encapsulation technology in improving PGPR efficacy. As a result, this review fills the gap and summarises the mechanism of PGPR as a biofertilizer for agricultural sustainability.

Introduction

There are various innovations and technological advancements that have been made in recent years to help with the growing challenges of food security and sustainable production in agriculture [1]. All these agricultural innovations are very important to meet the increasing food demand due to the increasing world population by using all the natural and synthetic resources. Nanotechnology can help by providing effective and long-term solutions to a variety of agricultural-related issues. Nanoparticles are of great scientific interest for bridging the gap between bulk materials and atomic or molecular structures. There has been continuous research going on for the last two decades in the field of nanotechnology, with the aim to utilize its applications in agriculture [2]. The major hurdle which comes up is to maintain the optimum growth conditions of nutrition and climate from lab to field. If the desired environment is not maintained, it can cause a major impact on the microbial count, which can affect the field result in a poor way. As a result, it is critical to develop innovative and effective methods for delivering PGPR in the field. Showing its major applications in not only biology, but also in chemistry, material science, physics and engineering, nanotechnology is a rapidly growing and developing field. Recently, due to its ability to improve disease detection, abiotic and biotic stress tolerance and prevention, nanotechnology has got a lot of attention [3]. When compared to traditional methods, nanomaterials can improve plant nutrient utilisation efficiency. Nanoparticles (NPs) with defined physicochemical properties can increase the yield of the crop, provide maximum nutrition to the soil and boost plant metabolism [4].

Carbon nanoparticles, metal nanoparticles, organic nanoparticles, and semiconductor nanoparticles are the most common types of nanoparticles [5]. Silver [6], titanium [7], zinc oxide [8], silica [9], carbon [10], boron [11, 12], gold [13] and zeolite [14] plant growth has shown to be aided by nanoparticles. The reason why nano-fertilizers have taken over the conventional chemical fertilizers is because of their ability to not cause the leaching and loss of nutrients after use and require a very small amount, reducing the risk of water and soil pollution.

Ideally, PGPR should have a good amount of rhizosphere efficiency, increase the capability of plant growth, should not have limited area of application and effectiveness, should not be harmful for the environment, should be able withstand other rhizobacteria, and should not be affected by heat, UV radiations, and various oxidising agents [15]. The reason for the need of a good PGPR biofertilizer to help with the ongoing demand of food production, one of the major key drivers, has been highlighted. The new addition of a technology named 'Nano-Encapsulation' has played a major role in the formulation of the recent PGPR biofertilizers.

Because of the low bioavailability, degradation of chemicals during storage, low stability, and sensitivity towards oxygen and UV light, it is very difficult to work with the plant derived compound. A variety of techniques have been developed to improve these properties. In this chapter, we would focus on nanotechnology and its various applications in the field of PGPR and agriculture. Nano-carrier systems can help improve the delivery of various bioactive molecules such as fatty acids, minerals, antioxidants, lycopene and vitamins. In encapsulation, there are one or more ingredients that are immobilised in a wall or matrix. The matrix can be a solid or a liquid phase, either heterogeneous or homogeneous [16].

PGPR (Plant Growth Promoting Rhizobacteria)

PGPR are the bacteria that live in the rhizosphere [17]. PGPR are bacteria which colonise the plant roots (rhizosphere) and help with the growth of the plant. It is the area with the maximum microbial activity in soil that leads to a pool of nutrients including the essential micro- and macronutrients. The microbial population of the rhizosphere is different from the surroundings because of the root exudates that are present, which acts as the source of nutrients for the growth of the microbes [18].

As Weller and Thomashow [19] showed in their research, that the narrow rhizosphere is a zone which is rich in nutrients when compared with the bulk soil; that was demonstrated by the number of bacteria present near the roots of the plant, that was in general, 10-100 times more than the bulk soil. A few microorganisms that colonize in the rhizosphere are actinomycetes, algae, bacteria, fungi and protozoa. On the other hand, the most abundant species in the rhizosphere are bacteria [20]. They are well known for their application to improve plant growth [21, 22]. The term PGPR was coined by Kloepper and Schroth [23] in order to describe these beneficial microbes, making way for further research into PGPR.

Depending on their interactions with the plants, these PGPRs can be divided as symbiotic bacteria that live inside the plants and exchange metabolites with them, and free-living rhizobacteria that lives outside the cells of the plants [24]. There are 2 ways in which the working mechanism of PGPR can be classified: Indirect and Direct. Under the direct mechanism comes rhizoremediation, plant stress control, biofertilization and root growth stimulation. On the other side, the rhizobacteria are indirectly involved in the promotion of plant growth by reducing the disease impact which includes induction of systemic resistance, competition for nutrients and niches, and antibiosis [25].

Application of PGPR to Enhance Plant Growth

In a variety of mechanisms, PGPR contributes in a good amount to the growth of the plant. Plant growth is promoted by PGPR in the following ways (i) production of the volatile organic compounds; (ii) the production of protection enzymes; (iii) plant growth regulators; (iv) the production of siderophores; (v) abiotic stress tolerance in plants, (vi) nutrient fixation for easy uptake by plants. The mode of action of various PGPR depends on the type of the host plant [26].

The environment of the soil causes a lot of stress as it influences the growth of the plant that is the major constraint for sustainable agriculture production. There are two categories in which these stresses are divided into: abiotic stress and biotic stress. The Biotic stress is caused due to the pathogens and pests in the plant that are insects, bacteria, nematodes, fungi, virus and so on, whereas, on the other hand, abiotic stress is caused due to the heavy metal content in the soil, nutrient deficiency, drought, temperature, salinity and so on.

Abiotic Stress Tolerance

These stresses are said to be the primary cause of the loss of agricultural yield. The severity of this abiotic stress is different, depending on the type of the soil (Deficiency of nutrition and hormone imbalance) and factors of the plant (i.e., physiological disorders such as abscission, disease susceptibility and so on) [27]. The response by the PGPR mechanism to abiotic stress has been studied previously. The unhealthy and toxic effects of cadmium pollution on the plant of barley may be attenuated by the ability of the bacteria to the cadmium ions from soil through binding mechanisms, therefore reducing the availability of the amount of cadmium in the soil [28].

Bacillus lentimorbus strain could increase the antioxidant capacity of carrot, lettuce and spinach parts that are edible while also helping in growth promotion. The results obtained are critical, particularly in terms of improving the nutrient content of these crops [29].

One more very important contribution of PGPR on the plants which are under abiotic stress conditions is the improved water status of leaves, particularly under drought or saline stress [30, 31]. *Pseudomonas aeruginosa* strain improved the growth of *Vigna radiata* (mung bean) plants under drought conditions [32]. Stomatal apertures help to determine the ability of the plants to use water for growth. The balance of the water content in the leaves and water uptake by the roots are done by the stomata. Under these drought conditions, plants inoculated with PGPR had higher stomatal conductance than the non-PGPR inoculated plants [30, 33].

Nutrients Available for the Plants

PGPR can increase the available nutrient concentration in the rhizosphere by fixing the nutrients and not allowing them to leach out [34]. Nitrogen is one of the most limited nutrients for plants due to its requirement for the synthesis of proteins and amino acids. Prokaryotes are the organisms that can convert the atmospheric nitrogen to the organic form so that the plant can easily use it [35, 36]. *Azospirillum* is one of the rarest examples of organisms that can fix nitrogen and is often associated with cereals in zones of temperature and is also seen to improve yields of rice crop [37].

Some phosphates can be solubilized by the PGPR [38], leading to an increase in soil availability of phosphate ions which can be taken up easily by the plants. The strain (2M4) of *Kocuriatur fanensis* was found to be a phosphate solubilizer after being isolated from the rhizosphere soil, IAA producer and also a siderophore producer [39]. The effect of PGPR on the nutrient uptake in rice and in experiment they used three PGPR strains: *P. fluorescens*, *P. putida* and *Pseudomonas* [40].

Regulators of Plant Growth

Also known as plant exogenous hormones, PGR (plant growth regulators) are synthetic substances that would mimic natural hormones of the plants. These can be used to control the growth of the plants and are a very crucial measure to increase the agricultural output. Phyto-stimulator is one of the most common modes of action for the growth promotion of PGPR. All the microorganisms that have the ability to produce or to alter the concentrations of the growth regulators such as GA, Ethylene, IAA and cytokines are classified as this. The given mechanism id the production of phytohormones which are plant hormones, such as cytokinin, auxin and GA [41, 42]. They are organic substances which are found in trace amount which would influence the plant (physiological, morphological and biochemical processes); and their synthesis is controlled tightly. Phytohormones are the plant growth regulators which are not synthesized naturally by the plants but they are exogenously synthesized through synthetic means and natural means.

Auxin is a critical molecule that regulates most plant processes directly or indirectly [43] as demonstrated by Ahmed and Hasnain [44], who found out that the property of producing auxins by *Bacillus* spp. had a good impact on the growing of *Solanum tuberosum*. One of the most well-known and effective as well as active auxins in plants is known as Indole-3-acetic acid (IAA) [45]. The regulations in the IAA affect a wide range of processes in the development of the plant and its growth [46, 47]. The low level of IAA stimulates the primary elongation of the roots. High level of IAA decreases the lengths of the roots, also increasing the formation of the root hair and stimulates the lateral root formation. As a result of the increased root surface area and length caused by bacterial IAA, there is good access for the soil nutrients for the plants. Gibberellin (GA), a phytohormone, is also a part of various processes like germination of the seeds and emergence, development of fruits and flowers, and the growth of stem and leaf [48]. Elongation of the shoot is the most major physiological effect of GA [49]. After inoculating the plant of tomato with gibberellins producing *Sphingomonas* sp. LK11 strain shows improvement in various characteristics of growth [50]. Division of cell, vascular differentiation and vascular cambium sensitivity as well as the proliferation of root hair are all stimulated by cytokinin in the plants [44, 51, and 52].

Another plant hormone which is known to regulate various processes like leaf abscission and fruit ripening is known as Ethylene [53]. When the concentration is high, defoliation is caused by ethylene along with cellular processes which inhibit the stems and the roots from growing and premature senescence [54]. This would all result in poor performance of the crop. Plants synthesised 1-aminocyclopropane-1-carboxylate (ACC) in response to the different conditions of stress from the environment such as flooding, pathogen infection, cold, and drought and in the presence of heavy metals, which is the precursor for ethylene [55]. In stress conditions, large levels of ethylene can affect certain processes in legumes, such as elongation of the roots or fixing of nitrogen [56] and can cause premature senescence [30] as shown in the figure.

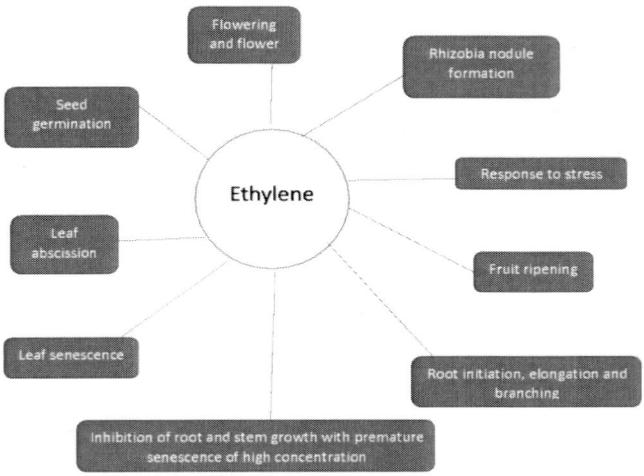

Figure 1. Effects of Phytohormone ethylene on different processes of development and growth of the plant.

Hormone Production

These are the chemical messengers who would affect the ability of the plant to respond to the surroundings. They are natural organic compounds which are effective when the concentration is very low and are mainly synthesized and also transported only in the specific parts of the plant. Also known as

Phytohormones, they influence physiological processes even when concentration is low. Various processes that are affected are growth, development and differentiation. Stomatal movement process may also be affected due to the hormones [57]. It is seen that every plant responds when there are two or more hormones acting in concert. They are also known as Plant Growth Regulator (derived from PGPR) because the hormones inhibit or stimulate the growth of the plant [58]. A few well-known hormones of the plant which may regulate the growth of the plant and also the development are: Cytokinins, Gibberellins, Auxins, Abscisic acid (ABA) and Ethylene [59, 60].

Production of Siderophores

One of the most abundant minerals on the surface of the earth is Iron, but it is not available to the plants in soil. It is mainly found in the form of Fe3+in nature, which is majorly insoluble; PGPR secretes siderophores in order to solve the problem. These siderophores are basically iron binding proteins that have low molecular weight which are involved in the process that would chelate ferric iron (Fe (iii)) in the environment. The microbial siderophores would provide Fe to the plants when it is scarce to boost their growth. The siderophore-producing strain which is the *Phyllobacterium* strain that promotes the quality and growth of strawberry [61].

Volatile Organic Compound Production

VOC, which is the volatile organic compound, which is produced by the PGPR plays a very important role in improving the growth of the plant and causes systemic resistance (ISR) to the pathogens. They are produced by a variety of bacterial species from different genera, which includes *Bacillus*, *Pseudomonas*, *Serratia*, *Arthrobacter*, and *Stenotrophomonas*. Two of the most well-known compounds: Acetoin and 2,3-butanediol, are known for the significant improvements in the growth of the plant [62]. The other PGPR strains would produce VOCs which can increase in the biomass of the plant, abiotic stress tolerance and disease resistance. Emission of VOC is the most common property of a number of microorganisms in the soil, although the quantity and the identity of these volatile compounds which are emitted are different between species [63, 64].

Role of Nano Particles for the Protection of Plants

Nanoparticles may be used for a large amount of plant protection applications, which includes detecting the pathogens (nano diagnostics), controlling pests against microbial pathogens, insect, fungi, and bacteria, controlling weed, pesticide remediation as well as induced resistance [65]. Nanoparticles (NPs) with defined physicochemical properties can boost plant metabolism, increasing the yield of crops and providing the necessary nutrients to the soil [66]. The most common types of these particles are carbon, metal, organic, and semiconductor nanoparticles [67]. The reason why these nano-fertilizers are more effective than the conventional chemical fertilisers is because they avoid causing leaching or loss of nutrients after the use and only required in a very less amount, reducing the risk of soil and water pollution., which reduces the risk of water and soil pollution [65]. The effects of various nanoparticles have been shown in Table 1.

Table 1. Effects of nanoparticles

Nano Particle	Method of synthesis	PGPR	Plant	Effect	Ref.
Silica	It is synthesized using rice husk ash by alkaline extraction which is followed by acid precipitation	*Bacillus megaterium, Bacillus brevis, Pseudomonas fluorescens, Azotobacter vinelandii*	Maize (*Zea mays*)	Leads to increase in the percentage of seed germination other than conventional silica	[68]
	Nanosilica synthesized with rice husk with the help of alkaline treatment later done by acid precipitation	-	Maize (*Zea mays*)	Increased bacterial biomass	[69]
Titania	Acidic Hydrolysis of titanium ethoxide, ($Ti(OC_2H_5)_4$) modified with triethanolamine, $N(C_2H_4OH)_3$	*B. thuringiensis AZP2, P. polymyxa A26*	Wheat (*Triticum aestivum cv. Stava*)	Enhanced performance of PGPR and their colonization	[70]
	Done by using the Sol-Gel approach which is a Captigel method and then applying TiBALDH precursors	*B. amyloliquefaciens subsp. plantarum UCMB5113*	Oilseed rape plants (*Brassica napus*)	Helped with the adhesion of the beneficial bacteria with the roots of the oilseed rape; it protects the plant from various fungal pathogen such as *Alternaria brassicae*	[71]

Table 1. (Continued)

Nano Particle	Method of synthesis	PGPR	Plant	Effect	Ref.
Silver	Chemical method using sodium borohydride (NaBH4)	*Pseudomonas fluorescence, Bacillus cereus*	Maize (*Zea mays*)	expand PGPR and induced raise in area, length and shoot of roots ratio of maize which is irrigated with the waste water from municipal	[72]
Gold	Biosynthesized by *B.subtilis SJ15*	*P. monteilii*	Cowpea *(Vigna unguiculata)*	Increasing the production of IAA in *P. monteilii*; plant probiotic effect is improved	[73]
Zeolite	Commercially purchased	*Bacillus spp.*	Maize (*Zea mays*)	Enhances height of the plant, area of the leaf, amount of leaves, total protein as well as chlorophyll	[74]
Zinc Oxide	-	*Brady rhizobium japonicum, Pseudomonas putida, Azospirillumlipoferum*	Soybean (*Glycine max L.*)	Enhances the height of the plant, amount of nodules present in each plant, grain weight and yield	[75]
Chitosan	-	*Bacillus sp.*	Maize (*Zea mays*)	Increased seed germination, height of the plant and area of the leaves	[76]

Application of Nanofarming in Agriculture

The most recent breakthrough in technology which has exhibited the most unique targets and characteristics with the most strength is nano particle engineering. Norio Taniguichi, who is a professor at the Tokyo University of Science, discovered the term "nanotechnology" in the year 1974 [77]. Although this term has been used in various disciplines, the thought that nanoparticles (NPs) may be helpful in the development of agriculture is a new technological invention which is yet in the initial stages of development [78]. Recent advances in the invention of nanomaterials of different shapes and size have led to a vast range of significance in agriculture, medicine, food processing, and environment science.

Additionally, as agriculture has to go through a number of and uncommon challenges for example less yield of crop because of biotic stress and abiotic stresses, such as pollution and deficiency in nutrients, the exposure of nanotechnology has shown realistic applications to enhance agriculture [79]. In recent years, the word precision farming or agriculture is used to address the achievement of wireless form of networking and the miniaturisation of the sensors for assessing, monitoring, and controlling the agricultural exercise. It is particularly related to the site-specific management of the crops with a vast range of agricultural pre-production and post-production features, which ranges from the horticultural crops to the crops in the fields [80].

A few scientific achievements are tissue engineering and, in the nanomaterial, based target delivery of Cas (CRISPR associated protein)/ CRISPR mRNA and sgRNA for the genetic modification of the crops [81-83]. Additionally, nanotechnology gives outstanding solutions to a lot of environmental problems. For example, the evolution of nano sensors has a wide range of applications in order to monitor the stress in the environment and improve plant disease resistance [84, 85]. As a result, these continuous upgrades in the field of nanotechnology, specifically for problem identification and the development of combined approaches for the sustainable growth of agriculture, has a wonderful potential to provide a wide range of equitable and social benefits.

Role of Nanoencapsulation in PGPR

Nowadays, the use of nanomaterials to carry micro and macronutrients to the plants is getting more in demand [86]. Nanoencapsulation can help to protect the crops from various pathogenic organisms, insects, and other kinds of pests [87]. By immobilising the two microorganisms: *Pantoea agglomerans* and *Burkholderia caribensis* in the nanofibers did not have any effect on their feasibility or other important properties [88]. The Immobilization of PGPR was attached to increase the germination of seed, and the length and dry weight of the roots of soybeans. While dealing with the applications of PGPR range, the major task is to maintain and keep up the count of the living bacteria. Nano materials along with the other nanofibers has the capability to be a hopeful eco-friendly factor for the development of the inoculums, where the PGPR utilizes its important growth promoting traits and the nanofibers secures the seeds and bacteria from the local abiotic stress.

Role of Nano-Fertilizers Along with PGPR

Recent studies have shown that conventional fertilisers can be exchanged with nano-fertilizers. They can not only enhance the utilization of the nutrients, reduce the toxicity in the soil, reduce the hopeful contrary force of overdosing, but also decrease application frequency [89]. The low-release nano fertilizer which would supply the nutrients PGPRs [90]. Release fertiliser systems can improve and magnify the seed vigour by speeding up the actions of the enzyme during germination [90].

With the use of engineered nanomaterials, with respect to sustainable agriculture, it has been shown a totally new way of the production of food which could overcome the uncertainty of the crop sector with finite resources which are available [91]. The revolution, known as green nanotechnology, has immediately transformed the canvas of global agriculture and nanomaterials as the nanofertilizers have provoked potential to meet the given global demand for food along with sustainable agriculture. Nanofertilizers could be the finest option for reducing micronutrient and macronutrient deficiency via better nutrient use ability and controlling the persistent problem of eutrophication [92]. Nanofertilizers are produced to manage the release of nutrients which are based on the need of the crop while also reducing differential losses.

Conventional nitrogen fertilisers, for example, have a large amount of loss from the soil because of evaporation, leaching, or even going down up to 50–70%, which would reduce the efficiency of the fertiliser and raise the cost of production [93]. On the other hand, nano formulations regarding nitrogenous fertiliser, synchronise fertilizer-N release with the crop consumption demand. Therefore, nanoformulations would not allow unwanted loss of nutrients through direct incorporation of crops, ignoring interaction of nutrients with water, soil, air, and microorganisms [94].

Advantages and Disadvantages of PGPR

Rhizobacteria are very important to maintain the fertility of soil and improve the growth and development of the plant. This development in the growth occurs via the difference of mechanisms [95]. For example, the production of Cyanide is noted to be a quality of certain species of *Pseudomonas*. Production of cyanide through bacteria is considered not only as a promoter of growth, but also as a growth inhibitor. Furthermore, cyanide is also said to act like a biocontrol agent against some pathogens of plant [96], but it might also have certain unwanted effects on the growth of the plant [97].

The production of auxin through PGPR could have certain likely and unlikely effects on the growth of the plan [57]. It is important to note that the concentration of auxin determines its effectiveness. For example, when the concentration is low, it enhances the growth of the plant, on the other hand, when the concentration is high, the growth of the root is inhibited [98]. Moreover, *Bradyrhizobium elkanii* produces rhizobitoxine, which has a dual effect. It can reduce the unwanted effects causing stress-induced ethylene production over nodulation since it is an inhibitor of the synthesis of ethylene [99]. On the contrary, Rhizobitoxine is divided as a plant toxin since it leads to foliar chlorosis in soybeans [100].

Conclusion

In the field of agriculture, nanotechnology is utilized basically to enhance the production of the crops while also improving the various systems of farming. The disclosure of engineered nanomaterials as well as the actions caused by them inside the context of sustainable agriculture has straight away changed the whole canvass of world agriculture regarding rapid growth, novelty, and enormity with the purpose to meet projected global demand for food. Protection of the environment from pollution is a very important trade aim in sustainable agriculture, and these materials ensure better control and conservation of plant inputs for production. Nanomaterials' ability encourages a latest green revolution with less risk of farming. Although, there are notable gaps in our understanding regarding the permissible limit, uptake capacity and ecotoxicity of different nanomaterials. As a conclusion, extra research is immediately required to understand the behaviour of changed inputs of agriculture, along with their working with macromolecules which are found in the living environments.

Nanoparticles together with PGPR have the ability to ease the way for a better future for agriculture. Interactions between these two can promote colonisation of roots, phytohormone and synthesis of secondary metabolites, and overall performance of rhizobacteria. PGPR nanoencapsulation and nano bio fertilizer along with PGPR are also currently rising at a very high speed. All these environmentally friendly compounds have a large amount of potential to change the traditional practices of fertiliser. PGPR nano-technology can be utilized in order to manage the stress of the plant as well as the disease in a less-input, sustainable, as well as environmentally friendly environment also known as the green way. Future innovative agricultural technologies will revitalise the agricultural sector.

As we have read the benefits of PGPR related to biofertilization, biocontrol, and bioremediation, all of these enhance the productivity of the crops and the functioning of the ecosystem, its promotion in the field of agriculture should be uplifted. With the continuous advancement of different technology and the development of successful research, the usage of PGPR will definitely become a reality and will also be instrumental to very important processes which ensure the stability and productivity of agro ecosystems, which then leads us to an ideal system of agriculture.

The importance and usage of nanotechnology in the field of agriculture should be extended and elaborated to decrease the environmental damage while meeting the global demand of crops. In the last few years, encouraging results and applications for the delivery of pesticides, fertilisers, and genetic material for the transformation of plants have been developed already. Based on this, a huge effort will be required in order to develop the details of nanotechnology with the bacteria that promote the growth of the plant. Discovering the distinctive nanomaterials required to incorporate these mutualistic bacteria could not be very easy, however, it is not impossible. That is why nanotechnology has all the necessary instruments to enhance the current biofertilizers which are used in order to support and enhance sustainable agriculture all around the world.

References

[1] Dwivedi, S., Saquib, Q., Al-Khedhairy, A. A., &Musarrat, J. (2016). Understanding the role of nanomaterials in agriculture. *Microbial Inoculants in Sustainable Agricultural Productivity*, 271-288. https://doi.org/10.1007/978-81-322-2644-4_17.

[2] Lv, M., Liu, Y., Geng, J., Kou, X., Xin, Z., & Yang, D. (2018). undefined. *Biosensors and Bioelectronics*, *106*, 122-128. https://doi.org/10.1016/j.bios.2018.01.049.

[3] Shalaby, T. A., Bayoumi, Y., Abdalla, N., Taha, H., Alshaal, T., Shehata, S., Amer, M., Domokos-Szabolcsy, É., & El-Ramady, H. (2016). Nanoparticles, soils, plants and sustainable agriculture. *Nanoscience in Food and Agriculture 1*, 283-312. https://doi.org/10.1007/978-3-319-39303-2_10.

[4] Siddiqui MH, Al-Whaibi MH, Mohammad F (2015) Nanotechnology and plant sciences: nanoparticles and their impact on plants. *Nanotechnol Plant Sci Nanopart Impact Plants:* 1–303.

[5] Buzea C, Pacheco I (2017) Nanomaterials and their classification. In: *EMR/ESR/EPR spectroscopy for characterization of nanomaterials.* Springer, New Delhi, pp 3–45.

[6] Yin, L., Colman, B. P., McGill, B. M., Wright, J. P., & Bernhardt, E. S. (2012). Effects of silver Nanoparticle exposure on germination and early growth of eleven wetland plants. *PLoS ONE*, *7*(10), e47674. .https://doi.org/10.1371/journal.pone.0047674.

[7] Abdel Latef, A. A., Srivastava, A. K., El-sadek, M. S., Kordrostami, M., & Tran, L. P. (2017). Titanium dioxide nanoparticles improve growth and enhance tolerance of broad bean plants under saline soil conditions. *Land Degradation & Development*, *29*(4), 1065-1073. https://doi.org/10.1002/ldr.2780.

[8] Laware, S. L., &Raskar, S. (2014). *Influence of Zinc Oxide Nanoparticles on Growth, Flowering and Seed Productivity in Onion,* ISSN: 2319-7706 Volume 3 Number 7(2014) pp.874-881.

[9] Rastogi, A., Tripathi, D. K., Yadav, S., Chauhan, D. K., Živčák, M., Ghorbanpour, M., El-Sheery, N. I., &Brestic, M. (2019). Application of silicon nanoparticles in agriculture. *3 Biotech*, *9*(3). https://doi.org/10.1007/s13205-019-1626-7.

[10] Mohamed, M. A., Hashim, A. F., Alghuthaymi, M. A., & Abd-Elsalam, K. A. (2018). Nano-carbon: Plant growth promotion and protection. *Nanotechnology in the Life Sciences*, 155-188. https://doi.org/10.1007/978-3-319-91161-8_7.

[11] Goudar, K.M., Geetha, K., Lingaraju, N., Shankar, A.G., &Raddy, R. (2018). Response of sunflower (Helianthus annuus L.) to nano boron nitride fertilization. *International Journal of Chemical Studies, 6*, 2624-2630.

[12] Shireen, F., Nawaz, M. A., Chen, C., Zhang, Q., Zheng, Z., Sohail, H., Sun, J., Cao, H., Huang, Y., &Bie, Z. (2018). Boron: Functions and Approaches to Enhance Its Availability in Plants for Sustainable Agriculture. *International journal of molecular sciences*, *19*(7), 1856. https://doi.org/10.3390/ijms19071856.

[13] Shukla, S. K., Kumar, R., Mishra, R. K., Pandey, A. K., Pathak, A., Zaidi, M. G., Srivastava, S., & Dikshit, A. (2015). Prediction and validation of gold nanoparticles (GNPs) on plant growth promoting rhizobacteria (PGPR): a step toward development of nano-biofertilizers. *Nanotechnology Reviews, 4*, 439 - 448.

[14] Yılmaz E., Sönmez İ., Demir H. (2014) Effects of zeolite on seedling quality and nutrient contents of cucumber plant (Cucumis sativus L. cv. Mostar F1) grown in different mixtures of growing media. *Commun Soil Sci Plant Anal* 45(21):2767–2777.

[15] Nakkeeran, S., Fernando, W. G., & Siddiqui, Z. A. (n.d.). Plant growth promoting Rhizobacteria formulations and its scope in commercialization for the management of pests and diseases. *PGPR: Biocontrol and Biofertilization*, 257-296. https://doi.org/10.1007/1-4020-4152-7_10.

[16] Bayraktar, O., Erdoğan, İ., Köse, M. D. and Kalmaz, G., (2017). Nanocarriers for plant-derived natural compounds. In *Nanostructures for Antimicrobial Therapy* (pp. 395-412). Elsevier.

[17] Ahmad, F., Ahmad, I., & Khan, M. (2008). Screening of free-living rhizospheric bacteria for their multiple plant growth promoting activities. *Microbiological Research*, *163*(2), 173-181. https://doi.org/10.1016/j.micres.2006.04.001.

[18] Burdman, S.; Jurkevitch, E.; Okon, Y. Recent advances in the use of plant growth promoting rhizobacteria (PGPR) in agriculture. *Microbial interactions in agriculture and forestry* (Volume II) 2000 pp.229-250 ref.135.

[19] Weller, D. M., & Thomashow, L. S. (n.d.). Current challenges in introducing beneficial microorganisms into the rhizosphere. *Molecular Ecology of Rhizosphere Microorganisms*, 1-18. https://doi.org/10.1002/9783527615810.ch1.

[20] Kaymak, H. C. (2010). Potential of PGPR in agricultural innovations. *Plant Growth and Health Promoting Bacteria*, 45-79. https://doi.org/10.1007/978-3-642-13612-2_3.

[21] Saharan, B. S., Nehra, V. (2011). Plant Growth Promoting Rhizobacteria: A Critical Review. *Life Sciences and Medicine Research,* Volume 2011: LSMR-21.

[22] Bhattacharyya, P. N., Jha, D. K. (2012) Plant growth-promoting rhizobacteria (PGPR): Emergence in agriculture. *Wood J. Microb. Biotechnol.* 2012, 28, 1327–1350.

[23] Kloepper J. W., Schroth M. N., 1978. Plant growth promoting rhizobacteria on radishes, In: *Proceedings of the 4th international conference on plant pathogenic bacteria, Angers, France* 1979. 879-882.

[24] Gray, E., & Smith, D. (2005). Intracellular and extracellular PGPR: Commonalities and distinctions in the plant–bacterium signaling processes. *Soil Biology and Biochemistry*, *37*(3), 395-412. https://doi.org/10.1016/j.soilbio.2004.08.030.

[25] Egamberdieva, D.; Lugtenberg, B. (2014) Use of Plant Growth-Promoting Rhizobacteria to Alleviate Salinity Stress in Plants. In *Use of Microbes for the Alleviation of Soil Stresses;* Springer: New York, NY, USA, 2014; Volume 1, pp. 73–96.

[26] García-Fraile, P.; Menéndez, E.; Rivas, R. 2015. Role of bacterial biofertilizers in agriculture and forestry. *AIMS Bioeng.* 2, 183–205.

[27] Nadeem, S. M.; Zahir, Z. A.; Naveed, M.; Ashraf, M. 2010. Microbial ACC-deaminase; prospects and applications for inducing salt tolerance in plants. *Crit. Rev Plant Sci.* 29, 360–393.

[28] Pishchik, V. N.; Vorobyev, N. J.; Chernyaeva, L. I.; Timofeeva, S. V.; Kazhemyakov, A. P.; Alexeev, Y. V. 2002. Experimental and mathematical simulation of plant growth promoting rhizobacteria and plant interaction under cadmium stress. *Plant Soil* 243, 173–186.

[29] Nautiyal, C. S.; Govindarajan, R.; Lavania, M.; Pushpangadan, P. 2008. Novel mechanisms of modulating natural antioxidants in functional foods: Involvement of plant growth promoting rhizobacteria NRRL B-30488. *J. Agric. Food Chem.* 56, 4474–4481.

[30] Ahmad, M.; Zahir, Z. A.; Khalid, M. 2013. Efficacy of Rhizobium and Pseudomonas strains to improve physiology, ionic balance and quality of mung bean under salt-affected conditions on farmer's fields. *Plant Physiol. Biochem.* 63, 170–176.

[31] Ahmed, A.; Hasnain, S. 2010. Auxin producing *Bacillus* sp.: Auxin quantification and effect on the growth *Solanum tuberosum. Pure Appl.* Chem. 82, 313–319.
[32] Sarma, R. K.; Saikia, R. R. 2014. Alleviation of drought stress in mung bean by strain *Pseudomonas aeruginosa* GGRK21. *Plant Soils* 377, 111–126.
[33] Naveed, M.; Hussain, M. B.; Zahir, Z. A.; Mitter, B.; Sessitsch, A. 2014. Drought stress amelioration in wheat through inoculation with *Burkholderiaphytofirmans* strain PsJN. *Plant Growth Regul.* 73, 121–131.
[34] Choudhary, D. K.; Sharma, K. P.; Gaur, R. K. 2011. Biotechnological perspectives of microbes in agro-ecosystems. *Biotechnol. Lett.* 33, 1905–1910.
[35] Lloret, L.; Martinez-Romero, E. 2005. Evolution and phylogeny of rhizobia. *Rev. Latinoam. Microbiol.* 47, 43–60.
[36] Raymond, J.; Siefert, J. L.; Staples, C. R. 2004. The natural history of nitrogen fixation. *Mol. Biol. Evol.* 21, 541–554.
[37] Tejera, N.; Lluch, C.; Martínez-Toledo, M. V. 2005. Isolation and characterization of *Azotobacter* and *Azospirillum* strains from the sugarcane rhizosphere. *Plant Soil* 270, 223–232.
[38] Wani, P. A.; Khan, M. S.; Zaidi, A. 2007. Synergistic effect of the inoculation with nitrogen-fixing and phosphate-solubilizing rhizobacteria on performance of field-grown chickpea. *J. Plant Nutr. Soil Sci.* 170, 283–287.
[39] Goswami, D.; Pithwa, S.; Dhandhukia, P.; Thakker, J. N. 2014. Delineating *Kocuriaturfanensis* 2M4 as a credible PGPR: A novel IAA producing bacteria isolated from saline desert. *J. Plant Interact.* 9, 566–576.
[40] Lavakush, Y. J.; Verma, J. P.; Jaiswal, D. K.; Kumar, A. 2014. Evaluation of PGPR and different concentration of phosphorus level on plant growth, yield and nutrient content of rice (Oryza sativa). *Ecol. Eng.* 62, 123–128.
[41] Lugtenberg, B. J.; Chin A-Woeng, T. F.; Bloemberg, G. V. 2002. Microbeeplant interactions: Principles and mechanisms. *Antonie Van Leeuwenhoek* 81, 373–383.
[42] Somers, E.; Vanderleyden, J.; Srinivasan, M. 2004. Rhizosphere bacterial signalling: A love parade beneath our feet. *Crit. Rev. Microbiol.* 30, 205–240.
[43] Tanimoto, E. 2005. Regulation and root growth by plant hormones-roles for auxins and gibberellins. *Crit. Rev. Plant Sci.* 24, 249–265.
[44] Aloni, R.; Aloni, E.; Langhans, M. 2006. Role of cytokinin and auxin in shaping root architecture: Regulating vascular differentiation, lateral root initiation, root apical dominance and root gravitropism. *Ann. Bot.* 97, 883–893.
[45] Hayat, R.; Ali, S.; Amara, U. Soil beneficial bacteria and their role in plant growth promotion: A review. *Ann. Microbiol.* 2010, 60, 579–598.
[46] Spaepen, S.; Vanderleyden, J.; Remans, R. Indole-3-acetic acid in microbial and microorganism-plant signaling. In *FEMS Microbiology Reviews;* Unden, F., Ed.; Blackwell Publishing Ltd.: New York, NY, USA, 2007; pp. 425–448.
[47] Vacheron, J.; Desbrosses, G.; Bouffaud, M.-L.; Touraine, B.; Moënne-Loccoz, Y.; Muller, D. 2013. Plant growth-promoting rhizobacteria and root system functioning. *Front. Plant Sci.* 4.
[48] Bottini, R.; Cassan, F.; Piccoli, P. 2004. Gibberellin production by bacteria and its involvement in plant growth promotion and yield increase. *Appl. Microbiol. Biotechnol.* 65, 497–503.

[49] Spaepen, S.; Vanderleyden, J. 2011. Auxin and plant-microbe interactions. *Cold Spring HarborPerspect. Biol.* 3.

[50] Khan, A. L.; Waqas, M.; Kang, S. M. 2014. Bacterial endophyte *Sphingomonas* sp. LK11 produces gibberellins and IAA and promotes tomato plant growth. *J. Microbiol.* 52, 689–695.

[51] Riefler, M.; Novak, O.; Strnad, M.; Schm€ulling, T. 2006. Arabidopsis cytokinin receptor mutants reveal functions in shoot growth, leaf senescence, seed size, germination, rootdevelopment, and cytokinin metabolism. *Plant Cell* 18, 40–54.

[52] Liu, F.; Xing, S.; Ma, H. 2013. Cytokinin-producing, plant growth-promoting rhizobacteria that confer resistance to drought stress in *Platycladusorientalis* container seedlings. *Appl. Microbiol. Biotechnol.* 97, 9155–9164.

[53] Reid, M. S. 1981. The role of ethylene in flower senescene. *Acta Hortic.* 261, 157–169.

[54] Li, Q.; Saleh-Lakha, S.; Glick, B. R. 2005. The effect of native and ACC deaminase containing *Azospirillum brasilense* Cdl843 on the rooting of carnation cuttings. *Can. J. Microbiol.* 51, 511–514.

[55] Glick, B. R. 2012. Plant growth promoting bacteria: Mechanisms and applications. *Scientifica* 2012.

[56] Jackson, M. B. 1991. Ethylene in root growth and development. In *The Plant Hormone Ethylene;* Matoo, A. K., Suttle, J. C., Eds.; CRC Press: Boca Raton, FL, USA, pp. 159–181.

[57] Davies, P., Ed.; 2013. *Plant Hormones: Physiology, Biochemistry and Molecular Biology;* Springer Science & Business Media: New York, NY, USA.

[58] Porcel, R.; Zamarreño, Á. M.; García-Mina, J. M.; Aroca, R. 2014. Involvement of plant endogenous ABA in *Bacillus megaterium* PGPR activity in tomato plants. *BMC Plant Biol*. 14, 36.

[59] Kende, H.; Zeevaart, J. 1997. The Five "Classical" Plant Hormones. *Plant Cell* 9, 1197–1210.

[60] Taiz, L.; Zeiger, E. 2002. Photosynthesis: Physiological and ecological considerations. *Plant Physiol.* 9, 172–174.

[61] Flores-Felix, J. D.; Silva, L. R.; Rivera, L. P. 2015. Plants probiotics as a tool to produce highly functional fruits: The case of Phyllobacterium and vitamin C in strawberries. *PLoS ONE* 10, e0122281.

[62] Ryu, C. M.; Farag, M. A.; Hu, C. H.; Reddy, M. S.; Wei, H. X.; Paré, P. W.; Kloepper, J. W. 2003. Bacterial volatiles promote growth in Arabidopsis. *Proc. Natl. Acad. Sci. USA,* 100, 4927–4932.

[63] Effmert, U.; Kalderás, J.; Warnke, R.; Piechulla, B. 2012. Volatile mediated interactions between bacteria and fungi in the soil. *J. Chem. Ecol.* 38, 665–703.

[64] Kanchiswamy, C. N.; Malnoy, M.; Maffei, M. E. 2015. Chemical diversity of microbial volatiles and their potential for plant growth and productivity. *Front. Plant Sci.* 6, 151.

[65] Worrall E., Hamid A., Mody K., Mitter N., Pappu H. 2018. Nanotechnology for plant disease management. *Agronomy.* 8:285. doi: 10.3390/agronomy8120285.

[66] Kumar, H.; Bajpai, V. K.; Dubey, R. C. 2010. Wilt disease management and enhancement of growth and yield of *Cajanus cajan* (L) var. Manak by bacterial combinations amended with chemical fertilizer. *Crop Protect.* 29, 591–598.

[67] Arora, N. K.; Khare, E.; Oh, J. H. 2008. Diverse mechanisms adopted by Pseudomonas fluorescent PGC2 during the inhibition of *Rhizoctonia solani* and Phytophthora capsici. *World J. Microbiol. Biotechnol.* 24, 581–585.

[68] Karunakaran, G., Suriyaprabha, R., Manivasakan, P., Yuvakkumar, R., Rajendran, V., Prabu, P., & Kannan, N. 2013. Effect of nanosilica and silicon sources on plant growth promoting rhizobacteria, soil nutrients and maize seed germination. *IET nanobiotechnology*, *7*(3), 70–77. https://doi.org/10.1049/iet-nbt.2012.0048.

[69] Rangaraj, S., Gopalu, K., Rathinam, Y., Periasamy, P., Venkatachalam, R., & Narayanasamy, K. 2014. Effect of silica nanoparticles on microbial biomass and silica availability in maize rhizosphere. *Biotechnology and applied biochemistry*, *61*(6), 668–675. https://doi.org/10.1002/bab.1191.

[70] Timmusk S., Seisenbaeva G., Behers L. 2018. Titania (TiO_2) nanoparticles enhance the performance of growth-promoting rhizobacteria. *Sci Rep.* Springer US:1–13.

[71] Palmqvist, N. G., Bejai, S., Meijer, J., Seisenbaeva, G. A., & Kessler, V. G. 2015. Nano titania aided clustering and adhesion of beneficial bacteria to plant roots to enhance crop growth and stress management. *Scientific reports*, *5*, 10146. https://doi.org/10.1038/srep10146.

[72] Khan N., Bano A. 2015. Role of plant growth promoting Rhizobacteria and Ag-nano particle in the bioremediation of heavy metals and maize growth under municipal wastewater irrigation. *Int J Phytoremediation* 6514(October).

[73] Panichikkal J., Thomas R., John J. C. 2019. Biogenic gold nanoparticle supplementation to plant beneficial *Pseudomonas monteilii* was found to enhance its plant probiotic effect. *Curr Microbiol.* Springer US 0(0):0.

[74] Khati, P., Parul, Bhatt, P., Nisha, Kumar, R., & Sharma, A. 2018. Effect of nanozeolite and plant growth promoting rhizobacteria on maize. *3 Biotech*, *8*(3), 141. https://doi.org/10.1007/s13205-018-1142-1

[75] Khoramdel RSSS 2016. Effects of Nano-zinc oxide and seed inoculation by plant growth promoting Rhizobacteria (PGPR) on yield, yield components and grain filling period of soybean (Glycine max L.). *Iran J Field Crop Res* 13(4):738–753.

[76] Khati, P., Chaudhary, P., Gangola, S., Bhatt, P., & Sharma, A. 2017. Nanochitosan supports growth of *Zea mays* and also maintains soil health following growth. *3 Biotech*, *7*(1), 81. https://doi.org/10.1007/s13205-017-0668-y.

[77] Khan M. R., Rizvi T. F. (2014) Nanotechnology: Scope and application in plant disease management. *Plant Pathol. J.* 2014;13:214–231. doi: 10.3923/ppj.2014.214.231.

[78] Gogos A., Knauer K., Bucheli T. D. (2012) Nanomaterials in plant protection and fertilization: Current state, foreseen applications, and research priorities. *J. Agril. Food Chem.* 2012;60:9781–9792. doi: 10.1021/jf302154y.

[79] Chen Y. W., Lee H. V., Juan J. C., (2016) Phang S.M. Production of new cellulose nanomaterial from red algae marine biomass Gelidium elegans. *Carbohydr. Polym.* 2016;151:1210–1219. doi: 10.1016/j.carbpol.2016.06.083.

[80] Dwivedi S., Saquib Q., Al-Khedhairy A. A., Musarrat J. (2016) Understanding the role of nanomaterials in agriculture. In: Singh D. P., Singh H. B., Prabha R., editors. *Microbial Inoculants in Sustainable Agricultural Productivity*. Springer; New Delhi, India: 2016. pp. 271–288.

[81] Ran Y., Liang Z., Gao C. (2017) Current and future editing reagent delivery systems for plant genome editing. *Sci. China Life Sci.* 2017;60:490–505. doi: 10.1007/s11427-017-9022-1.

[82] Miller J. B., Zhang S., Kos P., Xiong H., Zhou K., Perelman S. S., Zhu H., Siegwart D. J. (2017) Non-viral CRISPR/Cas gene editing in vitro and in vivo enabled by synthetic nanoparticle co-delivery of Cas9 mRNA and sgRNA. *Angew. Chem. Int.* Ed. 2017;56:1059–1063. doi: 10.1002/anie.201610209.

[83] Kim D. H., Gopal J., Sivanesan I. (2017) Nanomaterials in plant tissue culture: The disclosed and undisclosed. *RSC Adv.* 2017;7:36492–36505. doi: 10.1039/C7RA07025J.

[84] Afsharinejad A., Davy A., Jennings B., Brennan C. (2016) Performance analysis of plant monitoring nanosensor networks at THz frequencies. *IEEE Internet Things J.* 2016;3:59–69. doi: 10.1109/JIOT.2015.2463685.

[85] Kwak S.-Y., Wong M. H., Lew T. T. S., Bisker G., Lee M. A., Kaplan A., Dong J., Liu A. T., Koman V. B., Sinclair R. (2017) Nanosensor technology applied to living plant systems. *Annu. Rev. Anal. Chem.* 2017;10:113–140. doi: 10.1146/annurev-anchem-061516-045310.

[86] Jampílek J., Kráľová K. 2017. Nanomaterials for delivery of nutrients and growth-promoting compounds to plants. In: *Nanotechnology*. Springer, Singapore, pp 177–226.

[87] Castañeda, L. M., Genro, C., Roggia, I., Bender, S. D., Bender, R. J., & Pereira, C. N. 2014. Innovative Rice Seed Coating (*Oryza sativa*) with Polymer Nanofibres and Microparticles Using the Electrospinning Method. *Journal of Research Updates in Polymer Science, 3*, 33-39.

[88] De Gregorio P. R., Michavila G., Muller L. R., de Souza Borges C., Pomares M. F., de Sa E. L. S., Pereira C., Vincent P. A. 2017. Beneficial rhizobacteria immobilized in nanofibers for potential application as soybean seed bioinoculants. *PLoS One* 12(5):e0176930.

[89] Qureshi A., Singh D. K., Dwivedi S. 2018. Nano-fertilizers: a novel way for enhancing nutrient use efficiency and crop productivity. *Int J Curr Microbiol App Sci* 7(2):3325–3335.

[90] Mala, R., Selvaraj, R. C. A., Sundaram, V. B., Rajan, R. B. S. S., &Gurusamy, U. M. 2017. Evaluation of Nano Structured Slow Release Fertilizer on the Soil Fertility, Yield and Nutritional Profile of *Vigna radiata*. *Recent patents on nanotechnology*, *11*(1), 50–62. https://doi.org/10.2174/1872210510666160727093554.

[91] Godfray H. C. J., Beddington J. R., Crute I. R., Haddad L., Lawrence D., Muir J. F., Pretty J., 2010. Robinson S., Thomas S. M., Toulmin C. Food security: The challenge of feeding 9 billion people. *Science.* 327:812–818. doi: 10.1126/science.1185383.

[92] Shukla P., Chaurasia P., Younis K., Qadri O. S., Faridi S. A., Srivastava G. 2019. Nanotechnology in sustainable agriculture: Studies from seed priming to post-harvest management. *Nanotechnol. Environ. Eng.* 4:11. doi: 10.1007/s41204-019-0058-2.

[93] Yang H., Xu M., Koide R. T., Liu Q., Dai Y., Liu L., Bian X. 2016. Effects of ditch-buried straw return on water percolation, nitrogen leaching and crop yields in a rice-wheat rotation system. *J. Sci. Food Agric.* 96:1141–1149. doi: 10.1002/jsfa.7196.

[94] Panpatte D. G., Jhala Y. K., Shelat H. N., Vyas R. V. 2016. *Microbial Inoculants in Sustainable Agricultural Productivity.* Springer; New Delhi, India: Nanoparticles: The next generation technology for sustainable agriculture; pp. 289–300.

[95] Abdel-Aziz H. M. M., Hasaneen M. N. A., Omer A. M. 2016. Nano chitosan-NPK fertilizer enhances the growth and productivity of wheat plants grown in sandy soil. *Span. J. Agric. Res.* 14:17. doi: 10.5424/sjar/2016141-8205.

[96] Martinez-Viveros, O.; Jorquera, M. A.; Crowley, D. E.; Gajardo, G.; Mora, M. L. 2010. Mechanisms and practical considerations involved in plant growth promotion by rhizobacteria. *J. Soil Sci. Plant Nutr.* 10, 293–319.

[97] Bakker, A. W.; Schippers, P. 1987. Microbial cyanide production in the rhizosphere in relation to potato yield reduction and *Pseudomonas* spp.-mediated plant growth-stimulation. *Soil Biol. Biochem.* 19, 451–457.

[98] Xie, H.; Pasternak, J. J.; Glick, B. R. 1996. Isolation and characterization of mutants of the plant growth-promoting rhizobacterium *Pseudomonas putida* CR12–2 that overproduce indoleacetic acid. *Curr. Microbiol* 32, 67–71.

[99] Vijayan, R.; Palaniappan, P.; Tongmin, S. A.; Padmanaban, E.; Natesan, M. 2013. Rhizobitoxine enhances nodulation by inhibiting Ethylene synthesis of *Bradyrhizobiumelkanii* from Lespedeza species: Validation by homology modeling and molecular docking study. *World J. Pharm. Pharm. Sci.* 2, 4079–4094.

[100] Xiong, K.; Fuhrmann, J. J. 1996. Comparison of rhizobitoxine-induced inhibition of β-cystathionase from different bradyrhizobia and soybean genotypes. *Plant Soil* 186, 53–61.

Chapter 9

Biofertilizers for Crop Improvement

G. K. Chouhan[1]
D. Singh[1]
A. Mukherjee[1]
S. Singh[1]
A. P. A. Pereira[2]
and J. P. Verma[1,2,*]

[1]Institute of Environment and Sustainable Development, Banaras Hindu University, Varanasi, Uttar Pradesh, India
[2]Federal University of Ceará, Soil Science Department, Soil Microbiology Laboratory, Fortaleza, Ceará, Brazil

Abstract

The increasing world population is one of the significant threats to food security. Currently, the cultivable land for agriculture is gradually decreasing, and the soil is becoming infertile. The vigorous use of chemical fertilizers and pesticides is increasing continuously for higher crop productivity to meet the growing populations global food demand. Nowadays, chemical farming is causing many environmental problems that directly and indirectly affect human health. Therefore, organic agriculture (bio-fertilizer, biopesticide, Integrated Crop Management (ICM), Integrated pest management (IPM), crop rotation, green manuring, Farmyard manure (FYM), mixed cropping sequence, etc.) is the only alternative practice for saving our environment and producing quality food for the human being. This book chapter focuses on biofertilizers and their application for crop improvement. Bio-fertilizer

* Corresponding Author's Emails: verma_bhu@yahoo.co.in; jpv.iesd@bhu.ac.in.

In: Microbial Technologies and Their Applications
Editors: Santosh Kumar Karn and Anne Bhambri
ISBN: 979-8-88697-913-8

contains microorganism providing adequate nutrients to the host plant for proper growth and development. Living microorganisms have a specific function used to prepare biofertilizers to augment plant growth and reproduction. It has great potential to sustain long-term soil fertility and crop production. Nowadays, the application of liquid biofertilizer shows a very effective response in various crops, which has overcome the previous restrictions of biofertilizer applications. However, we need efficient and indigenous biofertilizers to enhance crop productivity, nutritional quality and soil health. These biofertilizers can replace chemical fertilizers. Biofertilizers are eco-friendly, cost-effective and socially acceptable for sustainable agricultural productivity.

Keywords: microorganisms, biofertilizers, biopesticide crop production, sustainable agricultural, PGPR, PSB, KSB, vermicompost

Introduction

The world population has increased and now it has reached 7.6 billion and only 1.35 billion people are contributed by India, which is growing daily. This increasing population impacts agricultural land and other resources to feed these populations. The 15th Indian Census population count shows 17.64% growth per decade, of which around 68.84% population belongs to rural areas. This growing population demands conventional farming to meet the demand for food, making the farmers depend upon chemical fertilizers and pesticides to increase agriculture productivity [1]. These chemical fertilizers are made up of harmful chemicals which cause various problems to plant, soil and human health, such as root weakening of plants, soil acidification, eutrophication of water bodies and various diseases in humans (Blue Baby Syndrome) [2-4]. Such chemicals have the remarkable ability to show an adverse impact on future generations. In this regard to being safe, there is a demand for an eco-friendly approach, such as biofertilizers that can play a significant role in sustainable agriculture (Table 1, Figure 1). Various countries are moving towards the use of biofertilizer alternatives to chemical fertilizers due to stringent regulations imposed on the usage of chemical fertilizers and replaced by the very effective alternative one that is biofertilizer [5]. Biofertilizer contains microorganisms that provide adequate nutrients to the host plant for proper growth and development. Living microorganisms have a specific function used to prepare biofertilizers to augment plant growth and reproduction when applied to soils, seeds, plant leaves or roots [6-10] (Table 1, Figure 1). These microbes reside in a narrow zone of soil around the root or interior of the plant and promote the plant's growth by solubilizing various complex nutrients and making them available to the host plant by their metabolic activities [11].

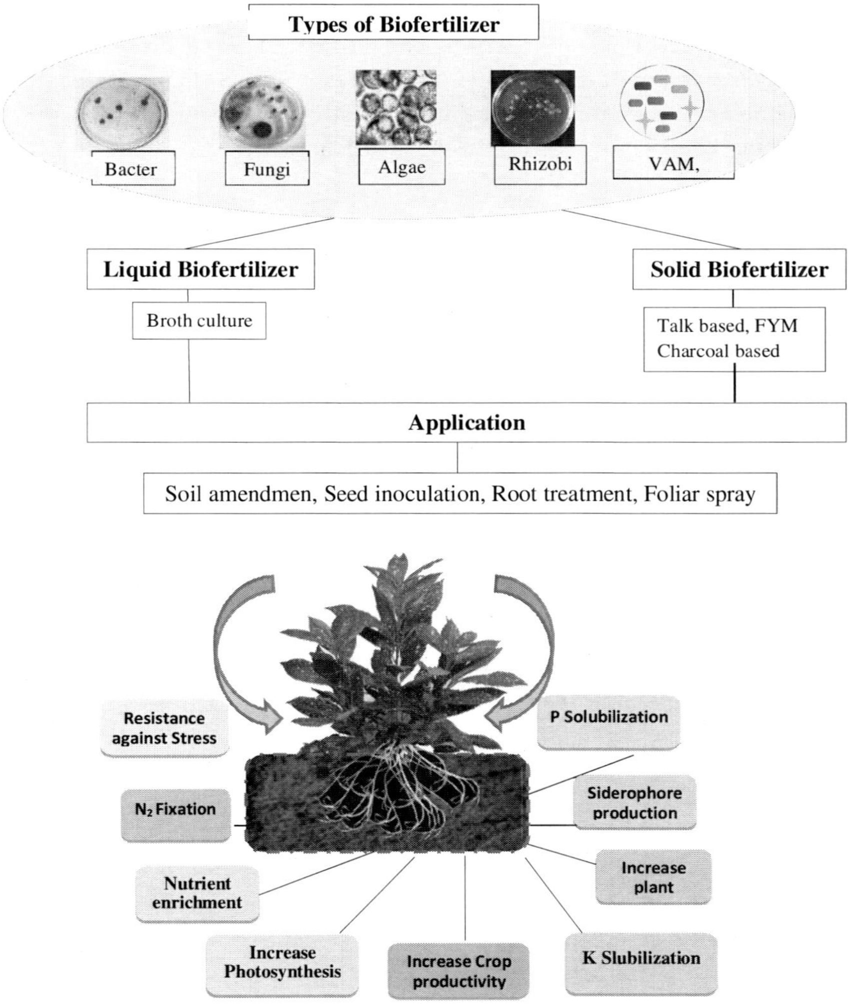

Figure 1. Benefits of biofertilizers to the plant.

Plants only utilize a meager amount of applied nutrients (10% - 40%) and the rest 60%-90% are insoluble or immobilized in soil. Biofertilizers are worked slowly and continue to release metabolic components such as N_2 fixation,

Phosphate solubilization, siderophore, Potassium solubilization, Zn solubilization, releasing of PGP (Plant growth-promoting) substance and biodegradation in the soil to convert these immobilize nutrients in soluble form and thus it forms an important component of integrated nutrient. In other words, biofertilizers are living natural beneficial microbes, including bacteria (*Azotobacter, Azospirilum, Rhizobium*), algae (Blue-green algae), Fungi (*Mycorrhizae*) alone or in the combined form and they augment the bioavailability of nutrients to the host plant. The role of biofertilizers in agriculture production is significant, particularly in the context of the increasing cost of chemical fertilizers as well as their hazardous effect on the plant, soil and human health.

Plant Growth-Promoting Rhizobacteria

PGPR are naturally occurring microbes that reside in the vicinity of roots in a narrow soil zone and provide safety and nutrients to host plants. These microbes show a symbiotic relationship with the plant. Plant roots secret important flavonoids in the vicinity of the root, attracting a variety of microbes that benefit the plant. In any situation, a plant cannot survive alone. They require the support of microbes for their proper growth and development and vice versa [12]. These microbes help the plant by easily absorbing nutrients and protecting against the diversified harmful microbes and unfavorable environmental conditions such as biotic and abiotic conditions [8, 13]. PGPR has several beneficial effects, such as efficient seed germination, increased plant height, seed weight, increased chlorophyll content, and increased nodulation in leguminous crops [6, 14, 15]. These microbes ensure the availability of certain essential macronutrients to the plant, such as Nitrogen, Phosphorus, Potassium, Sulfur and Iron. These microbes provide various bio-stimulator hormones and attract other bacterial and fungal microbes that enhance the plant's growth. Nowadays, many PGPR are identified and added to the natural flora of soil that covert the complex nutrient to the simple form is *Azotobacter, Azospirillum, Rhizobia, Enterobacter, Bacillus, Pseudomonas, Klebsiella, Trichoderma, Aspergillus* etc. (Table 1). Further, these microbes are categorized into subgroups such as Biostimulator, Bioprotectant and Biofertilizer. Bio-protectants include *Pseudomonas, Agrobacterium, Bacillus, Burkholderia* and *Bacillus.* These microbes suppress the growth of deleterious microbes that hinder the growth of plants.

Table 1. Group of microbes and their role as biofertilizers in plant growth and development

S. No.	Group of biofertilizer	Role in plants	Host plants	Reference
1	*Azotobacter*	It produces plant growth promoting substances like Nicotinic acid, Panthothenic acid, Biotin etc. It helps in better growth by secreting antibodies like Anisomycin against pathogens like *Alternaria, Fusarium, colletotrichum, Rhizotonia, Microfomina, Diplodia, Botryadiplodia* and *Cephalosporium*.	Maize, Sorghum, Wheat, Cotton, Rice and Potato.	[69, 70]
2	*Azotobacter*	Supplies 20-40 kg N/ha/year. Promote growth substance such as Gibberellic, IAA and vitamins.	Mustard, Sunflower, Banana, sugarcane, grapes, papaya, watermelon, chilly, tomato, ladyfinger and coconut.	[71-73]
3	Arbuscular mycorrhizal fungi (AMF)	Improves plant growth, improves nutrient acquisition, lessens the negative effects on plants and soil under the continuous cropping regime	American ginseng	[28]
4	*R. intraradices* and *F. mosseae*	Dual inoculation effective for promoting growth of faba bean grown in alkaline soils	Broad bean	[29]
5	*B. licheniformis* ASU, AMF (*A. bireticulata* and *G. caesaris*)	AMF and BFH in combination significantly increased nodule formation and N2-fixation, Growth promotion in sandy soil	Broad bean	[30]
6	*Azospirillum*	It helps in better growth of plant by secreting chemicals like Indole acetic acid, Indole latic acid and gibberellins	Rice	[74-77]
7	*Azospirilum*	Increase water and mineral uptake. Plant hormone production Enhance root growth.	Rice, Sugar, Millet, Wheat, Sorghum and Bajra	[78, 79]
8	AMF (*G. mosseae* or *G. intraradices*) and bacteria (*A. chroococcum, B. megaterium, B. mucilaginous*)	High biomass and seedling height	Mays	[31]

Table 1. (Continued)

S. No.	Group of biofertilizer	Role in plants	Host plants	Reference
9	Diazotrophic bacteria (*Bradyrhizobium*) and arbuscular mycorrhizal fungi (*Glomus etunicatum* and *Gigasporaalbida*)	Increased grain and shoot biomass yield and increased the available p and k	Black-eyed pea	[32]
10	Mycorrhizal fungi (*glomusmosae*) and phosphate solubilizing bacteria (*pseudomonas putida* strain p13 and *pantoeaag glomerans* strain p5)	Increase total chlorophyll content, increased leaf fresh weight	Aloe vera	[33]
11	Arbuscular mycorrhizae (*G. intraradices* and *G. mosseae*) and rhizobacteria (*Sinorhizobium meliloti* Rm1021, *B. subtilis* BS119 m, *Bradyrhizobium* sp. USDA 4438 and *P. fluorescens* WCS417 r)	Improve growth, nutritional status and essential oil production	Basil and Summer savory	[34]
12	*A. chroococcum* and *G. mosseae*	Improve plant growth, fruit yield, nutrition, metabolism and rhizosphere enzyme activities	Pomegranate	[80]
13	*Rhizobium* (symbiotic)	Fixes 200–300 kg N/ha/year. Increase yield upto 10-30%. Maintain soil fertility.	Pea, Pulses legumes, Cow pea, green gram, black gram, groundnuts and soyabeans etc.	[81, 82]
14	AMF species (*Pacispora franciscana*, *F. mosseae*, *F. geosporum*, *R. irregularis* and *G. tenebrosum* and beneficial bacteria	Improves grain yield and quality, higher mineral elements, polyphenol and flavonoid content	*Hordeumvulgare* ssp. nudum L.	[83]

S. No.	Group of biofertilizer	Role in plants	Host plants	Reference
15	Blue Green Algae	These microbes produce Nicotinic acid, Panthothenic acid, Folic acid to promote the proper growth of plants.	Rice	[62, 84]
16	Blue Green Algae	Fixes 20-40 kg N/ha/year. Promote growth subtances like gibberelic acid, Vitamins, and IAA.	Rice	[60]
17	AMF and *Thiobacillus*	Significant increases in N, P, K and S concentrations in soil, High nutrient concentrations	Onion L., Mays	[85]
18	Phosphate solubilizing micro-organisms	Group of there microbes secretes growth enhancing substances like biotin, Riboflavin etc.	Maize, wheat, sorghum, cotton and rice.	[86, 87]
19	Arbuscular mycorrhiza fungi and farmyard manure	Improve soil properties and increase grain yield, essential oil and α-linolenic acid under drought stress, Increased cation exchange capacity in soil	Common Purslane	[88]
20	PGPB (K. *pneumonia*) and arbuscular mycorrhizal fungi (*G. intraradices*)	Improved lant height, root length, fresh and dry weight under abiotic stress	Sweet And chili pepperand Prostrate knotweed	[89]
21	*G. mosseae,* *G. clarum,* *G. fasciculatum,* *G. intraradices,* *G. ambisporum, Gigasporagigantea, Acaulosporadenticulata,* *G. globiferum, Gigasporaalbida* and *G. pansiholus*	Increase in zinc, calcium, magnesium, manganese content, and seed oil content through endomycorrhizal fertigation method	Arachis	[90]
22	*Rhizobium* (*Mesorhizobium*) and mycorrhizal fungi (*F. mosseae*)	Significantly increased total yield by 43%	Chickpea	[91]
23	Mycorrhiza (*G. mosseae*) and growth promoting bacteria (*Azospirillum*)	Increased grain yield, biological yield, essential oil yield, chlorophyll a, chlorophyll b and total chlorophyll, and also reduced negative effects due to water deficit stress	Fennel	[27]

Table 1. (Continued)

S. No.	Group of biofertilizer	Role in plants	Host plants	Reference
24	Arbuscular mycorrhizal fungi (*Claroideo glomus etunicatum*), *Saccharomyces cerevisiae*, or *Trichoderma viride*	Promote production, protection and nutritive status under diseased conditions caused due to *Sclerotiumr olfsii*	Common bean	[16]
25	Mycorrhiza	It helps in production of Gibberellin and cytokinin which promote growth. They also show reduces growth of many pathogenic fungi like *Rhizoctonia, Fusarium, Phytophthora by secreting Diaretyrenitrile* and *Polyacetylene.*	Maize, sorghum, wheat, cotton, Rice	[26]
26	Arbuscular mycorrhizal fungi (symbiotic)	Increase root absorption area for nutrient access Fixes phosphate	Soybean, wheat, and corn	[92]
27	*Bacillus* spp.	Solubilize the phosphate and fix the nitrogen in soil Synthesis of growth hormones Production of antibiotics Increase crop yield	Many vegetables and fruits	[93, 94]
28	*Pseudomonas*	Production of siderophores and plant hormones Fixes phosphate Increase crop yield	Potato, reddish, sugar beat	[95]

Some of the Important Biofertilizers

Phosphate Solubilizing Micro-Organisms

Most of the phosphate in the soil is complex and cannot be absorbed by plants; hence do not promote plant growth as the plant only uses two forms (Monobasic and dibasic) of Phosphorus. Insoluble phosphate is mainly present in fixed forms such as phosphodiesters and inositol phosphate. Many rhizosphere microbes, such as Phosphorus solubilizing bacteria (PSB), can solubilize the complex phosphate to a simple soluble form and make them available to the host plant. PSB like *Pseudomonas putida, Bacillus megaterium, Flavobacterium, Micrococcus* and fungi like *Aspergillus, Fusarium, Sclerotium* and *Penicillium* are solubilized complexed phosphate and fixed Phosphorus and make them available to the host plant [16]. Approximately 20–25% of plants required phosphorus is fulfilled by PSB and PSF microbes. Thus, these microbes fulfil the need for chemical fertilizer and their manufacturing cost and help solubilize present unutilized Phosphorus in soil [17]. Among soil microbial populations, about 1-50% constitute PSB and about only 0.1 to 0.5% constitute PSF have the phosphate solubilizing ability [18]. About 20% of soil actinomycetes can solubilize insoluble phosphate and make it available to plants. Some microbes, such as *Streptomyces* and *Micromonospora*, can efficiently solubilize this complex phosphate. Some of the actinomycetes are capable of phosphate solubilization and are gaining popularity due to their ability to survive in different stress conditions such as drought and they produce plant stimulating hormones.

Azospirillum

These essential biofertilizers convert a notable amount of atmospheric nitrogen into ammonia and other soluble forms. In studies by [19], reported to fix around 20-40 kg of nitrogen per hectare in non-leguminous crops like sorghum and cotton. They also showed the role of *Azospirillum* in improving newborn leaves growth and increasing the shoot biomass of rice plants [19, 20]. Inoculation of *Azospirillum* in rice crops resulted in height increment (Figure 1). As per the studies done by [21], a seed of maize and sugarcane inoculated with *Azospirillum* showed a remarkable increase in plant height after 20 days. They also reported a maximum increment of 19.40 cm (i.e., 132.89%) observed after 80 days in the sugarcane plant.

Moreover, applying *Azospirillum* in combination with *Azotobacter* on paddy plants leads to a significant increase in values of many growth parameters like length of root and shoot, the weight of root, length of panicle etc. The percentage increase in the height of the plant was noted to be 8.04% at 10 days and 33.14% at 30 days in paddy crops [21]. *Azospirillum* application in plants helps improve their root system morphology, promotes vigorous growth in a shoot, and eventually leads to overall growth in the plant.

Rhizobium

These are some of the most successful diazotrophic microbes which belong to *the Rhizobiaceae* family. They are reported to fix around 300 kg of nitrogen per hectare [22]. They are significantly evolved to fix atmospheric nitrogen into a more usable form like ammonia. These microbes stay in symbiotic association with the root system of leguminous crops and perform a "give and take" relationship with them. Their agriculture usage has proved a notable biofertilizer because of their remarkable capacity to support plant growth. For its application in an agricultural field, seven bacteria have been combined. It is called a Cross-inoculating group and it contains such microbes which are quite distinct and particular in making symbiotic associations with legumes [10]. These microbes invade the root system of legume crops and form root nodules there. Then these microbes start converting unavailable nitrogen into soluble like ammonia. Strains like *Rhizobium leguminosarum* can produce many additional growth supporting chemicals like siderophores, indole acetic acid and help in solubilizing phosphate (Figure 1). Some studies have also reported their application in non-leguminous crops like *Lactuca Sativa* L. and *Daucus Carota* L. Application of rhizobium in these crops increased dry biomass of their root and shoot systems [23]. When the rhizobium inoculum was applied to three common been cultivars, clear improvement was observed in the chlorophyll content and percentage of protein. Application of this microbe in bean crops showed 28.02% increment in protein content and 1.3 mg/gr increment in chlorophyll content. It helped plants attain proper growth by directly improving the chlorophyll content of leaf, N-content, seed germination, and plant height [24]. These microbes produce substances like phytohormones, metabolites and also activate systemic resistance in plants against pathogenic microbes. In addition to these qualities, they can survive under abiotic stress, giving them broader applications [25].

Mycorrhizae as Biofertilizer

There have been many studies on mycorrhiza's role as biofertilizers. They have been used in addition to the plant-beneficial bacteria and thus play a synergistic effect [26]. In some studies, they have been used solely to improve plant growth and other plant and soil parameters. In a study [27, 28], arbuscular mycorrhizal fungi (AMF) were solely used to improve plant growth and nutrient acquisition and ameliorate adverse effects caused by continuous cropping regime on *Panax quinquefolius* L. Further, a study [29] revealed that dual inoculation of *Rhizophagus intraradices* and *Funneliformis mosseae* was effective in promoting growth for faba bean in alkaline soil conditions. [30] studied the combined effect of *Bacillus licheniformis* ASU, AMF (*Acaulospora bireticulata* and *Glomus caesaris*) and found a significant increase in the nodule formation, nitrogen fixation and growth of *Vicia faba* in sandy soil conditions. Wu et al. [31] observed high biomass and seedling height upon inoculating of AMF with *Azotobacter chroococcum*, *Bacillus megaterium* and *Bacillus mucilaginous*. Diazotrophic bacteria along with AMF also increased grain and shoot biomass yield in *Vigna unguiculata* [32]. An increase in the total chlorophyll content, chlorophyll content, fresh weight of the leaves in Aloe vera were found to be high when inoculated with AMF and phosphate solubilizing bacteria, *Pseudomonas putida* strain P13 and *Pantoea agglomerans* strain P5 [33]. Another study [34] revealed that AMF along with rhizobacteria (*Sinorhizobium meliloti, Bacillus subtilis, Bradyrhizobium* and *Pseudomonas fluorescens*) improves growth, nutritional status and essential oil production in *Ocimum basilicum* and *Satureja hortensis*. Considering the role played by AMF along with beneficial PGPR and PGPB it is imperative to say that the co-inoculation of AMF with PGPB or PGPR can play a phenomenal role not only in the promotion of plant growth but also in the amelioration of negative effects by different abiotic stresses.

Potassium Solubilizing Micro-Organisms

Potassium (K) is considered an essential nutrient for plants. It plays a vital role in the growth and development of plants. It participates in the various activities of plant metabolism. It activates multiple enzymes, enhances photosynthesis in a plant, helps in nitrogen uptake essential for protein synthesis, helps in sugar transport, and reduces respiration. Potassium is applied significantly in soil but unavailable to plant uptake. Only 1 to 2% of potassium present in the soil is available to the plants [35]. Bacteria solubilize the Potassium by releasing organic and inorganic acids, polysaccharides, chelation, and complexoysis [36-39]. Several microbial groups can solubilize Potassium, such as Bacteria, Fungi and Actinomycetes. Some examples of K-solubilizing microbes, such as *Bacillus* spp. (*Acidothiobacillus ferrooxidans, Paenibacillus* spp., *Bacillus mucilaginosus, B. edaphicus,* and *B. circulans*), *Arthrobacter* spp. and *Cladosporium* are well-known microbes involved in K solubilization [20, 37]. Two microbes, such as *B. edaphicus* and *B. mucilaginosus* are well-known microbes capable of solubilization and mobilization of potassium [37]. Inoculation of *B. mucilaginosus* in soil enhances groundnut oil content and biomass by 35.4% and 25%, respectively and it also increases the K and P availability [40]. Mainly two microbes such as *Bacillus licheniformis* and *Pseudomonas azotoformans* used in the rice field, show higher potassium solubility compared to other rhizosphere microbes [41]. *Enterobacter hormoechei* shows a very good ability to enhance potassium and chlorophyll content in cucumber by solubilizing Potassium in the rhizosphere region [42]. Some Fungi, such as *Aspergillus* spp. and *Penicillium* spp., also have the ability to solubilize Potassium by releasing organic and inorganic metabolites [43, 44]. Thus, the role of potassium solubilizer is significant in enhancing crop plants' productivity.

Zn Solubilization

Zn is an essential trace element for all crops. Zn is a critical metal element required for mostly almost all biological enzymatic activities. Zn element deficiency results in cellular stress due to the improper function of enzymes and altered rate of different metabolic reactions [45]. Approximately 2800 proteins require Zn elements as their structural and biological activity. Plants assimilate zinc as a Zn^{+2} form, which is needed for several different metabolic actions in the plant, including plant auxin hormones biosynthesis, photosynthesis, fertilization and inducing resistance to specific plant infectious pathogens [46]. Zinc Solubilizing Bacteria (ZSB) is an alternative way to convert the insoluble soil Zn into a solubilized form of Zn. During microbial growth, soil microbes release some organic acids mixed with the minerals in the soil and impound the zinc cations. The organic acid compounds like as keto-D-glutarate, formic acid, gluconic acid, 2-and 2,5-keto-derivatives propionic acid, lactic acid, citric acid, succinic acid, malic acid, oxalic acid in different conformations chelate Zn by the carboxyl and the hydroxyl groups and enhance the Zn solubility [47]. It helps to improve the mineral uptake process by the plant. Zn solubilizing ability of some microbes is reported. These are *Bacillus* sp. and *Pseudomonas* sp., *Pseudomonas fluorescens, Gluconacetobacter Klebsiella* sp. *Acinetobacter calcoaceticus, Bacillus cereus, B. aryabhattai, Pseudomonas taiwenensis, Serratia* sp., *Diazotrophicus. S*ome fungi are also able to solubilize the Zn in soil, such as *Penicillium luteum, Aspergillus niger, A. nomius, Beauveria caledonica* [46]. Microbes solubilized the Zn by producing organic acids such as formic acid, acetic acid, propionic acid, lactic acid, glycolic acid, fumaric and succinic acid.

Cyanobacteria

These are water-loving, multicolored microbes that may exist in a single-cell to multicellular form. They can live in both free and symbiotic forms. They are one of the most prominent sources of nitrogen to marine ecosystems. They also possess the capacity to thrive under diverse environmental conditions, from snow to hot springs [48]. These are among the most successful life forms on earth [49, 50], which played an immensely important role in making this planet more hospitable to present life forms. These organisms are the ones that bought oxygen on earth, contributing around 30% of the oxygen on earth annually [51, 52]. *Cyanobacteria* have economic importance as a biofertilizer due to their ability to fix atmospheric nitrogen into the soil even after growing in soil with no or negligible amount of nitrogen in it [53-58]. On top of this, these microbes can produce many biologically active secondary metabolites like hormones, an iron chelator, antifungal, antiviral compounds, etc., further improving their fertilizing capacity and giving them a significant therapeutic potential [49]. These microbes have application as biofertilizers in the rice field [59]. Blue-Green Alga is an example of prokaryotic cyanobacteria, quickly found growing in the Indian continent. These are Nitrogen-fixing photosynthetic autotrophs that help improve the overall growth of various plants and help in improving the quality of soil by reducing the intake of chemical fertilizers by farmers [22]. *Cyanobacteria* increase soil fertility by enhancing the level of vitamins like Vit-B12, Nitrogen content and improving the growth of plant-producing chemicals like Auxine, indole acetic acid, gibberellic acid etc. [60]. It has the capacity to fix nitrogen in both free and associated forms. Fern-like Azolla and plants like Cycads are some of its symbiotic partners. A heterocyst is a unique type of cell present in *cyanobacteria*. This cell contains enzymes like nitrogenase, which helps it to convert atmospheric nitrogen into a more usable form like ammonia [60]. As per [53, 60], *Cyanobacteria* fixes 20-30 kg N/ha annually. *Anabaena variabilis*, *Nostoc muscorum*, *Aulosira fertissima* and *Tolypothrix tenuis* are good biofertilizer cyanobacteria species. These are extensively used in south Asian countries, including India [60-63]. It was also reported that if *Cyanobacteria* is continuously inoculated for 3-4 cropping seasons in the same field, it is permanently established there [64].

Azolla

It is one of the eukaryotic organisms which is used as a biofertilizer. It belongs to a small group of ferns that can be found floating on the surface of water bodies like lakes, freshwater ponds etc. Due to its capacity to enrich the soil and promote plant growth, it is called as "Green Gold Mine." It performs multiple functions like nitrogen fixation, biofuel, compost, and bioaccumulation of heavy metals. This organism has found application in wastewater treatment and feeds for live stocks [65]. Its ability to multiply and produce high-nutrient biomass is an addition to its nitrogen-fixing quality, making it a prominent biofertilizers and livestock feed. It converts atmospheric nitrogen into a soluble form by forming a symbiotic relationship with *Cyanobacteria* [66]. On investigating the leaves of Azolla, they were found to contain 4-5% nitrogen of the plants dry weight. This nitrogen is further converted into a more usable form by Azolla. A study by Azolla was reported to fix about 20-40 kg of nitrogen per hectare in a paddy field in 20-25 days [67].

Methods of Application of Biofertilizers

Seed Treatment

This is the best method of using organic fertilizers. About 100 grams of jaggery or gum are boiled in 1 liter of water and it left for cooling. After that, 50 ml of PGPR and 50 grams of *Trichoderma* powder mixed well and made a solution. Mixed this solution with 20 kg seeds, so that each grain is coated with 10^8 colony forming units (CFU). Then seeds are dried in a shady place. Treated seeds are ready for sowing [22, 68].

Seedling Root Dip

This method is used for transplanted crops such as rice and vegetable crops (tomato, cauliflower, cabbage, onion, etc.). The roots of the plants are treated with bio-fertilizers. First, we had taken 1 kg of jaggery in 5 to 10 liters of water and boiled it. After that solution was left for cool down and then added 50 ml of PGPR and 50 grams of *Trichoderma* powder with proper mixing. This is liquid biofertilizers based medium. After this, the roots of 250 to 500 bunch of rice or vegetables are dipped in it for 10 to 15 min so that 10^8 colony forming units (CFU)/root layer is mounted on each plant root and then transplanted [22, 68].

Tuber Treatment Method

The tuber (potato and ginger) and stem (sugarcane) are treated with bio-fertilizers in liquid media. First, we had taken 1 kg of jaggery in 5 to 10 liters of water and bolide it. After that solution was left for cool down and then added 50 ml of PGPR and 50 grams of *Trichoderma* powder with proper mixing. This is liquid biofertilizers based medium. In this media, tubers and stem were dipped for 20 to 30 min to stick our microbial inoculum up to 10^8 Colony Forming Units (CFU). After that this is read for sowing [22, 68].

Soil Treatment

First, we had taken 1 kg of jaggery in 5 to 10 liters of water and boiled it. After that solution was left for cool down and then added 50 ml of PGPR and 50 grams of *Trichoderma* powder with proper mixing. This is liquid biofertilizers based medium. This media was mixed with 70 to 100 kilograms of soil or farmyard manure and kept it overnight. After that it was spread in the field and did ploughing [22, 68].

Advantages of Using Biofertilizers

Farmers Benefit

- The fertility of the land increases
- Increases the irrigation interval

- Biofertilizers are prepared in less time and are less costly than chemical fertilizers
- Increases the productivity of crops
- The use of organic fertilizers reduces the consumption of chemical fertilizers and increases the use efficiency of chemical fertilizers

Benefits of Soil

- Increases the number of beneficial bacteria and earthworms in the soil.
- Biofertilizers provide all the significant micronutrients required by plants
- The use of biofertilizers improves the fertility of the soil
- Insoluble phosphorus is converted into a soluble form by the activity of bacteria and is available in abundance to plants
- The effect of organic fertilizers is gradual, but soil fertility remains for a long time, so they should not be applied repeatedly in the field like chemical fertilizers.
- The water-holding capacity of the soil increases
- The evaporation of water from the soil decreases

Environmental Benefits

- The groundwater level rises
- There is a reduction in the pollution caused by water in the soil, food matter and land
- The use of waste in making food reduces disease
- There is a reduction in the cost of crop production and an increase in income.
- Organic fertilizers are best for human health and environmental protection, essential for sustainable farming

Constraints in Biofertilizer Technology

- Due to the lack of scientific knowledge and their continuous use among farmers, chemical technology is considered more modern and progressive
- Less experience in promoting and using traditional knowledge
- Lack of extension activities for agricultural extension and lack of detailed information for selling organic produce
- Getting less financial assistance from the government for agricultural expansion

Conclusion and Future Prospective

The approach of using life-based fertilizer to support our already degraded soil is the need of an hour. The usage of biofertilizers is an excellent practice to improve the quality of our soil in a more environmentally friendly manner. These biofertilizers not just provide nutritional support to the soil but also prevent its further degradation, by directly cutting down the usage of chemical-based fertilizers. These are pocket-friendly substitutes for the chemical. They improve the defense system of various plants and protect them from abiotic stresses like drought and acidity in a completely natural manner. This book chapter studies different microbes like *Azotobacter, Azospirillum, Rhizobium* and Blue-green algae used as biofertilizers. We understand the importance of putting an end to chemical-based fertilizers and using more sustainable and environmentally friendly approaches to feed the growing population. Further, this chapter talks about plant growth-promoting rhizobacteria, which shows a symbiotic relationship with plants and provide various substance for the proper growth of plants. They also secrete flavonoids to attract a variety of microbiomes suitable for plants. They can convert complex nutrients to a simple form for direct usage. *Rhizobium* bacteria show a prominent ability to fix a relatively higher amount of Nitrogen-fixing with leguminous Crops. This book chapter discusses even more biofertilizers, including Azolla, Cyanobacteria, Zn solubilizing bacteria, and Vermicomposting. These all seem to be brilliant substitutes for chemical fertilizers in many senses. But as far as the future is concerned, we need to be more willing to work in this area of agriculture. So that we can remove all the present constraints in this field and make it the most convenient method of improving soil quality and increasing crop production, hence, to fulfill the above requirements, we should develop such biofertilizers which are less expensive, easily available, easy to use and gives the same or better results as compared to chemical fertilizers. Moreover, to provide this field with holistic development, the government should invest in developing this technology and promote the production and usage of biofertilizers.

Acknowledgments

Authors are highly thankful to SERB (EEQ/2021/001083, SIR/2022/000626), DST (DST/INT/SL/P-31/2021, DST/SEED/SCSP/STI/2020/426/G) and Banaras Hindu University, IoE (6031) incentives grant for providing support for applied research in plant-microbe interaction research.

References

[1] Santos, V. B., Araújo, A. S., Leite, L. F., Nunes, L. A., & Melo, W. J., 2012. Soil microbial biomass and organic matter fractions during transition from conventional to organic farming systems. *Geoderma*, *170*, 227-231.

[2] Chun-Li, W., Shiuan-Yuh, C., & Chiu-Chung, Y., 2014. Present situation and future perspective of bio-fertilizer for environmentally friendly agriculture. *Annual Reports*, 1-5.

[3] Knobeloch, L., Salna, B., Hogan, A., Postle, J., & Anderson, H., 2000. Blue babies and nitrate-contaminated well water. *Environmental health perspectives*, *108*(7), 675-678.

[4] Youssef, M. M. A., & Eissa, M. F. M., 2014. Biofertilizers and their role in management of plant parasitic nematodes. A review. *Journal of Biotechnology and Pharmaceutical Research*, *5*(1), 1-6.

[5] Raja, N., 2013. Biopesticides and biofertilizers: ecofriendly sources for sustainable agriculture. *J Biofertil Biopestici*, *4*(1), 1-2.

[6] Chouhan, G. K., Jaiswal, D. K., Gaurav, A. K., Mukherjee, A. and Verma, J. P., 2021a. PGPM as a potential bioinoculant for enhancing crop productivity under sustainable agriculture. In *Biofertilizers* (pp. 221-237). Woodhead Publishing.

[7] Chouhan, G. K., Singh, S., Mukherjee, A., Gaurav, A. K., Lepcha, A., Kumari, S. and Verma, J. P., 2022. Exploration of Plant Growth-Promoting Rhizobacteria (PGPR) for Improving Productivity and Soil Fertility Under Sustainable Agricultural Practices. In *Re-visiting the Rhizosphere Eco-system for Agricultural Sustainability* (pp. 245-269). Springer, Singapore.

[8] Chouhan, G. K., Verma, J. P., Jaiswal, D. K., Mukherjee, A., Singh, S., de Araujo Pereira, A. P., Liu, H., Abd_Allah, E. F. and Singh, B. K., 2021b. Phytomicrobiome

for promoting sustainable agriculture and food security: Opportunities, challenges, and solutions. *Microbiological Research*, 248, p.126763.

[9] Khan, T. A., Mazid, M., & Mohammad, F., 2011. Journal of Industrial Research & Technology. *Journal of Industrial Research & Technology*, *1*(2), 135-146.

[10] Mukherjee, Arpan, Saurabh Singh, Anand Kumar Gaurav, Gowardhan Kumar Chouhan, Durgesh Kumar Jaiswal, Arthur Prudêncio de Araujo Pereira, Ajit Kumar Passari, Ahmed M. Abdel-Azeem, and Jay Prakash Verma. "Harnessing of phytomicrobiome for developing potential biostimulant consortium for enhancing the productivity of chickpea and soil health under sustainable agriculture." *Science of The Total Environment* 836 (2022): 155550.

[11] Malusá, E., Sas-Paszt, L., & Ciesielska, J., 2012. Technologies for beneficial microorganisms inocula used as biofertilizers. *The scientific world journal*, *2012*. doi:10.1100/2012/491206.

[12] Doyle, J. J., 1998. Phylogenetic perspectives on nodulation: evolving views of plants and symbiotic bacteria. *Trends in Plant Science*, *3*(12), 473-478.

[13] Saravanakumar, D., Vijayakumar, C., Kumar, N., & Samiyappan, R., 2007. PGPR-induced defense responses in the tea plant against blister blight disease. *Crop protection*, *26*(4), 556-565.

[14] Holzinger, A., Nagendra-Prasad, D., & Huys, G., 2011. Plant protection potential and ultrastructure of Bacillus subtilis strain 3A25. *Crop Protection*, *30*(6), 739-744.

[15] Tittabutr, P., Awaya, J. D., Li, Q. X., & Borthakur, D., 2008. The cloned 1-aminocyclopropane-1-carboxylate (ACC) deaminase gene from Sinorhizobium sp. strain BL3 in *Rhizobium* sp. strain TAL1145 promotes nodulation and growth of Leucaena leucocephala. *Systematic and Applied Microbiology*, *31*(2), 141-150.

[16] Abdelhafez, A. A., Eid, K. E., El-Abeid, S. E., Abbas, M. H., Ahmed, N., Mansour, R. R., Zou, G., Iqbal, J., Fahad, S., Elkelish, A. and Alamri, S., 2021. Application of soil biofertilizers to a clayey soil contaminated with *Sclerotium rolfsii* can promote production, protection and nutritive status of Phaseolus vulgaris. *Chemosphere, 271*, 129321.

[17] Chang, C. H., & Yang, S. S., 2009. Thermo-tolerant phosphate-solubilizing microbes for multi-functional biofertilizer preparation. *Bioresource Technology*, *100*(4), 1648-1658.

[18] Chen, Y. P., Rekha, P. D., Arun, A. B., Shen, F. T., Lai, W. A., & Young, C. C., 2006. Phosphate solubilizing bacteria from subtropical soil and their tricalcium phosphate solubilizing abilities. *Applied soil ecology*, *34*(1), 33-41.

[19] Isawa, T., Yasuda, M., Awazaki, H., Minamisawa, K., Shinozaki, S., & Nakashita, H., 2010. *Azospirillum* sp. strain B510 enhances rice growth and yield. *Microbes and environments*, *25*(1), 58-61.

[20] Nosheen, S., Ajmal, I., & Song, Y., 2021. Microbes as biofertilizers, a potential approach for sustainable crop production. *Sustainability*, *13*(4), 1868.

[21] Naing, K. M., & Zin, T., 2013. Effects of Nitrogen Fixing Bacteria, *Azotobacter* spp. and *Azospirillum* spp. on the Growth of Rice *Oryza sativa L* (Doctoral dissertation, MERAL Portal).

[22] Wahane, M. R., Meshram, N. A., More, S. S., & Khobragade, N. H., 2020. Biofertilizer and their role in sustainable agriculture-A. *The Pharma Innovation Journal* 2020; 9(7): 127-130.

[23] Flores-Félix, J. D., Menéndez, E., Rivera, L. P., Marcos-García, M., Martínez-Hidalgo, P., Mateos, P. F., Martínez-Molina, E., Velázquez, M. D. L. E., García-Fraile, P. and Rivas, R., 2013. Use of *Rhizobium leguminosarum* as a potential biofertilizer for *Lactuca sativa* and *Daucus carota* crops. *Journal of Plant Nutrition and Soil Science, 176(6),* pp. 876-882.

[24] Sara, S., Morad, M., & Reza, C. M., 2013. Effects of seed inoculation by *Rhizobium* strains on chlorophyll content and protein percentage in common bean cultivars (Phaseolus vulgaris L.). *International Journal of Biosciences (IJB)*, *3*(3), 1-8.

[25] Mabrouk, Y., Hemissi, I., Salem, I. B., Mejri, S., Saidi, M., & Belhadj, O., 2018. Potential of rhizobia in improving nitrogen fixation and yields of legumes. *Symbiosis*, *107*(73495).

[26] Šašek, V., & Musilek, V., 1968. Two antibiotic compounds from mycorrhizal basidiomycetes. *Folia microbiologica*, *13*(1), 43-45.

[27] Alipour, A., Rahimi, M. M., Hosseini, S. M. A., & Bahrani, A., 2021. Mycorrhizal fungi and growth-promoting bacteria improves fennel essential oil yield under water stress. *Industrial Crops and Products*, *170*, 113792.

[28] Liu, N., Shao, C., Sun, H., Liu, Z., Guan, Y., Wu, L., Zhang, L., Pan, X., Zhang, Z., Zhang, Y. and Zhang, B., 2020. Arbuscular mycorrhizal fungi biofertilizer improves American ginseng (*Panax quinquefolius* L.) growth under the continuous cropping regime. *Geoderma*, 363, p. 114155.

[29] Abd-Alla, M. H., El-Enany, A. W. E., Nafady, N. A., Khalaf, D. M., & Morsy, F. M., 2014. Synergistic interaction of *Rhizobium leguminosarum* bv. viciae and arbuscular mycorrhizal fungi as a plant growth promoting biofertilizers for faba bean (*Vicia faba* L.) in alkaline soil. *Microbiological research,* 169(1), 49-58.

[30] Nafady, N. A., Hassan, E. A., Abd-Alla, M. H., & Bagy, M. M. K., 2018. Effectiveness of eco-friendly arbuscular mycorrhizal fungi biofertilizer and bacterial feather hydrolysate in promoting growth of *Vicia faba* in sandy soil. *Biocatalysis and agricultural biotechnology*, *16*, 140-147.

[31] Wu, S. C., Cao, Z. H., Li, Z. G., Cheung, K. C., Wong, M. H., 2005. Effects of biofertilizer containing N-fixer, P and K solubilizers and AM fungi on maize growth: a greenhouse trial. *Geoderma*, 125, 155-166.

[32] Andrade, M. M. M., Stamford, N. P., Santos, C. E. R., Freitas, A. D. S., Sousa, C. A., & Junior, M. A. L., 2013. Effects of biofertilizer with diazotrophic bacteria and mycorrhizal fungi in soil attribute, cowpea nodulation yield and nutrient uptake in field conditions. *Scientia Horticulturae*, *162*, 374-379.

[33] Khajeeyan, R., Salehi, A., Dehnavi, M. M., Farajee, H., & Kohanmoo, M. A., 2019. Physiological and yield responses of Aloe vera plant to biofertilizers under different irrigation regimes. *Agricultural Water Management*, *225*, 105768.

[34] Khalediyan, N., Weisany, W., Schenk, P. M., 2021. Arbuscular mycorrhizae and rhizobacteria improve growth, nutritional status and essential oil production in *Ocimum basilicum* and *Satureja hortensis*. Industrial Crops and Products 160, 113163.

[35] Sparks, D. L., & Huang, P. M., 1985. Physical chemistry of soil potassium. *Potassium in agriculture*, 201-276.

[36] Ahmad, M., Nadeem, S. M., Naveed, M., & Zahir, Z. A., 2016. Potassium-solubilizing bacteria and their application in agriculture. Potassium solubilizing microorganisms for sustainable agriculture, 293-313.

[37] Etesami, H., Emami, S., & Alikhani, H. A., 2017. Potassium solubilizing bacteria (KSB): Mechanisms, promotion of plant growth, and future prospects A review. *Journal of soil science and plant nutrition*, *17*(4), 897-911.

[38] Meena, V. S., Maurya, B. R., Verma, J. P., Aeron, A., Kumar, A., Kim, K., & Bajpai, V. K., 2015. Potassium solubilizing rhizobacteria (KSR): isolation, identification, and K-release dynamics from waste mica. *Ecological Engineering*, *81*, 340-347.

[39] Sindhu, S. S., Parmar, P., Phour, M., & Sehrawat, A., 2016. Potassium-solubilizing microorganisms (KSMs) and its effect on plant growth improvement. In *Potassium solubilizing microorganisms for sustainable agriculture,* 171-185.

[40] Sugumaran, P., & Janarthanam, B., 2007. Solubilization of potassium containing minerals by bacteria and their effect on plant growth. *World Journal of Agricultural Sciences*, *3*(3), 350-355.

[41] Saha, M., Maurya, B. R., Meena, V. S., Bahadur, I., & Kumar, A., 2016. Identification and characterization of potassium solubilizing bacteria (KSB) from Indo-Gangetic Plains of India. *Biocatalysis and agricultural biotechnology*, *7*, 202-209.

[42] Prajapati, K., & Modi, H., 2016. Growth promoting effect of potassium solubilizing Enterobacter hormaechei (KSB-8) on cucumber (*Cucumis sativus*) under hydroponic conditions. *International Journal of Advanced Research in Biological Sciences*, *3*(5), 168-173.

[43] Kurjogi, M., Basavesha, K. N., & Savalgi, V. P., 2021. Impact of potassium solubilizing fungi as biopesticides and its role in crop improvement. In *Biocontrol Agents and Secondary Metabolites,* pp. 23-39.

[44] Xiafang, S., & Weiyi, H., 2002. Mechanism of potassium release from feldspar affected by the sprain Nbt of silicate bacterium. *Acta Pedologica Sinica*, *39*(6), 863-871.

[45] Fasim, F., Ahmed, N., Parsons, R., & Gadd, G. M., 2002. Solubilization of zinc salts by a bacterium isolated from the air environment of a tannery. *FEMS microbiology letters*, *213*(1), 1-6.

[46] Nitu, R., Rajinder, K., & Sukhminderjit, K., 2020. Zinc solubilizing bacteria to augment soil fertility–a comprehensive review. *International Journal of Agricultural Sciences and Veterinary Medicine*, *8*(1), 38-44.

[47] Saravanan, V. S., Kumar, M. R., & Sa, T. M. 2011. Microbial zinc solubilization and their role on plants. In *Bacteria in agrobiology: plant nutrient management,* 47-63.

[48] Whitton B. A., & Potts M., (eds) 2000. The ecology of cyanobacteria. Kluwer, Dordrecht. pp. 669.

[49] Gademann, K., & Portmann, C., 2008. Secondary metabolites from cyanobacteria: complex structures and powerful bioactivities. *Current Organic Chemistry*, *12*(4), 326-341.

[50] Schopf, J. W., 2000. The fossil record: tracing the roots of the cyanobacterial lineage. In *The ecology of cyanobacteria,* 13-35.

[51] DeRuyter, Y. S., 2008. Molecular structure of the photosynthetic apparatus. *The cyanobacteria, molecular biology, genomics and evolution*, 217-269.

[52] Karl, D., Michaels, A., Bergman, B., Capone, D., Carpenter, E., Letelier, R., Lipschultz, F., Paerl, H., Sigman, D. and Stal, L., 2002. Dinitrogen fixation in the world's oceans. *Biogeochemistry, 57(1)*, pp. 47-98.

[53] Issa, A. A., Abd-Alla, M. H., & Ohyama, T., 2014. Nitrogen fixing cyanobacteria: future prospect. *Advances in biology and ecology of nitrogen fixation*, *2*, 24-48.

[54] Ördög, V., Stirk, W. A., Lenobel, R., Bancířová, M., Strnad, M., Van Staden, J., Szigeti, J. and Németh, L., 2004. Screening microalgae for some potentially useful agricultural and pharmaceutical secondary metabolites. *Journal of applied phycology, 16(4),*309-314.

[55] Rosenberg, J. N., Oyler, G. A., Wilkinson, L., & Betenbaugh, M. J., 2008. A green light for engineered algae: redirecting metabolism to fuel a biotechnology revolution. *Current opinion in Biotechnology*, *19*(5), 430-436.

[56] Sielaff, H., Christiansen, G., & Schwecke, T., 2006. Natural products from cyanobacteria: Exploiting a new source for drug discovery. *IDrugs: the investigational drugs journal*, *9*(2), 119-127.

[57] Tan, L. T., 2010. Filamentous tropical marine cyanobacteria: a rich source of natural products for anticancer drug discovery. *Journal of Applied Phycology*, *22*(5), 659-676.

[58] Wase N. V., & Wright P. C., 2008 System biology of cyanobacterial secondary metabolite production and its role in drug discovery. *Expert Opin Drug Discov* 3:903–929

[59] González-López, J., Rodelas, B., Pozo, C., Salmerón-López, V., Martínez-Toledo, M. V., & Salmerón, V., 2005. Liberation of amino acids by heterotrophic nitrogen fixing bacteria. *Amino acids*, *28*(4), 363-367.

[60] Singh, J. S., Kumar, A., Rai, A. N., & Singh, D. P., 2016. Cyanobacteria: a precious bio-resource in agriculture, ecosystem, and environmental sustainability. *Frontiers in microbiology*, *7*, 529.

[61] Lumpkin, T. A., and Plucknett, D. L., 1982. Azolla as a Green Manure: Use and Management in Crop Production: Westview Tropical Agriculture Series No. 5. Boulder, Colo: Westview Press, 230.

[62] Venkataraman, G. S., & Neelakantan, S., 1967. Effect of the cellular constituents of the nitrogen-fixing blue-green alga, *Cylindrospermum muscicola*, on the root growth of rice plants. *The Journal of General and Applied Microbiology*, *13*(1), 53-61.

[63] Venkataraman, G. S., 1972. *Algal Biofertilizers and Rice Cultivation*. New Delhi: Today and tomorrow Printers and Publishers.

[64] Mishra, U., & Pabbi, S., 2004. Cyanobacteria: a potential biofertilizer for rice. *Resonance*, *9*(6), 6-10.

[65] Yadav, R. K., Abraham, G., Singh, Y. V., & Singh, P. K., 2014. Advancements in the utilization of Azolla-Anabaena system in relation to sustainable agricultural practices. In *Proc. Indian Natl. Sci. Acad* (80)2, 301-316.

[66] Bocchi, S., & Malgioglio, A., 2010. Azolla-Anabaena as a biofertilizer for rice paddy fields in the Po Valley, a temperate rice area in Northern Italy. *International Journal of Agronomy, 2010.*

[67] Setiawati, M. R., Damayani, M., Herdiyantoro, D., Suryatmana, P., Anggraini, D., & Khumairah, F. H. 2018. The application dosage of Azolla pinnata in fresh and powder form as organic fertilizer on soil chemical properties, growth and yield of rice plant. In *AIP conference proceedings* (Vol. 1927, No. 1, p. 030017). AIP Publishing LLC.

[68] Kumar, R., Kumawat, N., & Sahu, Y. K., 2017. Role of biofertilizers in agriculture. *Pop Kheti*, *5*(4), 63-66.

[69] Shende, S. T., Apte, R. G., & Singh, T., 1975. Multiple action of Azotobacter. In *Indian Journal of Genetics and Plant Breeding* (Vol. 35, No. 2, pp. 314-315). *Indian Agriculture res inst*, New Delhi Delhi-110012, India: Indian Soc Genet Plant Breed.

[70] Meshram, S. U., Pande, S. S., Shanware, A. S., Kamdi, R. R., & Tajane, V. S., 2004. Efficacy of biofertilizers integrated with chemical fertilizers in-vivo in soybean. *Biofertilizer Newsletter*, *12*(1), 7-10.

[71] Thomas, L., & Singh, I., 2019. Microbial biofertilizers: types and applications. In *Biofertilizers for sustainable agriculture and environment,* 1-19.

[72] Kumar Maurya, A., 2014. Performance of Bio-Fertilizers in Sustainable Agriculture Development. *Growth of Unorganised Sector in India*, p. 1.

[73] Jnawali, A. D., Ojha, R. B., & Marahatta, S., 2015. Role of Azotobacter in soil fertility and sustainability–a review. *Adv. Plants Agric. Res*, *2*(6), 1-5.

[74] Okon, Y., & Labandera-Gonzalez, C. A., 1994. Agronomic applications of Azospirillum: an evaluation of 20 years worldwide field inoculation. *Soil Biology and Biochemistry*, *26*(12), 1591-1601.

[75] Umali-Garcia, M., Hubbell, D. H., Gaskins, M. H., & Dazzo, F. B., 1980. Association of Azospirillum with grass roots. *Applied and Environmental Microbiology*, *39*(1), 219-226.

[76] Lin, W., Okon, Y., & Hardy, R. W., 1983. Enhanced mineral uptake by *Zea mays* and Sorghum bicolor roots inoculated with *Azospirillum brasilense*. *Applied and Environmental Microbiology*, *45*(6), 1775-1779.

[77] Tien, T. M., Gaskins, M. H., & Hubbell, D., 1979. Plant growth substances produced by *Azospirillum brasilense* and their effect on the growth of pearl millet (*Pennisetum americanum* L.). *Applied and environmental microbiology*, *37*(5), 1016-1024.

[78] Pathak, D. V., Kumar, M. and Rani, K., 2017. Biofertilizer application in horticultural crops. In Microorganisms for Green Revolution (pp. 215-227). Springer, Singapore.

[79] Okumura, R. S., Mariano, D. D. C., Dallacort, R., Albuquerque, A. D., Lobato, A. D. S., Guedes, E. S., Oliveiraneto, C. D., Conceição, H. D. and Alves, G. R., 2013. Azospirillum: a new and efficient alternative to biological nitrogen fixation in grasses. *J. Food Agric. Environ, 11,* pp. 1142-1146.

[80] Aseri, G. K., Jain, N., Panwar, J., Rao, A. V., & Meghwal, P. R., 2008. Biofertilizers improve plant growth, fruit yield, nutrition, metabolism and rhizosphere enzyme activities of pomegranate (*Punica granatum* L.) in Indian Thar Desert. *Scientia horticulturae*, *117*(2), 130-135.

[81] Brahmaprakash, G. P., & Sahu, P. K., 2012. Biofertilizers for sustainability. *Journal of the Indian Institute of Science*, *92*(1), 37-62.

[82] Zahran, H. H., 1999. Rhizobium-legume symbiosis and nitrogen fixation under severe conditions and in an arid climate. *Microbiology and molecular biology reviews*, 63(4), pp. 968-989.

[83] Jerbi, M., Labidi, S., Laruelle, F., Tisserant, B., Jeddi, F. B., & Sahraoui, A. L. H., 2022. Mycorrhizal biofertilization improves grain yield and quality of hulless Barley (*Hordeum vulgare* ssp. nudum L.) under water stress conditions. *Journal of Cereal Science*, *104*, 103436.

[84] Kaushik B. D., 1998 *Soil plant Microbe interaction in relation to Nutrient management.* Venus, New Delhi, pp 55–63.

[85] Mohamed, A. A., Eweda, W. E., Heggo, A. M., & Hassan, E. A., 2014. Effect of dual inoculation with arbuscular mycorrhizal fungi and sulphur-oxidising bacteria on onion (*Allium cepa* L.) and maize (*Zea mays* L.) grown in sandy soil under greenhouse conditions. *Annals of Agricultural Sciences*, *59*(1), 109-118.

[86] Lochhead, A. G., 1958. Soil bacteria and growth-promoting substances. *Bacteriological Reviews*, *22*(3), 145-153.

[87] Barea, J. M., Navarro, E., & Montoya, E., 1976. Production of plant growth regulators by rhizosphere phosphate-solubilizing bacteria. *Journal of Applied Bacteriology*, *40*(2), 129-134.

[88] Hosseinzadeh, M. H., Ghalavand, A., Boojar, M. M. A., Modarres-Sanavy, S. A. M., & Mokhtassi-Bidgoli, A., 2021. Application of manure and biofertilizer to improve soil properties and increase grain yield, essential oil and ω3 of purslane (*Portulaca oleracea* L.) under drought stress. *Soil and Tillage Research*, *205*, 104633.

[89] Rueda-Puente, E. O., Murillo-Amador, B., Castellanos-Cervantes, T., García-Hernández, J. L., Tarazòn-Herrera, M. A., Medina, S. M., & Barrera, L. E. G., 2010. Effects of plant growth promoting bacteria and mycorrhizal on *Capsicum annuum* L. var. aviculare ([Dierbach] D'Arcy and Eshbaugh) germination under stressing abiotic conditions. *Plant Physiology and Biochemistry*, *48*(8), 724-730.

[90] Pawar, P. B., Khadilkar, J. P., Kulkarni, M. V., & Melo, J. S., 2018. An approach to enhance nutritive quality of groundnut (*Arachis hypog*aea L.) seed oil through endo mycorrhizal fertigation. *Biocatalysis and Agricultural Biotechnology*, *14*, 18-22.

[91] Heydari, L., Bayat, H., & Gregory, A. S., 2021. Investigating the effect of inoculation of chickpea with rhizobium and mycorrhizal fungi (*Funneliformis mosseae*) on soil mechanical and physical behavior. *Geoderma,* 385, 114860.

[92] Kobae, Y., 2019. Dynamic phosphate uptake in arbuscular mycorrhizal roots under field conditions. *Frontiers in environmental Science*, 159.

[93] Leaungvutiviroj, C., Ruangphisarn, P., Hansanimitkul, P., Shinkawa, H., & Sasaki, K., 2010. Development of a new biofertilizer with a high capacity for N2 fixation, phosphate and potassium solubilization and auxin production. *Bioscience, biotechnology, and biochemistry*, *74*(5), 1098-1101.

[94] Radhakrishnan, R., Hashem, A., & Abd_Allah, E. F., 2017. Bacillus: A biological tool for crop improvement through bio-molecular changes in adverse environments. *Frontiers in physiology*, *8*, 667.

[95] Kloepper, J. W., Leong, J., Teintze, M., & Schroth, M. N., 1980. Enhanced plant growth by siderophores produced by plant growth-promoting rhizobacteria. *Nature*, *286*(5776), 885-886.

Chapter 10

Phytoplankton in Climate Change Regulation

H. A. S. N. Abeysiri[1,2]
K. R. V. Bandara[1,2]
K. A. Sirisena[3]
and P. M. Manage[1,2,*]

[1]Centre for Water Quality and Algae Research, Department of Zoology, University of Sri Jayewardenepura, Sri Lanka
[2]Faculty of Graduate Studies, University of Sri Jayewardenepura, Sri Lanka
[3]Department of Environmental Technology, Faculty of Technology, University of Colombo, Sri Lanka

Abstract

Phytoplankton is the foundation of aquatic food webs and is critical to the functioning and services of ecosystems all over the world. As a result of global climate change affected by increased wind speed, concentrations of greenhouse gas, precipitation, incoming solar radiation, and air temperature are expected to change (primarily CO_2). These critical climate variables have a direct influence on lake water quality by changing the flow and temperature balance. Periodic blooms of phytoplankton and a shift in the neutral trophic balance can emerge from high nutrient enrichment concentrations mixed with predictable climate variability. As a result, dissolved oxygen concentrations at low concentrations might fluctuate significantly. Algal production could also reach excessive levels. Climate change also affects phytoplankton size structure and species composition, favoring traits best altered to fluctuating climate conditions. Understanding the mechanisms that

* Corresponding Author's Email: pathmalal@sjp.ac.lk.

In: Microbial Technologies and Their Applications
Editors: Santosh Kumar Karn and Anne Bhambri
ISBN: 979-8-88697-913-8

govern phytoplankton adaptation to warming will improve the realism of models that describe the ecological and biogeochemical effects of climate change. Higher temperatures initially hampered population growth because respiration is more temperature-sensitive than photosynthesis, resulting in less carbon available for development. Tolerance to high temperatures evolved after 100 generations as a result of greater down-regulation of respiration in comparison to photosynthesis. By down-regulating respiration and allocating fixed carbon more efficiently to growth, phytoplankton overcame the metabolic constraint imposed by respiration's higher temperature sensitivity. The rapid evolution of phytoplankton carbon-use efficiency suggests a potentially broad mechanism for thermal adaptation. It implies that as temperatures rise, evolutionary responses in phytoplankton will alter biogeochemical cycles, thereby changing the structure and function of food webs. Phytoplankton has a significant evolutionary capacity in response to climate change.

Keywords: phytoplankton, climate change, water quality, environmental health, regulation

Introduction

Climate and Climate Change

Climate is the average weather in a particular location over a lengthy period. A climatic description includes the average temperature in different seasons, rainfall, and sunshine [1]. Climate change is defined as any long-term change in the statistics of climate variables such as temperature, precipitation, pressure, or wind [1]. Natural external forcing (variations in solar radiation, natural inner climate system processes) or human-induced climate change are both possible. Climate change is posing an increasing danger to biodiversity and ecosystems around the world [2]. Individual species and their interactions with other species and their ecosystems are affected by climate change, which influences the structure and function of ecosystems as well as the products and services that natural systems offer to society [2].

Impacts on Climate Change

As part of a natural cycle, the earth's climate is dynamic and changes regularly. The fact that current developments have been expedited due to man's activity concerns the world the most [3]. The causes of climate change can be separated into two categories: natural influences and human factors [4] (Figure 1).

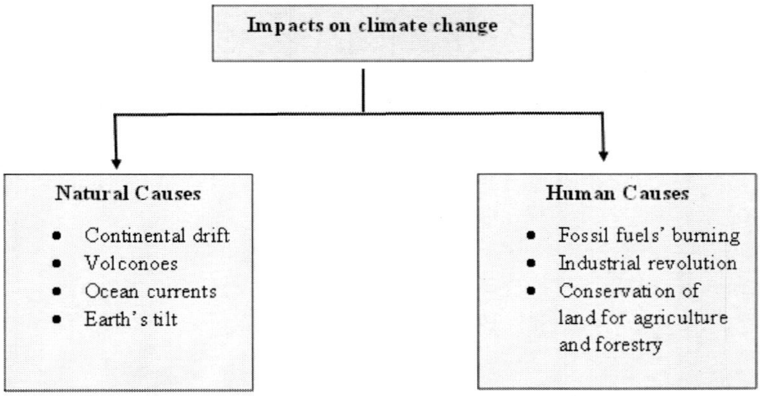

Figure 1. Impacts on climate change.

Continental drift changed the physical features of the landmass, their position, and the position of water bodies, all of which impacted the climate. The separation of landmasses changed the flow of ocean currents and winds, affecting climate [5]. When a volcano erupts, large amounts of sulfur dioxide (SO_2), water vapor, dust, and ash are released [6]. Climate change results from a progressive shift in the Earth's axis [7].

In the climate system, the seas play a significant role. They cover over 71% of the Earth's surface and absorb nearly twice as much solar radiation as the atmosphere or land surface. Ocean currents transmit nearly the same amount of heat across the globe as the atmosphere. Because the oceans are surrounded by land, heat is transported through the water via channels. Ocean currents are caused by winds pushing horizontally against the water surface. Some world places are more affected by ocean currents than others [8].

Climate change can be exacerbated by human activities such as fossil fuels and the conversion of land for forestry and agriculture. Since the beginning of the industrial revolution, human impacts on the climate system have increased considerably. These activities alter the ground surface and emit numerous compounds into the atmosphere, and have other environmental consequences [9]. The six principal greenhouse gases are carbon dioxide (CO_2), methane (CH_4), nitrous oxide (N_2O), hydrofluorocarbons (HFCs), perfluorocarbons (PFCs), and sulfur hexafluoride (SF_6). Water vapor is also a greenhouse gas [10].

Plankton

Plankton is microscopic organisms that float with the currents in the sea. Zooplankton acts as a medium between energy providers and consumers, transferring food energy to higher trophic levels [11]. Because they respond quickly to environmental changes, these creatures are important biological indicators of water quality and the trophic state of aquatic ecosystems [12]. Phytoplankton is the core of the aquatic food web. The density of phytoplankton as primary producers is directly related to the productivity of an aquatic environment. Phytoplankton, acts as an energy transducer, converting solar energy into chemical energy for food [12]. Phytoplankton, also known as microalgae, are algae that have chlorophyll and require sunlight to grow and survive, just like terrestrial plants [13].

Phytoplankton

Phytoplankton is typically microscopic and unicellular floating organisms, which are classified into the kingdoms Monera (Cyanobacteria) and Protista (Phyla: Chlorophyta, Rhodophyta, Dinoflagellata, Phaeophyta, Crysophyta, Bacillariophyta). They play an important role in primary production, nutrient cycling, and food webs and they account for a considerable portion of aquatic primary production [14]. They can be found in single-celled, colonial, or filamentous forms. Phytoplankton is non-swimmer. Diatoms (Bacillariophyceae) and dinoflagellates are two of the most prominent phytoplankton groups (Dinophyceae). Many studies have shown that diatoms, rather than dinoflagellates, dominate phytoplankton groups in marine and freshwater environments [15].

Climate Changes Impact on Phytoplankton and Eutrophication

Phytoplanktons are interrelated to yearly changes in consumption, water column stratification, accessibility of light and temperature [16]. These environmental factors may change influencing taxonomic composition and phytoplankton structure due to climate change. Effects on environmental parameters that could limit primary level production, most notably light and nutrients, can both directly and indirectly mediate phytoplankton responses.

Plant metabolism, which comprises both respiratory and photosynthetic processes, is directly influenced by temperature, whereas primary producers' metabolic rates are mostly controlled by photosynthesis [17]. Temperature has no effect on light-limited photosynthesis rates, however as temperature increases light saturated rates also increases, indirectly raising the light level at which light saturation begins. As a result, global warming is projected to enhance photoautotrophic production rates that are light-saturated but not light-limited. As a result, when sufficient resources are present, a temperature increase should result in higher biomass and plant growth rates [18]. Heterotrophic species' metabolism, on the other hand, is more temperature sensitive than photosynthesis rates [19]. As a result, warming must enhance herbivore consumption more than primary production.

Changes in thermal stratification patterns affect phytoplankton species composition such as growing season length and vertical mixing mechanisms [20] (Figure 2). Mixing of vertically is the most important factor influencing phytoplankton growth in the water column, due to variations in light and nutrition availability being frequently associated with mixing processes. Vertical mixing is greatly controlled by weather conditions in natural water. The tendency for stratification and suppression of mixing is produced by heat exchange processes and wind action [21]. Changes in meteorological forcing alter the balance between stratification and mixing [22, 23].

Turbulent diffusion and sedimentation of phytoplankton cells, which are the principal mechanisms producing non-motile cell vertical displacement, are directly affected by changes in duration of thermal stratification [24]. Velocities of phytoplankton sinking are increased non-linearly with size as a result of thermal stratification and vertical mixing [25], giving tiny and buoyant species [26].

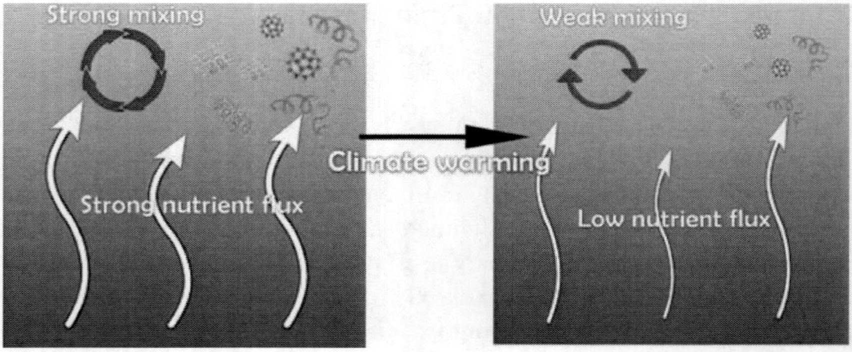

Figure 2. Effects of temperature on water column stratification (top arrows), nutrient redistribution (curvy arrows), and phytoplankton cell size and production.

Climate-phytoplankton interactions are complicated because other factors including density dependence, resource availability, and predation have a strong effect on the community's distribution, abundance, and size structure. Climate change affects plankton primarily through changes in seasonal dynamics, population size structure and species composition.

The Role of Phytoplankton in Atmospheric CO_2 Regulation

The use of fossil fuels has greatly raised atmospheric carbon dioxide (CO_2). CO_2 concentrations in the atmosphere, which had previously increased at a 1% yearly rate in the twentieth century, are now increasing at a 3% annual rate [27]. Rising CO_2 levels will have a significant impact on aquatic chemistry (carbonate ion levels and pH) [28]. The pH of aquatic water bodies is directly connected to dissolved inorganic carbon (DIC) such as CO_2, $H_2CO_3^-$, HCO_3^-.

The buffering capacity of marine environments limits inorganic DIC pH and speciation to a narrower range than freshwater ecosystems. Many lagoons are CO_2 supersaturated due to terrestrial C imports and sediment respiration [29]. Because of the substantial diel DIC decrease linked with eutrophic lakes' algal blooms, phytoplankton may become C-limited. Surface-dwelling cyanobacteria may have been fit over other phytoplankton when water column CO_2 concentrations are pushed down by thick blooms because they are closer to air CO_2, which can quickly seep into the surface waters and stimulate their growth [30]. Cyanobacteria have also been proven to benefit from low DIC environments. According to multiple studies [31], cyanobacteria outcompete eukaryotic algae in low CO_2, and high pH conditions. Other cyanobacteria, however, respond to higher CO_2 levels by raising carbon fixation, cell division rates, or both. There are several phylogenetic ways for phytoplankton to take up, convert or transport CO_2 and HCO_3^- [32].

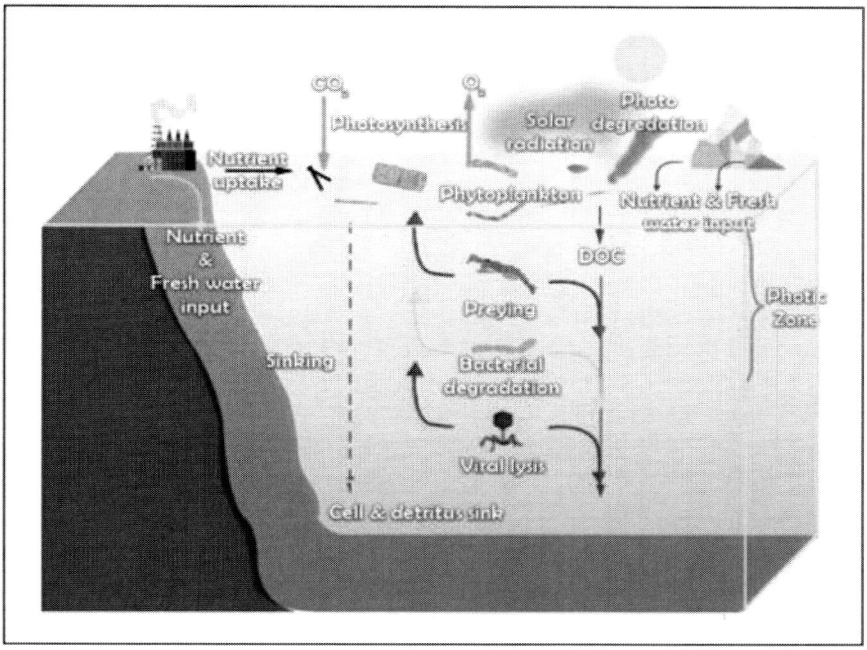

Figure 3. The role of phytoplankton in atmospheric CO_2 regulation.

Carbon-concentrating mechanisms (CCMs) are found in all eukaryotic cyanobacteria algae [33]. Cyanobacteria have established inorganic carbon absorption mechanisms. Its ribulose bisphosphate carboxylase-oxygenase (Rubisco) enzyme is divided into carboxysomes, which produce a high concentration of CO_2 around the Rubisco enzyme [34]. CCMs in cyanobacteria are more efficient at low CO_2 concentrations than those in other algae or higher plants [35], suggesting that they may be capable of growing in low CO_2 environments. Rises in CO_2 concentrations in the atmosphere affect individual cyanobacterial genera cell physiology and growth rates. Changing CO_2 levels may potentially have an impact on cyanobacterial community strain composition. One study used toxic and non-toxic cyanobacteria strains with high CO_2 availability to conduct competition experiments and found that the non-toxic strain had a competitive advantage [36].

The Significance of Phytoplankton Blooms for Environmental Health

Environmental health is the science of preventing human injury and illness and promoting well-being by identifying and assessing environmental sources and harmful agents, as well as limiting exposure to harmful physical, chemical, and biological agents in air, water, soil, food, and other environmental media or settings [37].

An algal bloom is a bloom of plankton algae that has grown out of control. A harmful algal bloom can result in significant financial losses for aquaculture, fisheries, and tourism businesses and significant environmental and human health consequences. A 'Red Tide' is a seawater discoloration caused by a toxic dinoflagellate bloom. *Karenia brevis* is a well-known red tide-forming dinoflagellate that was once thought to be endemic to the Gulf of Mexico and Florida's east coast, where red tides' were first recorded in 1844 [38]. Algal blooms can be brown, yellow, green, or milky in hue and red. In Japan's Seto Inland Sea, *Chattonella antiqua* produced multiple red tide outbreaks, killing 14 million farmed yellow-tail fish in 1972 [38].

On the west coast of the United States, [39] reported that important toxin-producing dinoflagellates are in the genus Alexandrium, which cause paralytic shellfish poisoning (PSP), and diatoms in the genus Pseudo-nitzschia, which make domoic acid, which causes domoic acid poisoning (DAP). On the California coast, PSP-producing blooms had been documented frequently. There were 22 hazardous episodes in California between 1962 and 1989, with *Alexandrium catenella* being the causal factor in many of them [39].

There were huge blooms of *Ceratium furca, C. divaricatum*, and *Prorocentrium micans*, lasting l-3 months between August and November, off the Washington coast in 1994 and 1995 [39].

Bloom formation in phytoplankton is a well-studied phenomenon that has become a major issue in surface waterbodies. Phytoplankton blooms are visible discolorations in water caused primarily by planktonic forms of cyanobacteria. The most common cyanobacterial genera are *Cylindrospermopsis, Dolicospermum, Mycrocystis, Nodularia* and *Trichodesmium* [40]. The formation of blooms is initiated by environmental conditions such as temperature, light, nutrient concentrations, and thermal stratification [41].

High nutrient concentrations, particularly phosphorus concentrations greater than 25-50 μg/L[1], water temperatures greater than 25°C, hydraulic retention times greater than 1 month, and stable water body stratification are common favorable conditions [41]. Some of the blooms that form are known to be hazardous, and such "harmful algal blooms" are large cyanobacterial blooms with negative effects such as decreased ecosystem stability, toxic compound production, and mortality. Toxins are produced by approximately 50% of cyanobacterial blooms [42] and are the cause of many toxicities in fish, mammals, and humans who consume water containing cyanobacteria.

Cyanotoxins are secondary metabolites produced by cyanobacteria that can belong to a variety of chemical groups. They can be peptides, alkaloids, or lipopolisacharides, and their toxicity classifies them as hepatotoxins, neurotoxins, cytotoxins, and skin and gastrointestinal irritants [43]. Endotoxins and exotoxins are both produced and have an effect either within the cell or after cell lysis. Despite the fact that there may be a variety of cyanotoxins in natural water bodies, the most commonly studied cyanotoxins are peptides such as microcystins and nodularins, alkaloids such as cylindrospermopsins, anatoxins, saxitoxins, and endotoxic lipopolysaccharides.

Consumption of cyanotoxin-contaminated water poses numerous health risks to humans and animals, causing cytotoxicity, hepatotoxicity, neurotoxicity, or dermatotoxicity [44, 45]. Approximately 75% of the water bodies in the country's CKDu endemic areas have reported the presence of cyanobacteria, raising the possibility that cyanobacteria and their toxins are one of the causes of CKDu [45]. It has also been observed that toxin-producing cyanobacterial species such as *M. aeruginosa, Cylindrospermopsis* sp., and *Anabaena* sp. are abundant in such Sri Lankan source water bodies [45].

The World Health Organization (WHO), is currently reviewing the health effects caused by cyanotoxins in water. The currently available data provides guideline value for Microcystins (MCs) in water, whereas, other cyanotoxins including Cylindrospermopsin (CYN), are under investigations. The guideline values have established considering both the free and cell-bound cyanotoxin concentrations and it is based on the life-time exposure that is required in order to implement remedial actions whenever the contamination exceeds this value in the short term. Except for WHO, Sri Lanka Standards Institute [42], have also established guideline values for MC and CYN for drinking water in Sri Lanka in 2013.

Direct and Indirect Effects of Climate Change on Water Treatment and Water Safety

There are two types of climate change concerns for the water reuse industry: effects of direct and indirect (Figure 4).

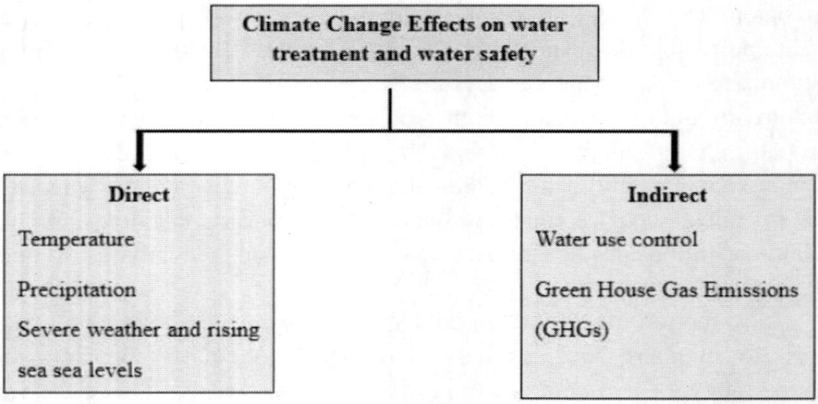

Figure 4. Impacts of direct and indirect of the climatic factors on water treatment and water safety.

Direct Impacts of the Climatic Factors

Temperature

The impact of climate change-induced changes in air temperature on the wastewater reuse process has received little attention [46]. Temperatures outside or beyond the ideal range impact biological processes, particularly temperature-sensitive nitrifying bacteria [47, 48]. Inspected the effects of changing air temperature on wastewater treatment efficiency. Higher temperatures were linked to a rise in the melting point numbers, they discovered. Melting points caused a stronger and colder influx, which reduced wastewater influent temperatures [48]. The increment in ambient temperature has received little attention in temperate zones due to it was usually assumed that a warmer climate would speed reaction kinetics and reduce energy consumption [46]. Warmer temperatures have been described to promote raw wastewater pipeline corrosion due to hydrogen sulfide production [49].

Precipitation

The most common effect of climate change on wastewater reclamation is increased rainfall intensity. In light of climate change provisions, hydraulic models were used to assess sewerage overfilling scenarios [50, 51]. Increasing pollutant concentrations, floatable materials, and main settling tanks limited wastewater treatment process performance at the beginning of the storm. The "first flush effect" occurred when rainwater carried debris and particles from the roads into the combined sewage system. Biological activities such as activated sludge, nitrification/denitrification, and sludge floc development were affected by the changes in influent components [52]. A shorter hydraulic retention duration also influenced the primary clarifier effluent quality [53]. The important national findings are:

- Denmark has developed new rainfall design and a consistent procedure for analyzing the impact of climate change on sewer systems.
- Norway has focused its efforts on mapping regional climate change scenarios and is now developing the implications for various sectors.
- Sweden recently presented a roadmap for actions related to climate change in urban areas.

Effects of Climate Change on Weather Changes and Sea Level Rises

The plants of treatment and reuse were usually positioned in low-lying locations or around the shore because most wastewater collection networks were gravitational. Inundation, storm surges, flooding, erosion, and saltwater intrusion are just a few of the dangers that come with rising sea levels [54]. Furthermore, increasing salinity concentrations can have a direct effect on the quality of recycled water as sea levels rise [55]. As sea levels rose as an outcome of climate change, flooding, hurricanes, storms, cyclones, and thunderstorms were expected to become more severe [54].

Indirect Impacts of the Climatic Factors

Water Use Control

Water reduction programs were implemented as part of the effort to adjust to climate change. In terms of resource conservation, it was helpful. The volume of wastewater discharged to transmission systems, on the other hand, was reduced in proportion. As a result, the strength of the wastewater increased, speeding up the corrosion of the transport network [56].

Green House Gasses (GHGs) Emission

While water reuse attempted to reduce greenhouse gas emissions (GHGs) [57]. A wastewater reclamation plant produced two GHG sources. Offsite GHGs (CO_2, CH_4, and N_2O) are connected with energy demand, chemical manufacture, and transportation, whereas onsite GHGs (CO_2, CH_4, and N_2O) are associated with wastewater and sludge treatment operations [58]. According to [59], wastewater treatment and reclamation account for 56% of the water industry's GHG emissions.

How Does Phytoplankton Contribute to Global Climate Regulation?

When the thermal structure of the water column changes, the phytoplankton community reorganizes, favoring small-sized algal cells and species that can control their buoyancy [60].

Deviations in primary producers will have an impact on primary production as well as the biogeochemical cycle of carbon and other elements. Additionally, cyanobacteria or dinoflagellate blooms have a major effect on ecosystem water quality, trophic transfer, and fish production. Phytoplankton changeability is the main driver of biogeochemical changeability and yearly variations in fish recruitment [61]. A deeper understanding of phytoplankton's intrinsic natural variability is necessary for assessing the degree of global change's effect on aquatic ecosystem functioning. Predicting how compositional changes spread up to higher trophic levels and the judgement of these changes is a significant challenge [62].

The Importance of Phytoplankton

Phytoplankton is extremely important to live on earth. They impact us in several ways. Phytoplankton is the main source of a food chain that contributes to the world's most important fish stocks. Thus, they are answerable for formulating a biological community and regulating the food web [63, 64]. Phytoplankton acts as an important component of the marine ecosystem as it releases oxygen during photosynthesis and supports the energy exchange process. In turn, these microscopic organisms offer food for upper trophic levels in the food web. For example, mussels (oysters, clams, scallops) consume phytoplankton almost exclusively for their food. They play a vital role in mitigating climate change and global warming and reducing global CO_2 emissions [65].

Phytoplankton is important because it can produce a variety of highly toxic compounds bio-toxins. Some compounds can be released into the surrounding water while others are reserved in the phytoplankton, and can transform through the food web leading to bioaccumulation and biomagnification in higher trophic level species. In most cases, fish and seafood do not appear to be affected by these potent compounds, but organisms such as marine mammals and humans, can become ill or even die.

Cyanobacteria act as primary producers which make their own food from sunlight apart from plants and a number of other organisms. Primary producers are organisms that produce biomass from inorganic substances in an ecosystem. They are called autotrophs because they are almost photo-synthetically active and are the base of every food chain [66]. Many cyanobacteria is exceptional as they are autotrophic fix atmospheric nitrogen [67]. The significant adaptations of cyanobacteria are structural changes in their cells. A heterocyst is a specialized cell found in most cyanobacteria that fix nitrogen aerobically [66].

Spirulina plantensis and *Aphanizomenonflos-aquae*, both cyanobacteria, have been used as high-nutritive dietary supplements. It also contains a wide range of anti-inflammatory and medicinal minerals [68].

Many natural compounds originating from cyanobacteria have been shown to be pharmacologically significant as possible chemotherapeutic agents [69, 70]. Antioxidant, anticancer, neuroprotective, anti-inflammatory, hepatoprotective, and hypocholesterolemic effects have all been found in phycocyanin [71]. Phycocyanin is often used as a natural food colorant in chewing gum, ice cream, soft drinks, soft candies, jellies, and cosmetics [72]. Because of their specific fluorescent properties, phycobiliproteins are utilized in fluorescent immunoassays, diagnostic fluorescence microscopy, biomedical research, and protein markers for electrophoretic methods [73, 74]. Nostocarboline, a cholinesterase inhibitor manufactured by Nostoc, exhibits an effect similar to galanthamine, a medicine licensed to treat Alzheimer's disease [75].

Cyanobacteria are common in wastewater treatment plants [76]. Biodegradation and biosorption of heavy metals can be achieved by *Nostoc,*

Synechococcus, *Oscillatoria*, and *Nodularia* species isolated from pharmaceutical and textile industrial effluents [77]. A few cyanobacteria species have recently been explored for biofuel production due to their ability to convert solar energy to chemical energy, the most efficient of all living organisms [78].

Destruction of Phytoplankton in the Environment

Anthropological Implications

Phytoplankton is the primary producer in aquatic ecosystems. Their carbon dioxide sequestration and biomass production equal that of all terrestrial plants combined. A variety of environmental factors affect phytoplankton productivity, the majority of which are changing dramatically as a outcome of human-caused global warming. The vast majority of these interactions are either additive or synergistic in nature.

Phytoplankton is also exposed to higher amounts of solar UV-B radiation due to ozone depletion in the stratosphere. Eutrophication is caused by the transfer of sediments and dissolved organic matter into coastal waters, limiting UV penetration. These environmental factors have been shown to influence physiological and ecological processes.

Ocean acidification changes the chemistry of saltwater as atmospheric CO_2 levels rise, lowering calcification in phytoplankton, macro-algae, and some zoological taxa. Warming oceans affect phytoplankton bloom species composition, favoring harmful prokaryotic and eukaryotic species. Pollution from automobiles is harming phytoplankton.

Land Use

Changes in biodiversity loss and ecosystem function are caused mainly by human-induced land use and climate change [79, 80]. One of the processes through which land use and climate affect community and ecosystem characteristics are altering biogeochemical cycles [81].

Agriculture

On a global scale, fertilizer has resulted in a two-fold increase in N and a four-fold increase in the amount of P available for primary production [82]. In contrast to more disturbed agricultural or urban landscapes, unbroken woodland landscapes often produce lower nitrogen (N) and phosphorus (P) pools [81, 83].

Surface Water Runoff

Climate change has the possible to exacerbate erosion, loss of organic matter, salinization, loss of soil biodiversity, landslides, and flooding [84]. Heavy rains caused a lot of soil and sand input, which influenced plankton growth by changing the ambient environmental conditions. Heavy rains reduced pH, conductivity, transparency, light availability, and chlorophyll concentrations while increasing nitrate concentrations and turbidity [85].

The Effects of Global Warming on Phytoplankton Succession and Trophic Interactions

Phytoplankton Succession

Seasonal phytoplankton succession is a community phenomenon driven by processes that regulate the population dynamics of a variety of diverse producers and consumers.

In particular, temperature-related alterations in plankton phenology have been reported in various ecosystems; however, other factors such as biotic relationships sometimes make it difficult to extract a pure climate signal [86]. In the Western Scheldt Estuary, earlier bloom onset has coincided with rising temperatures over the last 30 years, indicating bloom timing shifts [87].

Trophic Interactions

Climate change can affect species in a variety of methods, from individual species traits to effects of population-level that affect how interactions of predator-prey play out.

Changes in plant traits can mediate predator-prey interactions. Climate change possibly affects plant quality by changing nutrient content, increasing protein:carbohydrate ratios [88], and changing concentrations of plant secondary compounds [89]. Changes in herbivore foraging behavior and nutrient content can have an impact on interactions of predator-prey. Plant phenology, distribution, and abundance changes are also likely to mediate interactions of predator-prey [90, 91].

Temperature and other abiotic factors related to climate change can change species traits such as development time [92], oviposition rates, and specific behavioral traits [93, 94]. Climate change has the potential to change the interactions of species by changing the interspecific efficiency and intraspecific interactions (Figure 5).

Production of phytoplankton biomass and carbon dioxide sequestration equal that of all terrestrial plants combined. The environmental factors influence phytoplankton productivity, the majority of which are changing dramatically as an outcome of human-caused global warming [88]. The vast majority of these interactions are either additive or synergistic in nature.

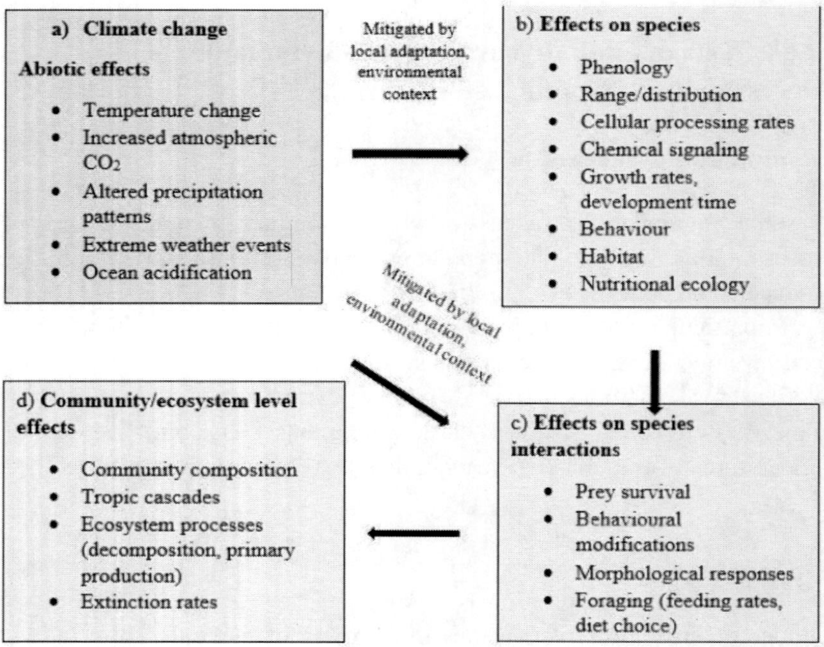

Figure 5. Climate change and tropic interactions.

Eutrophication is caused by the transfer of dissolved organic matter and sediments into coastal waters, limiting UV penetration. These environmental factors have been shown to influence physiological and ecological processes. As atmospheric CO_2 levels rise, ocean acidification alters the chemistry of saltwater, lowering calcification in macroalgae, phytoplankton, and some zoological taxa. The makeup of organisms alters due to global warming, favoring harmful eukaryotic and prokaryotic phytoplankton blooms. Phytoplankton is being harmed by pollution from crude oil spills, heavy metals, persistent organic pollutants, and home and industrial wastewaters, aggravated by solar UV radiation [95].

There have been few comprehensive studies of the effects of multiple stressors. The increasing pollution of the environment is another anthropogenic stressor affecting phytoplankton. Coastal and freshwater habitats are particularly vulnerable. However, sewage does not appear to harm phytoplankton even at high concentrations. Despite the river's high contaminant levels found high concentrations of phytoplankton and zooplankton [96].

Climate Change and Microbes

Change of climate affects the structure and functionality of terrestrial microbial communities in both direct and indirect ways. A list of the consequences of climate change on microorganisms is available. These include death and disturbance, metabolic activity is heavily influenced both directly and indirectly, diversity, biomass reduction, and composition lead to disappearance with effects of negative or positive on physiology and greenhouse gas emissions, and biomass reduction, diversity, and composition lead to extinction with negative or positive effects on greenhouse gas emissions and physiology. As the temperature increases, microbial community structures alter, and processes including methanogenesis, respiration, and fermentation speed up [97].

Heat waves, wildfires, intense storms, floods, distinction, natural catastrophes, excessive heat, poor air quality, drought, and the growth and establishment of diseases are all examples of climate change's impact on abiotic and biotic components [98].

Way Forward to Rich Phytoplankton Diversity

As a primary producer, phytoplankton significantly impacts advanced trophic levels. Phytoplankton development and diversity may be influenced by seasonal temperature variations and rising water temperatures, among other causes. The temperature effects and variety on phytoplankton growth were studied by [99].

Enlarged species variety enhanced rates of phytoplankton growth and phosphorus content at all temperatures, with the most significant values at 18 °C. Over yielding was detected at nearly every temperature range, and in most cases, complementing features could be used to explain it. The proportion of cyanobacteria in communities increased as the temperature climbed.

Conclusion

Some ecological activities, such as biogeochemical cycling, primary production, energy transfer across the food web, and plankton community structure, have been altered due to climate change. Changes in primary producers have impacted primary production and carbon and other elements' biogeochemical cycling. Furthermore, cyanobacteria bloom substantially impacts trophic ecosystem transfer, water quality changes, and intensive aquaculture changes. The number of physical differences and species' ability to acclimate to changing environmental conditions will considerably impact food-web dynamics as the future climate warms and becomes more variable.

References

[1] Ali, S., Liu, Y., Ishaq, M., Shah, T., Ilyas, A. and Din, I. U. 2017. Climate change and its impact on the yield of major food crops: Evidence from Pakistan. *Foods,* 6(6), p. 39.

[2] Díaz, F. P., Latorre, C., Carrasco-Puga, G., Wood, J. R., Wilmshurst, J. M., Soto, D. C., Cole, T. L. and Gutiérrez, R. A. 2019. Multiscale climate change impacts on plant diversity in the Atacama Desert. *Global Change Biology,* 25(5), pp. 1733-1745.

[3] Reddy, P. P. 2015. Causes of climate change. In *Climate Resilient Agriculture for Ensuring Food Security* (pp. 17-26). Springer, New Delhi.

[4] Byg, A. and Salick, J. 2009. Local perspectives on a global phenomenon—climate change in Eastern Tibetan villages. *Global environmental change*, 19(2), pp.156-166.

[5] Khan, M. Z. A. 2012. Climate change: Cause & effect. *Journal of Environment and Earth Science*, 2(4), pp. 48-53.

[6] Moxnes, E. D., Kristiansen, N. I., Stohl, A., Clarisse, L., Durant, A., Weber, K. and Vogel, A. 2014. Separation of ash and sulfur dioxide during the 2011 Grímsvötn eruption. *Journal of Geophysical Research: Atmospheres*, 119(12), pp. 7477-7501.

[7] Wanner, H., Beer, J., Bütikofer, J., Crowley, T. J., Cubasch, U., Flückiger, J., Goosse, H., Grosjean, M., Joos, F., Kaplan, J. O. and Küttel, M. 2008. Mid-to Late Holocene climate change: an overview. *Quaternary Science Reviews*, 27(19-20), pp. 1791-1828.

[8] Neumann, G. 2014. *Ocean currents*. Elsevier.

[9] Abas, N., Kalair, A. and Khan, N. 2015. Review of fossil fuels and future energy technologies. *Futures*, 69, pp. 31-49.

[10] Ritchie, H., Roser, M. and Rosado, P., 2020. CO_2 and greenhouse gas emissions. *Our world in data*.

[11] Gajbhiye, S. N. 2002. *Zooplankton-Study methods, importance and significant observations.*

[12] Jakhar, P. 2013. Role of phytoplankton and zooplankton as health indicators of aquatic ecosystem: A review. *International Journal of Innovation Research Study*, 2(12), pp. 489-500.

[13] Metting, F. B. 1996. Biodiversity and application of microalgae. *Journal of industrial microbiology*, 17(5), pp. 477-489.

[14] Dawes, C. J. 1998. *Marine botany*. John Wiley & Sons.

[15] Al-Kandari, M., Al-Yamani, F. and Al-Rifaie, K. 2009. Marine phytoplankton atlas of Kuwait's waters. *Kuwait institute for scientific research*, 351.

[16] Cloern, J. E., 1996. Phytoplankton bloom dynamics in coastal ecosystems: a review with some general lessons from sustained investigation of San Francisco Bay, California. *Reviews of Geophysics* 34(2): 127–168.

[17] Dewar, R. C., B. E. Bedlyn & R. E. McMurtrie. 1999. Acclimation of the respiration/photosynthesis ratio to temperature: insights from a model. *Global Change Biology* 5: 615–622.

[18] Padilla-Gamino, J. L. & R. C. Carpenter. 2007. Seasonal acclimatization of *Asparagopsistaxiformis* (Rhodophyta) from different biogeographic regions. *Limnology and Oceanography* 52: 833–842.

[19] Paerl, H. W. and Huisman, J. 2009. Climate change: a catalyst for global expansion of harmful cyanobacterial blooms. *Environmental microbiology reports*, 1(1), pp. 27-37.

[20] Lopez-Urrutia, A., E. San Martin, R. P. Harris & X. Irigoien. 2006. Scaling the metabolic balance of the oceans. *Proceedings of the National Academy of Sciences of the United States of America* 103: 8739–8744.

[21] Smol, J. P., Wolfe, A. P., Birks, H. J. B., Douglas, M. S., Jones, V. J., Korhola, A., Pienitz, R., Rühland, K., Sorvari, S., Antoniades, D. and Brooks, S. J. 2005. Climate-

driven regime shifts in the biological communities of arctic lakes. *Proceedings of the National Academy of Sciences*, 102(12), pp. 4397-4402.

[22] Wetzel, R. G. 2001. *Limnology: lake and river ecosystems.* gulf professional publishing.

[23] Boyd, P. W. & S. C. Doney. 2002. Modelling regional responses by marine pelagic ecosystems to global climate change. *Geophysical Research Letters* 29(16), pp. 53–56.

[24] Livingstone, D. M. 2003. Impact of secular climate change on the thermal structure of a large temperate central European lake. *Climate Change* 57(1–2): 205–225.

[25] Huisman, J., N. N. P. Thi, D. M. Karl & B. Sommeijer. 2006. Reduced mixing generates oscillations and chaos in the oceanic deep chlorophyll maximum. *Nature* 439(7074): 322–325.

[26] Winder, M. and Sommer, U. 2012. Phytoplankton response to a changing climate. *Hydrobiologia*, 698(1), pp. 5-16.

[27] Strecker, A. L., Cobb, T. P. and Vinebrooke, R. D. 2004. Effects of experimental greenhouse warming on phytoplankton and zooplankton communities in fishless alpine ponds. *Limnology and Oceanography*, 49(4), pp. 1182-1190.

[28] Füssel, H. M. 2009. An updated assessment of the risks from climate change based on research published since the IPCC Fourth Assessment Report. *Climatic change*, 97(3), pp. 469-482.

[29] Cao, L. and Caldeira, K. 2008. Atmospheric CO_2 stabilization and ocean acidification. *Geophysical Research Letters*, 35(19).

[30] Maberly, S. C. 1996. Diel, episodic and seasonal changes in pH and concentrations of inorganic carbon in a productive lake. *Freshwater Biology,* 35(3), pp. 579-598.

[31] Paerl, H. W. and Huisman, J. 2009. Climate change: a catalyst for global expansion of harmful cyanobacterial blooms. *Environmental microbiology reports,* 1(1), pp. 27-37.

[32] Oliver, R. L. and Ganf, G. G. 2000. Freshwater blooms. In *The ecology of cyanobacteria* (pp. 149-194). Springer, Dordrecht.

[33] Reinfelder, J. R. 2011. Carbon concentrating mechanisms in eukaryotic marine phytoplankton. *Annual review of marine science*, 3, pp. 291-315.

[34] Giordano, M., Beardall, J. and Raven, J. A. 2005. CO_2 concentrating mechanisms in algae: mechanisms, environmental modulation, and evolution. *Annu. Rev. Plant Biol.*, 56, pp. 99-131.

[35] Badger, M. R., Hanson, D. and Price, G. D. 2002. Evolution and diversity of CO_2 concentrating mechanisms in cyanobacteria. *Functional Plant Biology,* 29(3), pp.161-173.

[36] Badger, M. R. and Price, G. D. 2003. CO_2 concentrating mechanisms in cyanobacteria: molecular components, their diversity and evolution. *Journal of experimental botany*, 54(383), pp. 609-622.

[37] Van de Waal, D. B., Verspagen, J. M., Finke, J. F., Vournazou, V., Immers, A. K., Kardinaal, W. E. A., Tonk, L., Becker, S., Van Donk, E., Visser, P. M. and Huisman, J. 2011. Reversal in competitive dominance of a toxic versus non-toxic cyanobacterium in response to rising CO2. *The ISME journal*, 5(9), pp.1438-1450.

[38] Battersby, S. ed., 2016. *Clay's handbook of environmental health.* Routledge.

[39] Hallegraeff, G. M. 2003. Harmful algal blooms: a global overview. *Manual on harmful marine microalgae*, 33, pp. 1-22.

[40] Horner, R. A., Garrison, D. L. and Plumley, F. G. 1997. Harmful algal blooms and red tide problems on the US west coast. *Limnology and Oceanography*, 42(5part2), pp. 1076-1088.

[41] Huisman, J., Codd, G. A., Paerl, H. W., Ibelings, B. W., Verspagen, J. M. and Visser, P. M. 2018. Cyanobacterial blooms. *Nature Reviews Microbiology*, 16(8), pp. 471-483.

[42] Zhao, Y., Yu, R. C., Kong, F. Z., Wei, C. J., Liu, Z., Geng, H. X., Dai, L., Zhou, Z. X., Zhang, Q. C. and Zhou, M. J. 2019. Distribution patterns of picosized and nanosized phytoplankton assemblages in the East China Sea and the Yellow Sea: implications on the impacts of Kuroshio intrusion. *Journal of Geophysical Research: Oceans*, 124(2), pp. 1262-1276.

[43] World Health Organization, 2015. World health statistics 2015. *World Health Organization.*

[44] Cires, S. and Ballot, A. 2016. Are view of the phylogeny, ecology and toxin production of bloom-forming *Aphanizomenon* spp. and related species within the Nostocales (cyanobacteria). *Harmful Algae*, 54, pp. 21-43.

[45] Metcalf, J. S., Banack, S. A., Powell, J. T., Tymm, F. J., Murch, S. J., Brand, L. E. and Cox, P. A. 2018. Public health responses to toxic cyanobacterial blooms: perspectives from the2016 Florida event. *Water Policy*, 20(5), pp.919-932.

[46] Manage, P. 2019. Cyanotoxins: A hidden cause of Chronic Kidney Disease of unknown etiology (CKDu) in Sri Lanka-A review. *Sri Lanka J. Aquat. Sci,* 24(1), pp. 1-10.

[47] Metcalf & Eddy, Tchobanoglous G, Stensel HD, Burton F. 2002. *Wastewater engineering: treatment and reuse.* New York: McGraw-Hill.

[48] Eckenfelder W, Wesley J. *Industrial water pollution control.* Boston: McGraw-Hill; 2000.

[49] Plósz B. G., Liltved H., Ratnaweera H. 2009. Climate change impacts on activated sludge wastewater treatment: a case study from Norway. *Water Sci Technol;* 60:533–41.

[50] KWL. 2008. *Vulnerability of Vancouver sewerage area infrastructure to climate change*. Vancouver: Kerr Wood Leidal Associates Ltd.

[51] Mark O., Svensson G., König A, Linde J. J. 2008. Analyses and Adaptation of Climate Change Impacts on Urban Drainage Systems. In: Mark, et al., editors. *11th International Conference on Urban Drainage*. UK: Edinburgh, Scotland.

[52] Semadeni-Davies, A., Hernebring, C., Svensson, G. and Gustafsson, L. G. 2008. The impacts of climate change and urbanisation on drainage in Helsingborg, Sweden: Suburban stormwater. *Journal of hydrology*, 350(1-2), pp. 114-125.

[53] Wilén, B. M., Lumley, D., Mattsson, A. and Mino, T. 2006. Rain events and their effect on effluent quality studied at a full scale activated sludge treatment plant. *Water science and technology,* 54(10), pp. 201-208.

[54] Schütze, M., Campisano, A., Colas, H., Schilling, W. and Vanrolleghem, P. A. 2002. Real-time control of urban wastewater systems-where do we stand today? In *Global Solutions for Urban Drainage* (pp. 1-17).

[55] Parry M. L., Canziani O. F., Palutikof J. P., van der Linden P. J, Hanson C. E. 2007. Impacts, Adaptation and Vulnerability. In: Parry ML, editor. *IPCC Fourth Assessment Report: Climate Change 2007.* United Kingdom: Cambridge University Press, Cambridge.

[56] Howe, C. 2007. *Implications of potential climate change for Melbourne's water resources.*

[57] Larsen T. A. 2011. Redesigning wastewater infrastructure to improve resource efficiency. *Water Sci Technol*;63:2535–41.

[58] Hardisty P. E., Sivapalan M., Humphries R. 2013. Determining a sustainable and economically optimal wastewater treatment and discharge strategy. *J Environ Manag;*114:285–92.

[59] BaniShahabadi M., Yerushalmi L., Haghighat F. 2009. Impact of process design on greenhouse gas (GHG) generation by wastewater treatment plants. *Water Res,* 43:2679–87.

[60] Sweetapple C., Fu G. and Butler D. 2014. Multi-objective optimization of wastewater treatment plant control to reduce greenhouse gas emissions. *Water Res,* 55, pp. 52–62.

[61] Huisman, J., J. Sharples, J. M. Stroom, P. M. Visser, W. E. A. Kardinaal, J. M. H. Verspagen & B. Sommeijer, 2004. Changes in turbulent mixing shift competition for light between phytoplankton species. *Ecology* 85: 2960–2970.

[62] Behrenfeld, M. J., R. T. O'Malley, D. A. Siegel, C. R. McClain, J. L. Sarmiento, G. C. Feldman, A. J. Milligan, P. Falkowski, R. M. Letelier & E. S. Boss, 2006. Climate-driven trends in contemporary ocean productivity. *Nature* 444: 752–755.

[63] Platt, T., C. Fuentes-Yaco & K. T. Frank, 2003. Spring algal bloom and larval fish survival. *Nature* 423: 398–399.

[64] Field, C. B., Behrenfeld, M. J., Randerson, J. T. and Falkowski, P., 1998. Primary production of the biosphere: integrating terrestrial and oceanic components. *Science,* 281(5374), pp. 237-240.

[65] Vajravelu, K., Prasad, K. V., Ng, C. O. and Vaidya, H., 2017. MHD squeeze flow and heat transfer of a nanofluid between parallel disks with variable fluid properties and transpiration. *International Journal of Mechanical and Materials Engineering,* 12(1), pp. 1-14.

[66] Kumar, C. S. and Perumal, P., 2012. Studies on phytoplankton characteristics in Ayyampattinam coast, India. *Journal of Environmental Biology*, 33(3), p. 585.

[67] Kulasooriya, S. A., 2011. Cyanobacteria: pioneers of planet earth. *Ceylon Journal of Science (Bio. Sci.),* 40(2), pp. 71-88.

[68] Francis, T. B., Scheuerell, M. D., Brodeur, R. D., Levin, P. S., Ruzicka, J. J., Tolimieri, N. and Peterson, W. T., 2012. Climate shifts the interaction web of a marine plankton community. *Global Change Biology*, 18(8), pp. 2498-2508.

[69] Kulshreshtha, A., Jarouliya, U., Bhadauriya, P., Prasad, G. B. K. S. and Bisen, P. S., 2008. Spirulina in health care management. *Current pharmaceutical biotechnology*, 9(5), pp. 400-405.

[70] Skulberg, O. M., 2000. Microalgae as a source of bioactive molecules–experience from cyanophyte research. *Journal of Applied Phycology*, 12(3), pp. 341-348.

[71] Singh, U. B., Ahluwalia, A. S., Sharma, C., Jindal, R. and Thakur, R. K., 2013. Planktonic indicators: A promising tool for monitoring water quality (early-warning signals). *Ecology, Environment and Conservation*, 19(3), pp. 793-800.

[72] Nagaoka, S., Shimizu, K., Kaneko, H., Shibayama, F., Morikawa, K., Kanamaru, Y., Otsuka, A., Hirahashi, T. and Kato, T., 2005. A novel protein C-phycocyanin plays a crucial role in the hypocholesterolemic action of *Spirulina platensis* concentrate in rats. *The Journal of nutrition*, 135(10), pp. 2425-2430.

[73] Sekar, S. and Chandramohan, M., 2008. Phycobiliproteins as a commodity: trends in applied research, patents and commercialization. *Journal of Applied Phycology*, 20(2), pp. 113-136.

[74] Glazer, A. N., 1994. Phycobiliproteins—a family of valuable, widely used fluorophores. *Journal of Applied Phycology,* 6(2), pp. 105-112.

[75] Aráoz, R., Lebert, M. and Häder, D. P., 1998. Electrophoretic applications of phycobiliproteins. *Electrophoresis,* 19(2), pp. 215-219.

[76] Becher, P. G., Beuchat, J., Gademann, K. and Jüttner, F., 2005. Nostocarboline: isolation and synthesis of a new cholinesterase inhibitor from Nostoc 78-12A. *Journal of natural products*, 68(12), pp. 1793-1795.

[77] Lincoln, E. P., Wilkie, A. C. and French, B. T., 1996. Cyanobacterial process for renovating dairy wastewater. *Biomass and Bioenergy*, 10(1), pp. 63-68.

[78] Dubey, S. K., Dubey, J., Mehra, S., Tiwari, P. and Bishwas, A. J., 2011. Potential use of cyanobacterial species in bioremediation of industrial effluents. *African journal of Biotechnology*, 10(7), pp. 1125-1132.

[79] Nozzi, N. E., Oliver, J. W. and Atsumi, S., 2013. Cyanobacteria as a platform for biofuel production. *Frontiers in bioengineering and biotechnology*, 1, p. 7.

[80] Sala, O. E., Stuart Chapin, F. I. I. I., Armesto, J. J., Berlow, E., Bloomfield, J., Dirzo, R., Huber-Sanwald, E., Huenneke, L. F., Jackson, R. B., Kinzig, A. and Leemans, R., 2000. Global biodiversity scenarios for the year 2100. *Science,* 287(5459), pp. 1770-1774.

[81] DeFries, R. S., Foley, J. A. and Asner, G. P., 2004. Land-use choices: Balancing human needs and ecosystem function. *Frontiers in Ecology and the Environment,* 2(5), pp. 249-257.

[82] Howarth, R., Shindell, D., Santoro, R., Ingraffea, A., Phillips, N. and Townsend-Small, A. 2012. Methane emissions from natural gas systems. *National Climate Assessment,* pp. 2011-0003.

[83] Falkowski, P., Scholes, R. J., Boyle, E. E. A., Canadell, J., Canfield, D., Elser, J., Gruber, N., Hibbard, K., Högberg, P., Linder, S. and Mackenzie, F. T. 2000. The global carbon cycle: a test of our knowledge of earth as a system. *Science,* 290(5490), pp. 291-296.

[84] MacDonald, A. M., Bonsor, H. C., Dochartaigh, B. É. Ó. and Taylor, R. G. 2012. Quantitative maps of groundwater resources in Africa. *Environmental Research Letters,* 7(2), p.024009.

[85] Poff, N. L. 2002. Ecological response to and management of increased flooding caused by climate change. Philosophical Transactions of the Royal Society of London. *Series A: Mathematical, Physical and Engineering Sciences*, 360(1796), pp. 1497-1510.

[86] D'Croz, L., Rosario, J. B. D. and Gondola, P. 2005. The effect of fresh water runoff on the distribution of dissolved inorganic nutrients and plankton in the Bocas del Toro Archipelago, Caribbean Panama. *Caribbean Journal of Science*.

[87] Winder, M. and Schindler, D. E. 2004b. Climatic effects on the phenology of lake processes. *Global Change Biology*, 10(11), pp. 1844-1856.

[88] Kromkamp, J. C. and Van Engeland, T. 2010. Changes in phytoplankton biomass in the Western Scheldt Estuary during the period 1978–2006. *Estuaries and Coasts,* 33(2), pp. 270-285.

[89] Zvereva, E. L. and Kozlov, M. V. 2006. Consequences of simultaneous elevation of carbon dioxide and temperature for plant–herbivore interactions: a meta analysis. *Global Change Biology,* 12(1), pp. 27-41.

[90] Dell, A. I., Pawar, S. and Savage, V. M. 2014. Temperature dependence of trophic interactions are driven by asymmetry of species responses and foraging strategy. *Journal of Animal Ecology,* 83(1), pp.70-84.

[91] Ainsworth, E. A. and Long, S. P. 2005. What have we learned from 15 years of free-air CO_2 enrichment (FACE)? A meta-analytic review of the responses of photosynthesis, canopy properties and plant production to rising CO_2. *New phytologist,* 165(2), pp. 351-372.

[92] Cleland, E. E., Chuine, I., Menzel, A., Mooney, H. A. and Schwartz, M. D. 2007. Shifting plant phenology in response to global change. *Trends in ecology & evolution*, 22(7), pp. 357-365.

[93] Culler, L. E., Ayres, M. P. and Virginia, R. A. 2015. In a warmer Arctic, mosquitoes avoid increased mortality from predators by growing faster. *Proceedings of the Royal Society B: Biological Sciences*, 282(1815), p. 20151549.

[94] Bodlah, I., Rasheed, M. T., Fareen, A. G., Ajmal, M. S. and Bodlah, M. A. 2016. First record of two new species of genus Tetraponera (Hymenoptera: Pseudomyrmecinae: Formicidae) from Pakistan. *Journal of Entomology and Zoology Studies*, 4(4), pp. 1028-1030.

[95] Gunderson, A. R. and Leal, M. 2016. A conceptual framework for understanding thermal constraints on ectotherm activity with implications for predicting responses to global change. *Ecology Letters,* 19(2), pp. 111-120.

[96] Rieuwerts, J., 2017. *The elements of environmental pollution*. Routledge.

[97] Canovas, S., Picot, B., Casellas, C., Zulkifi, H., Dubois, A. and Bontoux, J., 1996. Seasonal development of phytoplankton and zooplankton in a high-rate algal pond. *Water Science and Technology,* 33(7), pp. 199-206.

[98] Panswad, T., Doungchai, A. and Anotai, J., 2003. Temperature effect on microbial community of enhanced biological phosphorus removal system. *Water Research,* 37(2), pp. 409-415.

[99] D'Alpaos, A. 2011. The mutual influence of biotic and abiotic components on the long-term ecomorphodynamic evolution of salt-marsh ecosystems. *Geomorphology*, 126(3-4), pp. 269-278.

[100] Schabhüttl, S., Hingsamer, P., Weigelhofer, G., Hein, T., Weigert, A. and Striebel, M. 2013. Temperature and species richness effects in phytoplankton communities. *Oecologia*, 171(2), pp. 527-536.

Chapter 11

Nitrite, Nitrate, Ammonia and Phosphate Pollution and Their Remediation

F. S. Idroos[1]
P. A. K. C. Wijerathna[1,2]
H. A. S. N. Abeysiri[1,2]
and P. M. Manage[1,2,*]
[1]Centre for Water Quality and Algae Research, Department of Zoology, University of Sri Jayewardenepura, Sri Lanka
[2]Faculty of Graduate Studies,University of Sri Jayewardenepura, Gangodawila, Nugegoda, Sri Lanka

Abstract

Pollutants in aqueous solutions have negative environmental, economic, and societal consequences. In rural and urban regions, pollution caused by nitrite, nitrate, ammonia, and phosphorus is a significant concern. Nitrate is one of the most prevalent contaminants found in groundwater. Increased use of artificial fertilizers, waste disposal, and changes in land use are the primary drivers of the gradual increase in nitrate levels in groundwater supplies. Methemoglobinemia is caused by the presence of nitrate in drinking water. Furthermore, nitrate and nitrite levels in water and food have increased cancer risk. In commercial fertilizers, Ammonia is a critical component. In an aquatic environment, Ammonia and ammonia-related chemicals are abundantly available. Ammonia's biological effects in humans are dose-dependent and influenced by ambient concentrations. Phosphate ion discharge also leads to eutrophication in water, with significant environmental consequences.

* Corresponding Author's Email: pathmalal@sjp.ac.lk.

In: Microbial Technologies and Their Applications
Editors: Santosh Kumar Karn and Anne Bhambri
ISBN: 979-8-88697-913-8

Algal blooms are triggered when there is too much phosphorus in the water. Algal blooms eventually release poisons that can kill wildlife. Algae, fungus, and bacteria are among the microorganisms that may convert nitrate ions into organic material. The detoxification process that releases Ammonia involves several microbial enzymes, including nitrate and nitrite reductases. Microorganisms including algae, bacteria, and fungi have been employed successfully in bioremediation. Bacteria and fungi use a several ways to extract nitrogen and phosphorus. These processes include adsorption, biosorption, bioaccumulation, mineralization, ammonia adsorption, nitrogen fixation, ammonia volatilization, nitrification, and immobilization. These bacteria transform inorganic pollutants into organic matter through an assimilatory reduction mechanism. As a result, microbe-mediated biological treatment could be recommended as a viable option for treating Nitrite, Nitrate, Ammonia, and Phosphate-contaminated waters with minimum environmental impact.

Keywords: nitrate, nitrite, ammonia, phosphate, pollution, bioremediation

Introduction

Water is a limited precious commodity. Almost 70% of the Earth's crust is submerged in water making it a valuable natural resource. Freshwater accounts for only 2.66% of total world water resources, including groundwater, lakes, rivers, polar ice, and glaciers. Only 0.6% of freshwater is suitable for human consumption [1]. Hence, careful management of water resources, and effective wastewater treatment is crucial.

Unregulated ways of disposing household, agricultural and industrial waste into freshwater bodies have resulted in high levels of Nitrate, nitrite, and phosphates in surface waters [2]. Nutrient enrichment of lowland water bodies is of worldwide concern [3]. Consumption of Nitrogen fertilizer has expanded dramatically over the world in recent years, resulting in high nitrate and nitrite levels in drinking water and rivers [4]. Nitrogen enters the water environment in agricultural areas in three ways: drainage, leaching, and flow [5].

Presence of excess nitrates and phosphates in water is a critical environmental concern as they act as main nutrients for microorganisms [6]. Excessive release of Nitrate and phosphate compounds to the water bodies results in uncontrolled growth of certain microorganisms especially Cyanobacteria [3-7].

Agriculture methods like fertilizer and cultivation, solid waste management, and leachate percolation are linked to leaching of nitrates and nitrites into soil. In some arid and semiarid environments, irrigation increases nitrate accumulation in the soil and water evaporation [8].

Accumulation of nitrates in land surfaces results its penetration into deep layers depends on the environment [9]. Nitrifies bacteria available in the soil convert the nitrogenous compound in chemical fertilizers to nitrate. Negatively charged Nitrates could easily reach groundwater [10]. However, the majority of nitrogenous fertilizers are not absorbed. Hence this will be causing problems with both underground and surface water bodies. Plants use 50% of nitrogen fertilizers added to the soil and about 2-20% loss by evaporation from the soil. About 15-25% combine with organic compounds in the soil, and residual 2-10% interact with surface and groundwater [10-11].

The pollution of ground water by means of industrialization and urbanization has progressively developed many harmful environmental consequences. The nitrate problem in groundwater should be examined in a worldwide context [12]. In Europe 22% of cultivated areas, the nitrate concentration of ground water concentration ($\geq$11.3 mg/L) above which greater than the environmental guidelines. Municipal and industrial wastes may contribute to large quantities of nitrate in groundwater in highly urbanized areas [13-14]. One of the critical sources of elevated nitrate in groundwater has been the excessive use of nitrogenous fertilizers in agriculture. Apart from nitrate, nitrogen is administered in the forms of ammonium (NH_4^+) and amide (NH_2^-). Both ammonium and amide produce nitrate in the soil system through mineralization, which occurs quickly in tropical and subtropical soils [15]. These produced nitrates are highly soluble in water. Hence nitrate is easily leached into the subsurface layers and eventually into the groundwater [16].

The pollution of water bodies by phosphate released via many sources is a serious environmental issue. Phosphorus is a plant nutrient in which excessive amounts can contribute to significant phytoplankton development. Septic systems, leaky infrastructure, landfills, and industrial activities are examples of agricultural and non-agricultural sources contributing to phosphate accumulation in water bodies [17]. The conveyance of phosphorous fertilizer with surface water flow may cause phosphate levels to rise in drinking water and rivers.

Furthermore, the water pollution caused by Nitrite, Nitrate, Ammonia and Phosphate is a critical concern to most of the coastal regions in the world. Coastal zones are the world's most critical aquatic environments, with more significant ecological and economic importance [17]. However, human activities like the unauthorized immediate discharge of untreated effluent water from different industries and releasing domestic and industrial solid wastes may be a critical threat to marine ecosystems [18]. Except the direct effects, the synergistic effects of those contaminants on the marine ecosystem and their potential risk to living beings may be very high [19]. Thus, proper regulations and a sustainable management of these contaminants are essential for the world.

Chemical Nature of Inorganic Pollutants: Nitrites, Nitrates, Ammonia and Phosphates

Ammonium (NH_4^+) and nitrate are the most common forms of inorganic nitrogen (N) (NO_3^-). Nitrite (NO_2^-) may also be present, although the amount is usually too tiny to detect until neutral or alkaline soils are fertilized with NH_4^+or NH_4^+-forming fertilizers [20]. Several other inorganic N compounds have been hypothesized as intermediates during microbial N transformations in soils, including hydroxylamine (NH_2OH), hyponitrous acid ($H_2N_2O_2$), and nitramide (NH_2NO_2). However, these molecules are thermodynamically unstable and have not been identified in soil [20].

Phosphorus is the second most commonly used chemical element in agricultural production. Phosphorus is a highly reactive element with no pure form [21]. Phosphorus is available primarily as orthophosphate in both aquatic and terrestrial ecosystems. Orthophosphate is found in water bodies predominantly as $H_2PO_4^-$ in acidic conditions and HPO_4^{2-} in alkaline conditions.

In surface sediments, phosphorus was primarily composed as inorganic phosphorous forms such as bound phosphorus to iron and calcium aluminium (Fe-P, Ca-P, Al-P) and dissolved forms in water [22]. Inorganic phosphorous are the most abundant phosphorus species in sediments, accounting for approximately 85% of total phosphorus. Phosphorus bonded to iron is the most prevalent form of inorganic phosphorus, accounting for more than 40% of total phosphorus. Aluminum and calcium coupled phosphorus represented less than 20% of total phosphorus [22].

Point and Non-Point Sources of Inorganic Pollutants

Sewage effluent, weathering, natural soil nitrate, atmospheric N deposition, septic systems, soil organic matter, livestock waste, nitrogenous fertilizers, and wastewater treatment plants are sources of organic nitrogen to water. Mineralization, hydrolysis, and bacterial nitrification convert these inorganic nitrogen sources to nitrate. Hence, nitrate is the most abundant form of nitrogen in water sources. Nitrite is quickly oxidized to nitrate because nitrite is chemically unstable, but in a reducing environment, it can concentrate to a limited extent in ground and surface water [21].

Nitrate can reach the water sources mainly through agricultural processes like excessive use of chemical fertilizers, direct disposal of wastewater to the water bodies, and sewage water systems with nitrogenous excretory materials. For ecological nitrogen balance, the flow of dissolved nitrogen from top layers to groundwater sources has been troublesome. Even though soluble organic forms of nitrogen have been detected in rain and drainage waters for many years, they have yet to be identified as the dominant form of nitrogen in agricultural soils, most likely due to analytical issues. Wooded watersheds have been identified as a critical nitrogen supplier to surface water.

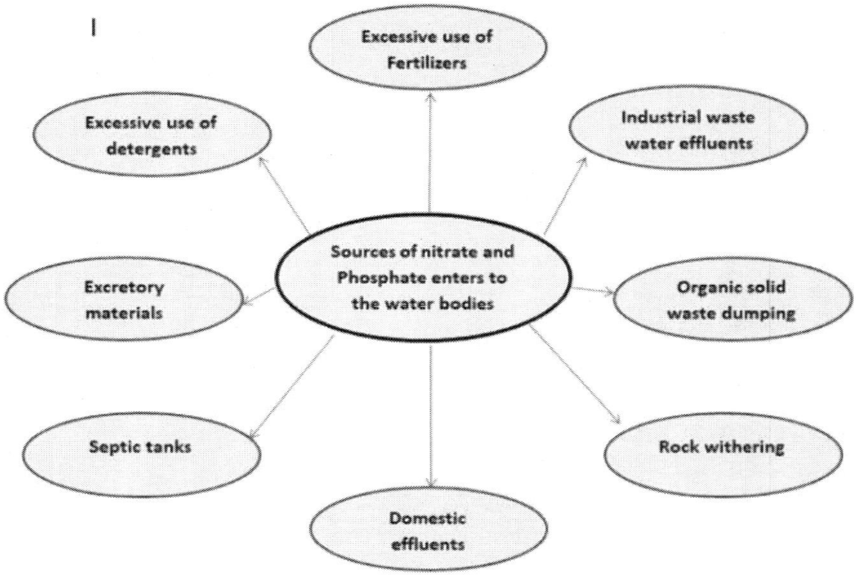

Figure 1. Different point and non-point sources of inorganic pollutants.

Phosphorus is an important element for animals and plants; without it, there would be no life. Phosphorus can reach to eco systems by several ways including both anthropogenic and natural processes. Mineral and rock weathering, effluent water and sewage stocks (point sources), drainage from farmlands, soil degradation, and leaching from grasslands (diffuse source) are some of the major sources of phosphorous which enters to the water bodies. Moreover, excessive fertilizing in agriculture lead to phosphorus imbalances in the soil, most of which are transferred to aquatic ecosystems [23].

Another important source of P in aquatic habitats is detergents. Although some governments have banned phosphate-based detergents, they are still commonly used in others. The phosphorus deposit in the soil causes an increase in the soil solution. Phosphates can also be released when phosphate-containing organic pesticides decompose. As a result, they may emerge as particles in the bodies of aquatic organisms. Water sources may include organic P and phosphate coupled with microscopic sediment particles.

Bioavailability of Inorganic Pollutant and Their Role in Aquatic Ecosystems

Nitrate is one of the most stable active anions in water owing to its characters such as proclivity for disintegration and lack of interaction with the water matrix. Nitrogen is the most abundant chemical element in the earth's atmosphere (almost 80%), and it is also found in many critical biomolecules such as amino acids and nucleotides [24]. After carbon, oxygen, and hydrogen, nitrate is the fourth most common chemical element in living tissues. Ammonium (NH_4^+), nitrite (NO_2^-), and nitrate (NO_3^-) are the most common ionic forms of dissolved inorganic nitrogen in aquatic ecosystems (NO_3^-). Aquatic animals can readily absorb these inorganic nitrogenous compounds from the surrounding water. Unionized ammonia is the most dangerous, whereas ammonium and nitrate ions are the least harmful [25]. Unionized ammonia is the most toxic type, followed by ammonium and nitrate ions. Nitrogen-fixing macrophytes, algae, and bacteria can remove ammonium (NH_4^+), nitrite (NO_2^-), and nitrate (NO_3^-) from water [26].

In some cases, inorganic nitrogen contamination can cause toxic algae to grow. Inorganic nitrogen pollution can impact aquatic species' ability to survive, prosper, and reproduce due to the direct toxicity of inorganic nitrogenous chemicals. Inorganic nitrogen pollution of ground and surface rivers can have severe health and economic repercussions [26].

Phosphorus is a macronutrient required by all living things, and it is returned to the soil and converted into inorganic phosphate by biological and chemical processes in natural ecosystems.

When inorganic phosphorus was unavailable, most organic forms of dissolved phosphorus may be employed as a supplemental P source [27]. The organic P in these humic–Al–Fe complexes had a very poor bioavailability to algae, indicating that the humic compounds' P fractions will not aid algal development. This is because nitrates and phosphates are essential to plant nutrients that promote growth [28].

For plant development and cattle production, phosp play music you can do it with myhorus is an essential macronutrient. Phosphate absorption and bioavailability in plants used for animal feed and human consumption are becoming more of a focus for researchers [29]. Furthermore, because phosphorus's harmful effects on the aquatic environment, an immediate need for modification of feeding regime (eutrophication). In these settings, the enzyme phytase, which can liberate bioavailable phosphate from one of the critical components in the phosphate cycle, is crucial [30].

Nitrite, Nitrate, Ammonia and Phosphate as Limiting Factor for Cyanobacteria and Algae Growth

Eutrophication of aquatic ecosystems is caused by excessive nutrient enrichment, primarily phosphate and nitrogen. The primary way of nutrient enrichment in marine settings is the direct disposal of waste water without proper treatment [31]. Nutrient loading in coastal waterways can be exacerbated by losing coastal and riverine wetlands, which retain and maintain nutrients. Nitrogen is commonly associated with coastal system impairment whereas Phosphorus is frequently the primary cause of impairment in freshwater systems [32].

Eutrophication occurs in freshwater habitats at Nitrate and Phosphate concentrations of 1–2 mgL-1 and 0.1 mg L-1, respectively, or below the concentration of P [33]. Increased nutrient loads encourage the growth of phytoplankton, algae, gelatinous creatures, and other aquatic plants, resulting in harmful algal blooms and dead zones in coastal and freshwater bodies.

Eutrophication's effects on coastal and inland waters [34].

Increasing phytoplankton biomass:

- Rises in gelatinous zooplankton blooms of coastal habitats.
- Enhance the abundance of epiphytic and benthic microorganisms.
- Growth of rooted macroalgae and macrophytes.
- Changes of water quality parameters.
- Reduction in seeming aesthetic value of the water body.

Potential Environmental and Human Health Risk of Nitrites, Nitrates, Ammonia and Phosphates

Nitrate is a chemical that is commonly considered to be non-toxic. The transformation of nitrate into nitrite, nitric oxide, and N-nitroso substances has a negative effect on public health.

NO_3^- is converted to NO_2^- inside the human body, causing methaemoglobinemia by compromising hemoglobin's ability to absorb oxygen [35].

Nitrite ions rejoin with haemoglobin in red blood cells to form methaemoglobin (MetHb), which binds oxygen and blocks its transport by altering the nature of iron in haemoglobin from Fe^{2+} to Fe^{3+}. MetHb levels in human blood are normal at 1–3 percent, and MetHb concentrations of 10% or higher are associated with reduced oxygen transport [36]. A methaemoglobin concentration of more than 10% in infants can cause cyanosis, also known as "blue-baby syndrome." High nitrate ingesting has been connected to acute respiratory disorders and different defects at the birth of infants, cancer (ovarian and liver), reproductive problems and disorders of central nervous system [35-40].

Phosphate is not a hazardous form to living beings if they are not available in extremely high concentrations. An exceptionally high phosphorus level could cause gastrointestinal issues. The presence of phosphorus in freshwater can have a secondary effect on public health due to its participation in eutrophication. Some cyanobacteria species develop secondary metabolites known as cyanotoxins under suitable nutritional and light conditions. The most common are *Anabaena, Cylindrospermopsis, Aphanizomenon, Microcystis*, genera that produce toxins [40-43].

These toxic materials can have health consequences on human such as dermatotoxicity, hepatotoxicity [44-47], neurotoxicity and nephrotoxicity [44-

49] as well as general inhibition of protein synthesis. Number of research studies have been conducted to investigate the impact of Cyanotoxins to systems, such as reproductive, excretory, gastrointestinal and cardiac systems [50].

Fatigue, weakening of muscle functions, headaches, Vision problems, fever, problems of gastrointestinal tract, vomiting and some allergic conditions of cyanotoxity [51]. Seizures, respiratory arrest, liver failure, and death may result in extreme cases [51]. HABs can also have an indirect impact on humans by disrupting freshwater and marine ecosystems, as well as food sources resulting from them.

Nitrite, Nitrate, Ammonia and Phosphate Removal Methods

Nutrients such as Phosphate and Nitrates contaminate aquatic environment by means of activities such as tanning, pharmaceuticals, fertilizer, and mining. The immediate release of wastewater has severe consequences for the ecosystem, including eutrophication [52]. Phosphates and nitrates are the primary nutrients responsible for eutrophication. Consequently, the effective treatment of industrial wastewater effluents is the most crucial obligation for environmental sustainability [53]. Chemical, physical, and biological treatments are the three main types of nitrate and phosphate treatment techniques.

Chemical Methods

The application of various chemicals can be used to remove dissolved nitrate and phosphate ions. Iron, aluminum, lime, calcium, and magnesium could be precipitate phosphorus from wastewater [54]. The reagent ions and the soluble salts in wastewater can react with precipitation to produce highly distributed phosphate clot deposits [55].

Aluminum hydroxide is an excellent adsorbent for the precipitation of orthophosphate and condensed phosphate [56]. A higher concentration of Phosphate could be precipitated using minute amount of magnesium salts in wastewater prior to introducing to the anaerobic sludge digester, which causes a significant reduction in suspended particles and chemical oxygen demand.

Precipitation Calcium–phosphorus is another effective low-cost method for the removal of phosphorus [57]. Using calcite as a seeding element to deposit phosphate via immediate precipitation of calcium phosphate ($Ca_5(PO_4)_3OH$) is a straight forward process [57].

Physical Methods

Adsorption systems are the most crucial approach for wastewater treatment owing to their convenience, cost-effectiveness, and simplicity. Various types of adsorbing materials are used in the adsorption process to remove numerous anions like fluoride, nitrate by effluent water [58].

For nitrate removal, powdered activated carbon and carbon nanotubes are excellent absorbents. Carbon nanotubes remove nitrates more effectively than powdered activated carbon, following chemical activation with $ZnCl_2$, conventional granular activated carbon (GAC) removed nitrates [59]. In an electrical furnace, bamboo powder charcoal (BPC) was heated to 900 degrees Celsius [60]. The clays are aluminosilicates that are hydrous. Fine-grained clay minerals, quartz, carbonate, and metal oxides make up these clay-based adsorbents. They're utilized to eliminate nitrates in wastewater [61]. Employing agricultural waste as an effective adsorbent material to absorb nitrate is a great way to help the environment by reducing trash. It is a terrific strategy that can be used again and again.

Nanotechnology

Nanotechnology is one of the most influential and extensively used ways to eliminate nitrate and phosphate from wastewater. This technique employs a variety of nanomaterials, including nanoscale zero-valent iron. Nanoparticles such as cobalt, nickel, carbon/titanium oxide nanoparticles, carbon nanotubes, and different nanofibers might be used in this treatment process. Graphene nanocomposites are also utilized to remove phosphate and nitrate [61, 62].

According to various research, palladium and copper as graphene catalysts and iron as reductants result in nitrate and nitrogen removal efficacy of 82% and 66%, respectively [75]. Magnetic graphene nanoparticles remove 86.4% of nitrate at the same time under optimal pH3, temperature 50°C, contact period 60 minutes, adsorbent dosage 2 g/l, and starting concentration 50 mg/l [63].

Biological Approaches

Denitrification, which is carried out by autotrophic and heterotrophic bacteria, is the most common biological method for eliminating nitrate from wastewater [64]. Due to the carbon requirements for the growth and development of denitrification bacteria, mixotrophic denitrification is appropriate for most wastewater treatment systems [58]. Heterotrophic bacteria denitrification in anaerobic or anoxic environments, such as *Pseudomonas*, *Micrococcus*, *Achromobacter*, and *Bacillus* [52].

Denitrifying bacteria use nitrate as a terminal electron acceptor from the respiratory process in biological denitrification [58]. The microorganisms that cause denitrification can be found all around the world. Denitrifying bacteria transform dangerous inorganic nitrogen molecules into innocuous nitrogen gas.

The biological method removes the tremendous amount of phosphorus from wastewater, allowing activated sludge to realize its full potential. Only 20% to 40% of the phosphorous may be eliminated [55]. Increasing the amount of sludge required to remove 50% of the phosphorus is necessary. There are several ways to get rid of phosphorus. Phostrip, anaerobic–oxic (A/O), and extended anaerobic sludge contact are the three procedures [65]. The amount of sludge required to remove 50% of the phosphorus must be increased. Phosphorus can be removed using a variety of methods [65].

Further research has discovered that *Cynobacterium* and *Phormidium bohneri* in residential water can eliminate both N and P by growing for 50 and 75 hours, respectively [66]. Different micro and macro algal species have the greater potential to reduce nitrogen and phosphate from effluent water. *Chlorella/ vulgaris* and *Scenedesmus/dimorphus* are two microalgae that may remove up to 55% phosphate from pig-forming effluents. *Spirulina platensis*may eliminate NO_3^-, PO_4^{3-}, NH_3 in artificial effluent water [66].

Bioremediation as an Affective Technique to Advocate Inorganic Pollution

In developed countries, inorganic pollution of water bodies is seen as a significant environmental issue [67]. As a result, various physicochemical processes for treating these pollutants have been devised, including ozonation, adsorption, membrane processing, flocculation, solvent extraction, and so on

[68]. However, due to their high costs, non-specific character, and possibility for extra pollution, the continued relevance of these technologies is questioned. As a result, there has been a surge in interest in bioremediation, or environmentally friendly bio-based remedies [67].

Bioremediation is usually viewed as a low-cost, ecologically friendly solution gaining traction in environmental treatment. Biological agents are used in bioremediation to improve hazardous pollutant detoxification and breakdown, either by intracellular accumulation or enzymatic transformation into less or non-toxic compounds [69]. The type of biological agent utilized for waste treatment significantly impacts bioremediation procedures. Bacteria (*Bacillus* sp., *Pseudomonas* sp., etc.), fungi (*Aspergilus niger*, *Rhizopus arrhizus*, etc.), macro and microalgae species (*Chlorella vulgaris*, *Anabaena variabilis*, etc.), cyanobacteria, and other aquatic plants are all considered to be important bioremediation factors [70]. Bioattenuation, biostimulation, and bioaugmentation are all examples of bioremediation [81]. Members of the genera *Pseudomonas*, *Rhizobium*, *Cupriavidus*, *Shewanella*, *Ensifer*, and *Thermomonas* converted nitrate to nitrite and dinitrogen. Furthermore, *Bacillus* sp. and *Proteus* sp. isolated from Sri Lankan agricultural fields reduced nitrate levels, according to [71].

These biological agents break down and detoxify environmental contaminants through intracellular accumulation processes or enzymatic transformation pathways. The rate of pollutant conversion in bioremediation is determined by the substrate's chemical structure, hydrophobicity, and polarity [72]. Furthermore, various environmental factors such as temperature, pH, redox state, and specific soil features have an impact on biological deterioration [73]. However, the activity of these biological agents for biodegradation may be limited by the biotic and abiotic stress conditions present in the waste-water such as xenobiotic and heavy metals [74]. Therefore, instead of the cells, the application of broad-spectrum enzymes has become the most efficient bioremediation tool.

Bioremediation Methods of Inorganic Chemicals

Microbial Removal (Bacteria, Cyanobacteria, Algae)

Microorganisms possess the capacity to break down and detoxify contaminants in the environment due to their natural transformation mechanisms [76]. Owing to their efficient enzymatic pathways, microbes are capable of eliminating or modifying various contaminants, Microbes degrade many contaminants in the natural environment. However, genetically engineered microorganisms and microalgae are be used to improve their degradation effectiveness at a quicker pace in a shorter time period [77].

Several research has been conducted on microbial bioremediation of inorganic contaminants [78-80, 82]. Investigated the bio-transformation capacity of several bacterial species and contaminants, with the *Bacillus licheniformis* releasing the most significant percentage of copper-93% from treated wood [83]. Furthermore, [84] investigated wastewater bioremediation's potential with two combined *Bacillus* species. The results demonstrated that the removal efficiencies of biological oxygen demand (BOD) and chemical oxygen demand (COD) were also significantly higher, at 66.66% and 66.67%, respectively, with better metal removal efficiency.

Bioremediation of inorganic pollutants using genetically engineered bacteria is a new method seen as an environmentally acceptable and effective way to remove harmful metal contaminations [85, 86]. Genetically modified bacteria have a better degradative capability and have been successfully used to degrade various contaminants under controlled settings [86]. *Ralstonia eutropha*, *Pseudomonas putida*, *Mycobacterium marinum*, *Escherichia coli*, *Sphingomonas desiccabilis*, *Bacillus idriensis*, and other bacteria have genes introduced into their genomes that enable them to bioremediate harmful metal compounds in contaminated environments [87].

Phycoremediation is the treatment of wastewater with macro and microalgae to remove organic and inorganic contaminants [88]. Microalgae are a vast collection of autotrophic microorganisms that include about 50,000 species that live in various settings, including freshwater, brackish water, ocean, and wastewater [89]. Proteins, carbohydrates, lipids, fatty acids, pigments, vitamins, and enzymes are found in their cells and may be helpful to humans. They can produce far more biomass (per unit land area) than higher plants, making them an essential source of biofuel [90].

In wastewater systems, microalgae use nitrogen (as nitrates, nitrites, and ammonium) and phosphates as essential elements for algal development [88]. Microalgae are able to use inorganic nitrogen and phosphate for growth. Further they are able to remove heavy metals and some harmful organic chemicals. Hence, microalgal cultures are an attractive alternative to tertiary treatments [91]. Microalgae adsorb and break down persistent chemicals such as hydrocarbons, heavy metals (HMs), antibiotics, and chlorinated compounds and the common organic and inorganic pollutants such as nitrates, phosphates, and ammonium. Many *Chlorella* and *Scenedesmus* microalgal species have been found to collect polysaccharides extracellularly, encapsulating their cells [92].

Only a few research have focused on nutrient removal utilizing *Chlorella pyrenoidosa* and *Scenedesmus*. According to the study, *Chlorella pyrenoidosa* can remove% of ammonium nitrogen, 96% of phosphate, and 80% of nitrates from raw samples, demonstrating the micro algae's ability to cure environmental toxins [93, 94]. Investigated the mechanism and effectiveness of nitrogen and phosphorus removal in wastewater using several laboratories-grown algae cultures. In the Pilot Project, total nitrogen, NH_4-N, and Phosphate-P removal efficiency ranged from 75.3% to 87.9%, 100%, and 50.6%, respectively.

Phytoremediation

Phytoremediation is a low-cost, green solution that is also environmentally benign. Traditional physicochemical cleansing methods, which are costly, have advantages over phytoremediation. Traditional approaches modify soil microorganisms and change soil properties [95, 96]. Phytoremediation is a type of bioremediation that uses green plants and microorganisms to remove toxic metals from damaged environments [97]. It employs a range of strategies to remove metal pollutants from contaminated areas, including phytoextraction, phytostabilization, phytodegradation, phytostimulation, phytovolatilization, and rhizofiltration [98].

The majority of the research focuses on heavy metal phytoremediation from the contaminated matrix [99]. Plant roots that have mingled with the water provide a habitat for microorganisms and a filter system for toxins [100].

Eichhornia crassipes is a floating herb with a long stem and air bladders often employed in phytoremediation. This plant thrives in a variety of nutritional and temperature environments. Because of their metabolic activities, which reduce the phytotoxicity of organic pollutants, *E. crassipes*' phytoremediation attribute in combination with endophytic bacteria has evolved in recent years [101]. *Pseudomonas*, *Bacillus*, and *Enterobacter* strains that dwell in association or freely in wastewater bodies can break down the principal chemicals such as hydrocarbons, alkanes, and many pesticides (polychlorinated biphenyls) [102].

Adsorption of Microorganisms

Biosorption is a physicochemical process. The mechanisms involved in this physiologically irrelevant activity include absorption, adsorption, ion exchange, surface complexation, and precipitation. Biosorption biotechnology is widely used for pollutant removal and recovery from the solution because of its simplicity, the comparable operation to conventional ion exchange technology, apparent efficiency, and availability of biomass and waste bio-products. Biosorption is extracting or recovering organic and inorganic compounds from a solution using biological materials such as living or dead microbes and their components, seaweeds, plant materials, industrial and agricultural wastes, and natural residues. The majority of biosorption research has been done on microbial systems, primarily using bacteria, microalgae, fungi, and hazardous metals and radionuclides such as uranium and thorium. According to many researchers, Biosorption is a subtype of adsorption in which the sorbent is a biological matrix [103].

Microbial Fuel Cell Technology

Microbial fuel cells (MFCs) are bio-electrochemical systems that use the power of respiring bacteria to transform organic substrates directly into electricity. MFC could be a promising alternative to conventional activated sludge process-based treatment systems in energy consumption and sludge removal [104]. MFCs have significant advantages over typical treatment systems regarding energy, environmental, economic, and operational benefits.

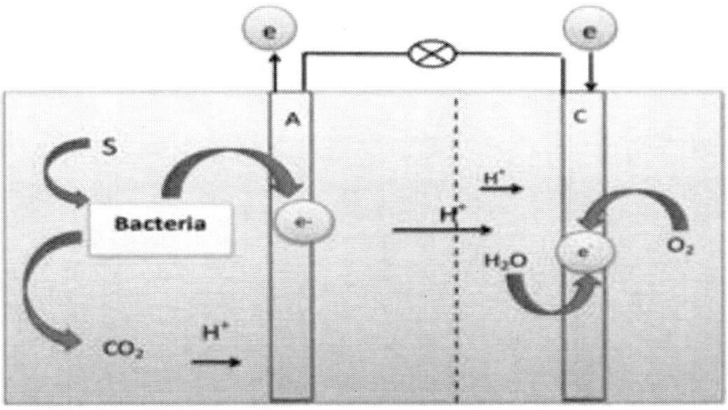

Figure 2. Microbial fuel cell technology.

MFCs are an environmentally favorable technology because they can create electricity directly from organic matter in wastewater without separating, purifying, or converting the energy products. They can function at low temperatures [105]. According to the manufacturer, MFCs can generate up to 1.43 kWh/m^3 from primary sludge and 1.8 kWh/m^3 from treated effluent [106].

Bioreactor Technology

Phosphate and nitrate nutrients are overlayed in streams by wastewater from diverse businesses such as tanning, pharmaceuticals, fertilizers, and mining. The release of untreated wastewater into the water system and the environment is a big problem. A Membrane Bio-Reactor (MBR) is an activated sludge process which could be used to minimize effects of wastewater. MBR can be used in aerobic and anaerobic conditions under an efficient filtration process that can hold tiny particles. It is a type of Activated sludge Process.

Despite the wide variation in LFL quality, it is used as a very effective treatment method to treat leachate via membrane separation of high biomass concentration despite the high biomass concentration [107]. Conventional MBR processes consist of two essential components: the bioreactor and the membrane module, which are responsible for the biodegradation and filtration processes [108]. In most cases, anaerobic reactors are used to treat young LFL containing a high organic load that has been discharged from landfill sites.

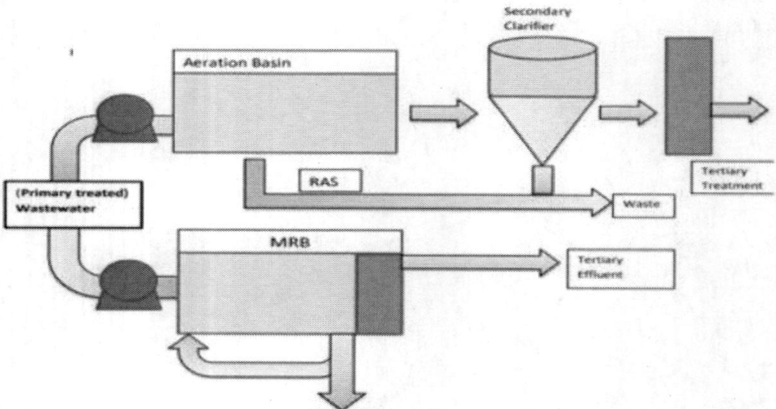

Figure 3. Membrane bio reactor technology.

Using *Saccharomyces cerevisiae* as a novel inoculum and extracellular polymeric substances for an MBR with LFL, [109] achieved removal efficiencies of 68% for COD, color, NH_3, and humic substances removal, and 79% for NH_3 and humic substances removal while lowering the membrane fouling potential by a factor of two. [110] developed a completely autotrophic nitrogen removal system with a fluidized bed granular activated carbon added membrane reactor to biodegrade organic compounds [108]. recorded that powdered activated carbon could remove micropollutants from LFL wastewater. It has been discovered that the process of micropollutant removal has been improved by 10% to 40% due to the new technology. The bacterial community has become more diverse due to the addition of powdered activated carbon, which has had a positive impact on the environment.

Hence, in conclusion Nitrates, Nitrites and Phosphate which are released to the environment could be effectively treated using an appropriate biological method. However, the selection of an appropriate method requires expertise knowledge and skill.

References

[1] Mohammed, T. M. A. 2015. *A Comparative Study of Khartoum State Water and Zamzam Water*.

[2] Sethunga. H. and Manage, P. M. 2010. Contamination status of algae toxins microcystins in some selected water bodies in Sri Lanka, *International Environmental Symposium*, pp. 25.

[3] Allan, J. D., Castillo, M. M., and Capps, K. A. 2021. *Stream ecology: structure and function of running waters*. Springer Nature.

[4] Ahmed, M., Rauf, M., Mukhtar, Z., and Saeed, N. A. 2017. Excessive use of nitrogenous fertilizers: an unawareness causing serious threats to environment and human health. *Environmental Science and Pollution Research*, 24(35), pp. 26983-26987.

[5] Wang, M., Chen, S., Jia, X. and Chen, L. 2021. Concept and types of bioremediation. In *Handbook of Bioremediation* (pp. 3-8). Academic Press.

[6] Janati, W., Benmrid, B., Elhaissoufi, W., Zeroual, Y., Nasielski, J., and Bargaz, A. 2021. Will Phosphate Bio-Solubilization Stimulate Biological Nitrogen Fixation in Grain Legumes? *Frontiers in Agronomy*, 3, p.637196.

[7] Idroos, S. F. and Manage, P. M. 2014. Seasonal occurrence of Microcystin-LR with respect to physico-chemical aspects of Beira lake water. *International Journal of Multidisciplinary Studies*, *1*(2).

[8] Bortolini, L., Maucieri, C., and Borin, M. 2018. A tool for the evaluation of irrigation water quality in the arid and semi-arid regions. *Agronomy*, 8(2), p. 23.

[9] Ashitha, A., Rakhimol, K. R., and Mathew, J. 2021. Fate of the conventional fertilizers in environment. In *Controlled Release Fertilizers for Sustainable Agriculture* (pp. 25-39). Academic Press.

[10] Khan, M. N., Mobin, M., Abbas, Z. K., and Alamri, S. A. 2018. Fertilizers and their contaminants in soils, surface and groundwater. *Encyclopedia of the Anthropocene*, 5, pp. 225-240.

[11] Chaudhary, I. J., Neeraj, A., Siddiqui, M. A., and Singh, V. 2020. Nutrient management technologies and the role of organic matrix-based slow-release biofertilizers for agricultural sustainability: A review. *Agricultural Reviews*, 41(1), pp. 1-13.

[12] Rahman, A., Mondal, N. C. and Tiwari, K. K. 2021. Anthropogenic nitrate in groundwater and its health risks in the view of background concentration in a semi arid area of Rajasthan, India. *Scientific reports*, 11(1), pp. 1-13.

[13] Adelana, S. M. A. 2005. Nitrate health effects. *Water encyclopedia*, 4, pp. 30-42.

[14] Nejatijahromi, Z., Nassery, H. R., Hosono, T., Nakhaei, M., Alijani, F. and Okumura, A. 2019. Groundwater nitrate contamination in an area using urban wastewaters for agricultural irrigation under arid climate condition, southeast of Tehran, Iran. *Agricultural Water Management*, 221, pp. 397-414.

[15] Ghaly, A. E., and Ramakrishnan, V. V. 2015. Nitrogen sources and cycling in the ecosystem and its role in air, water and soil pollution: A critical review. *Journal of Pollution Effects & Control*, pp. 1-26.

[16] Lin, Z. Z., Chen, H. M. and Yang, H. 2020. The potential pollution risk of groundwater by a ceramic permeable brick paving system. *Water, Air, & Soil Pollution*, 231(7), pp. 1-16.

[17] Sarvajayakesavalu, S., Lu, Y., Withers, P. J., Pavinato, P. S., Pan, G. and Chareonsudjai, P. 2018. Phosphorus recovery: a need for an integrated approach. *Ecosystem health and sustainability*, 4(2), pp. 48-57.

[18] Bashir, I., Lone, F. A., Bhat, R. A., Mir, S. A., Dar, Z. A., and Dar, S. A. 2020. Concerns and threats of contamination on aquatic ecosystems. In *Bioremediation and biotechnology* (pp. 1-26). Springer, Cham.

[19] Cabral, H., Fonseca, V., Sousa, T., and Costa Leal, M. 2019. Synergistic effects of climate change and marine pollution: An overlooked interaction in coastal and estuarine areas. *International journal of environmental research and public health*, 16(15), p. 2737.

[20] Mulvaney, R. L. 1996. Nitrogen—inorganic forms. *Methods of soil analysis: Part 3 Chemical methods*, *5*, 1123-1184.

[21] Nieder, R., Benbi, D. K. and Reichl, F. X. 2018. Reactive water-soluble forms of nitrogen and phosphorus and their impacts on environment and human health. In *Soil components and human health* (pp. 223-255). Springer, Dordrecht.

[22] Xiang, S. L., & Zhou, W. B. (2011). Phosphorus forms and distribution in the sediments of Poyang Lake, China. *International Journal of Sediment Research*, *26*(2), 230-238.

[23] Bouwman, A, F., Beusen, A. H. W. and Billen, G. 2009. Human alteration of the global nitrogen and phosphorus soil balances for the period 1970–2050. *Glob Biogeochem Cycles* 23:GB0A04.

[24] Gligorovski, S., Strekowski, R., Barbati, S., and Vione, D. 2015. Environmental implications of hydroxyl radicals (• OH). *Chemical Reviews*, 115(24), pp. 13051-13092.

[25] Shefat, S. H. T., Chowdhury, M. A., Haque, F., Hasan, J., Salam, M. A. and Shaha, D. C. 2020. Assessment of physico-chemical properties of the pasur river estuarine water. *Annals of Bangladesh Agriculture*, 24(1), pp. 1-16.

[26] Prakash, A. and Khanam, S. 2021. Nitrogen Pollution Threat to Mariculture and Other Aquatic Ecosystems: An Overview. *Journal of Pharmacy and Pharmacology*, 9, pp. 428-433.

[27] Tian, J., Ge, F., Zhang, D., Deng, S. and Liu, X., 2021. Roles of phosphate solubilizing microorganisms from managing soil phosphorus deficiency to mediating biogeochemical P cycle. *Biology*, 10(2), p. 158.

[28] Miyasaka, S. C., Hue, N. V. and Dunn, M. A. 2016. 16 Aluminum. *Plant Nutrition*, p. 439.

[29] Zamparas, M., Kapsalis, V. C., Kanteraki, A. E., Vardoulakis, E., Kyriakopoulos, G. L., Drosos, M. and Kalavrouziotis, I. K. 2019. Novel composite materials as P-adsorption agents and their potential application as fertilizers. *Glob Nest J*, 21(1), pp. 48-57.

[30] Morales, G. A., Marquez, L., Hernández, A. J. and Moyano, F. J. 2016. Chapter 9 Phytase effects on protein and phosphorus bioavailability in fish diets. In *Phytate destruction-consequences for precision animal nutrition* (pp. 1-8). Wageningen Academic Publishers.

[31] Rodellas, V., Garcia-Orellana, J., Masqué, P., Feldman, M. and Weinstein, Y. 2015. Submarine groundwater discharge as a major source of nutrients to the Mediterranean Sea. *Proceedings of the National Academy of Sciences*, *112*(13), pp. 3926-3930.

[32] Howarth, R. W. and Marino, R. 2006. Nitrogen as the limiting nutrient for eutrophication in coastal marine ecosystems: evolving views over three decades. *Limnology and oceanography*, *51*(1part2), pp. 364-376.

[33] Grizzetti, B., Bouraoui, F., Billen, G., van Grinsven, H., Cardoso, A. C., Thieu, V. and Garnier, J. 2011. Nitrogen processes in aquatic ecosystems. In: edited by, Sutton, M. A., Howard, C. M., Erisman, J. W., Billen, G., Bleeker, A., Grennfelt, P., van Grinsven, H., and Grizzetti, B. *The European nitrogen assessment*. Cambridge University Press, Cambridge, pp 379–404.

[34] Smith, V. H. 2003. Eutrophication of freshwater and marine ecosystems: a global problem. *Environ Sci Pollut Res Int* 10:126–139.

[35] Kross, B. C., Ayebo, A. D. and Fuortes, L. J., 1992. Methemoglobinemia: nitrate toxicity in rural America. *Am Fam Physician* 46(1):183–188.

[36] Walker, R. 1990. Nitrates, nitrites and N-nitrosocompounds: a review of the occurrence in food and diet and the toxicological implications. *Food Addit Contam* 7:717–768.

[37] Fewtrell, L. 2004. Drinking-water nitrate, methemoglobinemia, and global burden of disease: a discussion. *Environ Health Perspect* 112:1371–1374.

[38] Tsezou, A., Kitsiou-Tzeli Galla, A., Gourgiotis, D., Mitrou, S., Molybdas, P., and Sinaniotis, C., 1996. High nitrate content in drinking water: cytogenetic effects in exposed children. *Arch Environ Health* 51:458–461.

[39] Mueller, B. A., Newton, K., Holly, E. A. and Preston-Martin, S. 2001. Residential water source and the risk of childhood brain tumors. *Environ Health Perspect* 109:551–556.

[40] Kramer, M. H., Herwaldt, B. L. and Craun, G. F. 1996. Surveillance of waterborne-disease outbreaks United States, 1993–1994. *CDCP Surveill Summ MMWR 45:1–33.*

[41] Idroos, S. F. and Manage, P. M. 2014. Seasonal occurrence of Microcystin-LR with respect to physico-chemical aspects of Beira lake water. *International Journal of Multidisciplinary Studies*, *1*(2).

[42] Carmichael, W. W. 2001. Health effects of toxin-producing cyanobacteria: "The CyanoHABs." *Hum Ecol Risk Assess* 7:1393–1407.

[43] Manage, P. M., Yasawardna S. G. and Wedage W. S. 2009c. Hepatotoxic effects of *Microcystis aeruginosa* (PCC7820) on Wistar Rats, *Golden Jubilee Special Issue of the Vidyodaya Journal*, pp. 23-26.

[44] Abeysiri H. A. S. N., Manage S. P., Wanigasuriya K., Suresh T. S. and Manage P. M. 2021e, Hepatotoxicity of Microcystin-LR in Wistar rats, pp. 31, *26th International Forestry and Environment Symposium*, University of Sri Jayewardenepura.

[45] Abeysiri H. A. S. N., Wanigasuriya K., Suresh T. S., Beneragama D. H. and Manage P. M. 2021b, Nephrotoxicity of Microcystin-LR in Wistar rats, *8th International Symposium on Water Quality and Human Health: Challenges Ahead*, University of Peredeniya, Sri Lanka.

[46] Abeysiri H. A. S. N., Wanigasuriya K., Suresh T. S., Beneragama D. H. and Manage P. M. 2021a, Hepatotoxicity of Cylindrospermopsin in Wistar Rats, *8th International Conference on Multidisciplinary Approaches*, University of Sri Jayewardenepura, Sri Lanka.

[47] Abeysiri H. A. S. N., Wanigasuriya K., Suresh T. S., Beneragama D. H. and Manage P. M. 2021c, Renal-toxicity of Cylindrospermopsin in Wistar rats *The Premier Postgraduate Research Congress of Sri Lanka*, University of Peredeniya, Sri Lanka.

[48] Abeysiri, H. A. S. N., Dilrukshi, K. T., Abinaiyan, I., Manage, S. P., Wanigasuriya, K., Suresh, T. S. and Manage, P. M. *In-vivo* Toxicity Study of Cylindrospermopsin on Wistar Rats. In *Proceedings of the 24th International Forestry and Environment Symposium*, Negambo, Sri Lanka, 11 – 12 October 2019; Page 55.

[49] Abeysiri H. A. S. N., Dilrukshi K. T., Wanigasuriya K., Perera I. C. and Manage P. M. 2021d. Cytotoxic effect of Microcystin-LR, Microcystin-RR, Nodularin and Cylindrospermopsin on (Vero-89) kidney cell line, pp. 30, *26th International Forestry and Environment Symposium*, University of Sri Jayewardenepura.

[50] US EPA (US Environmental Protection Agency)., 2014. Cyanobacteria and cyanotoxins: information for drinking water systems. EPA-810F11001, *Office of Water*, Washington DC.

[51] Volterra, L., Boualam, M., Menesguen, A., Duguet, J., Duchemin, J., Bonnefoy, X. 2002. Eutrophication and health. World Health Organization Regional Office for Europe. *European Communities*. Available from http://europa.eu.int.

[52] Velusamy, K., Periyasamy, S., Kumar, P. S., Vo, D. V. N., Sindhu, J., Sneka, D. and Subhashini, B. 2021. Advanced techniques to remove phosphates and nitrates from waters: a review. *Environmental Chemistry Letters*, 19(4), pp. 3165-3180.

[53] Mustafa, H. M. and Hayder, G. 2021. Recent studies on applications of aquatic weed plants in phytoremediation of wastewater: A review article. *Ain Shams Engineering Journal*, 12(1), pp. 355-365.

[54] Bacelo, H., Pintor, A. M., Santos, S. C., Boaventura, R. A., and Botelho, C. M. 2020. Performance and prospects of different adsorbents for phosphorus uptake and recovery from water. *Chemical Engineering Journal*, 381, p. 122566.

[55] Ruzhitskaya, O. and Gogina, E., 2017. Methods for removing of phosphates from wastewater. In *MATEC Web of Conferences* (Vol. 106, p. 07006). *EDP Sciences*.

[56] Manikam, M. K., Halim, A. A., Hanafiah, M. M. and Krishnamoorthy, R. R. 2019. Removal of ammonia nitrogen, nitrate, phosphorus and COD from sewage wastewater using palm oil boiler ash composite adsorbent. *Desalin. Water Treat*, 149, pp. 23-30.

[57] Yin, H., Yan, X. and Gu, X. 2017. Evaluation of thermally-modified calcium-rich attapulgite as a low-cost substrate for rapid phosphorus removal in constructed wetlands. *Water Research*, 115, pp. 329-338.

[58] Yang, Y., Chen, Z., Wang, X., Zheng, L. and Gu, X. 2017. Partial nitrification performance and mechanism of zeolite biological aerated filter for ammonium wastewater treatment. *Bioresource Technology*, *241*, pp. 473-481.

[59] Zhang, J., Lü, F., Shao, L. and He, P. 2014. The use of biochar-amended composting to improve the humification and degradation of sewage sludge. *Bioresource technology*, *168*, pp. 252-258.

[60] Battas, A., Gaidoumi, A. E., Ksakas, A. and Kherbeche, A. 2019. Adsorption study for the removal of nitrate from water using local clay. *The Scientific World Journal*, 2019.

[61] Siddeeg, S. M., Tahoon, M. A., Alsaiari, N. S., Shabbir, M. and Rebah, F. B. 2021. Application of functionalized nanomaterials as effective adsorbents for the removal of heavy metals from wastewater: a review. *Current Analytical Chemistry*, 17(1), pp. 4-22.

[62] Yun, Y., Li, Z., Chen, Y. H., Saino, M., Cheng, S. and Zheng, L. 2016. Reduction of nitrate in secondary effluent of wastewater treatment plants by Fe0 reductant and pd–cu/graphene catalyst. *Water, Air, & Soil Pollution*, 227(4), pp. 1-10.

[63] Kumar, M., Seth, A., Singh, A. K., Rajput, M.S. and Sikandar, M. 2021. Remediation strategies for heavy metals contaminated ecosystem: A review. *Environmental and Sustainability Indicators*, 12, p. 100155.

[64] Rezvani, F., Sarrafzadeh, M. H., Ebrahimi, S. and Oh, H. M. 2019. Nitrate removal from drinking water with a focus on biological methods: a review. *Environmental Science and Pollution Research*, 26(2), pp. 1124-1141.

[65] Mohammadi, E., Daraei, H., Ghanbari, R., Athar, S. D., Zandsalimi, Y., Ziaee, A., Maleki, A., and Yetilmezsoy, K. 2019. Synthesis of carboxylated chitosan modified

with ferromagnetic nanoparticles for adsorptive removal of fluoride, nitrate, and phosphate anions from aqueous solutions. *Journal of Molecular Liquids*, 273, pp. 116-124.

[66] Velusamy, K., Periyasamy, S., Kumar, P. S., Vo, D. V. N., Sindhu, J., Sneka, D. and Subhashini, B. 2021. Advanced techniques to remove phosphates and nitrates from waters: a review. *Environmental Chemistry Letters*, 19(4), pp. 3165-3180.

[67] Rezania, S., Ponraj, M., Talaiekhozani, A., Mohamad, S. E., Din, M. F. M., Taib, S. M., Sabbagh, F. and Sairan, F. M. 2015. Perspectives of phytoremediation using water hyacinth for removal of heavy metals, organic and inorganic pollutants in wastewater. *Journal of environmental management*, 163, pp. 125-133.

[68] Villegas, L. G. C., Mashhadi, N., Chen, M., Mukherjee, D., Taylor, K. E. and Biswas, N. 2016. A short review of techniques for phenol removal from wastewater. *Current Pollution Reports*, 2(3), pp. 157-167.

[69] Wang, S., Wei, S., Liang, H., Zheng, W., Li, X., Hu, C., Currell, M. J., Zhou, F. and Min, L. 2019. Nitrogen stock and leaching rates in a thick vadose zone below areas of long-term nitrogen fertilizer application in the North China Plain: a future groundwater quality threat. *Journal of Hydrology*, 576, pp. 28-40.

[70] Roy, U., Manna, S., Sengupta, S., Das, P., Datta, S., Mukhopadhyay, A. and Bhowal, A., 2018. Dye removal using microbial biosorbents. *Green adsorbents for pollutant removal*, pp. 253-280.

[71] Gunasekara, N. D., Idroos, F. S. and Pathmalal, M. M., 2019. Microbial Degradation of Nitrate: *Put Microbes to Work*.

[72] Zhang, H., Yuan, X., Xiong, T., Wang, H. and Jiang, L. 2020. Bioremediation of co-contaminated soil with heavy metals and pesticides: Influence factors, mechanisms and evaluation methods. *Chemical Engineering Journal*, 398, p. 125657.

[73] Roy, U., Manna, S., Sengupta, S., Das, P., Datta, S., Mukhopadhyay, A. and Bhowal, A. 2018. Dye removal using microbial biosorbents. *Green adsorbents for pollutant removal*, pp. 253-280.

[74] Usman, K., Al-Ghouti, M. A. and Abu-Dieyeh, M. H. 2018. Phytoremediation: Halophytes as promising heavy metal hyperaccumulators. *Heavy Met*.

[75] Yun, Y., Li, Z., Chen, Y. H., Saino, M., Cheng, S. and Zheng, L. 2016. Reduction of nitrate in secondary effluent of wastewater treatment plants by Fe0 reductant and pd–cu/graphene catalyst. *Water, Air, & Soil Pollution*, 227(4), pp. 1-10.

[76] Ayangbenro, A. S., and Babalola, O. O. 2017. A new strategy for heavy metal polluted environments: a review of microbial biosorbents. *International journal of environmental research and public health*, 14(1), p. 94.

[77] Lalrintluangi, I. F., Kumar, V. and Bisarya, D. 2020. *Biodegradation of pesticides: A review*. Biodegradation, 8(11).

[78] Xing, D., Magdouli, S., Zhang, J. and Koubaa, A. 2020. Microbial remediation for the removal of inorganic contaminants from treated wood: Recent trends and challenges. *Chemosphere*, 258, p. 127429.

[79] Gong, X., Huang, D., Liu, Y., Peng, Z., Zeng, G., Xu, P., Cheng, M., Wang, R., and Wan, J. 2018. Remediation of contaminated soils by biotechnology with

nanomaterials: bio-behavior, applications, and perspectives. *Critical reviews in biotechnology*, 38(3), pp. 455-468.

[80] Idroos, S. F. and Manage, P. M. 2014. Seasonal occurrence of Microcystin-LR with respect to physico-chemical aspects of Beira lake water. *International Journal of Multidisciplinary Studies*, *1*(2).

[81] Kumar, M., Seth, A., Singh, A. K., Rajput, M. S. and Sikandar, M. 2021. Remediation strategies for heavy metals contaminated ecosystem: A review. *Environmental and Sustainability Indicators*, 12, p. 100155.

[82] Sannigrahi, P., Ragauskas, A. J. and Tuskan, G. A. 2010. Poplar as a feedstock for biofuels: a review of compositional characteristics. *Biofuels, Bioproducts and Biorefining*, *4*(2), pp. 209-226.

[83] Kumar, V. and Chandra, R. 2017. Bacteria-assisted phytoremediation of industrial waste pollutants and ecorestoration. In *Phytoremediation of environmental pollutants* (pp. 159-200). CRC Press.

[84] Mahfooz, Y., Yasar, A., Islam, Q. U., Rasheed, R., Naeem, U., and Mukhtar, S. 2021. Field testing phytoremediation of organic and inorganic pollutants of sewage drain by bacteria assisted water hyacinth. *International Journal of Phytoremediation*, 23(2), pp. 139-150.

[85] Verma, S. and Kuila, A. 2019. Bioremediation of heavy metals by microbial process. *Environmental Technology & Innovation*, 14, p. 100369.

[86] Dixit, R., Malaviya, D., Pandiyan, K., Singh, U. B., Sahu, A., Shukla, R., Singh, B. P., Rai, J. P., Sharma, P. K., Lade, H., and Paul, D. 2015. Bioremediation of heavy metals from soil and aquatic environment: *an overview of principles and criteria of fundamental processes*. Sustainability, 7(2), pp. 2189-2212.

[87] Saxena, G., Kishor, R., Saratale, G. D. and Bharagava, R. N. 2020. Genetically modified organisms (GMOs) and their potential in environmental management: constraints, prospects, and challenges. In *Bioremediation of industrial waste for environmental safety* (pp. 1-19). Springer, Singapore.

[88] Shahid, A., Malik, S., Zhu, H., Xu, J., Nawaz, M. Z., Nawaz, S., Alam, M. A. and Mehmood, M. A. 2020. Cultivating microalgae in wastewater for biomass production, pollutant removal, and atmospheric carbon mitigation; a review. *Science of the Total Environment*, 704, p. 135303.

[89] Ahmad, A., Buang, A. and Bhat, A. H. 2016. Renewable and sustainable bioenergy production from microalgal co-cultivation with palm oil mill effluent (POME): A review. *Renewable and Sustainable Energy Reviews*, 65, pp. 214-234.

[90] Demarco, M., de Moraes, J. O., Matos, Â. P., Derner, R. B., de Farias Neves, F., and Tribuzi, G. 2022. Digestibility, bioaccessibility and bioactivity of compounds from algae. *Trends in Food Science & Technology*.

[91] Randrianarison, G. and Ashraf, M. A. 2017. Microalgae: a potential plant for energy production. *Geology, Ecology, and Landscapes*, 1(2), pp. 104-120.

[92] Ummalyma, S. B., Pandey, A., Sukumaran, R. K. and Sahoo, D. 2018. Bioremediation by microalgae: Current and emerging trends for effluents treatments for value addition of waste streams. In *Biosynthetic technology and environmental challenges* (pp. 355-375). Springer, Singapore.

[93] Lekshmi, B., Joseph, R. S., Jose, A., Abinandan, S. and Shanthakumar, S. 2015. Studies on reduction of inorganic pollutants from wastewater by Chlorella pyrenoidosa and Scenedesmus abundans. *Alexandria Engineering Journal*, 54(4), pp. 1291-1296.

[94] Koju, A., Mennerich, A., and von Campenhausen Joost, I. 2013. Nutrient removal of waste water using Algae Culture and its applicability in Nepal. In *During two expert seminars in year 2012 the chair of "Water Resources Management" of the University Siegen has focused on both sides of the water story, on the water abstraction and sustainable water supply (seminar on "Wassergewinnung", Hof, Germany) on the one hand and on water pollution through diverse uses (seminar on "Water pollution by agriculture," Santa Maria, Brazil) on the other hand.* (p. 123).

[95] Malik, J. A. ed. 2021. *Handbook of Research on Microbial Remediation and Microbial Biotechnology for Sustainable Soil*. IGI Global.

[96] Bharagava, R. N., Saxena, G., and Mulla, S. I. 2020. Introduction to industrial wastes containing organic and inorganic pollutants and bioremediation approaches for environmental management. In *Bioremediation of industrial waste for environmental safety* (pp. 1-18). Springer, Singapore.

[97] Bharagava, R. N., Saxena, G., and Mulla, S. I. 2020. Introduction to industrial wastes containing organic and inorganic pollutants and bioremediation approaches for environmental management. In *Bioremediation of industrial waste for environmental safety* (pp. 1-18). Springer, Singapore.

[98] Chirakkara, R. A., Cameselle, C., and Reddy, K. R. 2016. Assessing the applicability of phytoremediation of soils with mixed organic and heavy metal contaminants. *Reviews in Environmental Science and Bio/Technology*, 15(2), pp. 299-326.

[99] Sarwar, N., Imran, M., Shaheen, M. R., Ishaque, W., Kamran, M. A., Matloob, A., Rehim, A. and Hussain, S. 2017. Phytoremediation strategies for soils contaminated with heavy metals: modifications and future perspectives. *Chemosphere*, 171, pp. 710-721.

[100] Rehman, K., Imran, A., Amin, I. and Afzal, M. 2018. Inoculation with bacteria in floating treatment wetlands positively modulates the phytoremediation of oil field wastewater. *Journal of hazardous materials*, 349, pp .242-251.

[101] Chen, Z., Zheng, Y., Ding, C., Ren, X., Yuan, J., Sun, F., and Li, Y. 2017. Integrated metagenomics and molecular ecological network analysis of bacterial community composition during the phytoremediation of cadmium-contaminated soils by bioenergy crops. *Ecotoxicology and environmental safety*, 145, pp. 111-118.

[102] Hussain, A., Ali, S., Rizwan, M., ur Rehman, M. Z., Javed, M. R., Imran, M., Chatha, S. A. S., and Nazir, R. 2018. Zinc oxide nanoparticles alter the wheat physiological response and reduce the cadmium uptake by plants. *Environmental Pollution*, 242, pp. 1518-1526.

[103] Michalak, I., Chojnacka, K. and Witek-Krowiak, A. 2013. State of the art for the biosorption process—a review. *Applied biochemistry and biotechnology*, *170*(6), pp. 1389-1416.

[104] Imanthi, K. P. A., Idroos, F. S., and Manage, P. M. 2019 December. Development of a Third Generation Biofuel Source using Cyanobacteria and Landfill Leachate. In *Proceedings of the 6th International Conference on Multidisciplinary Approaches* (iCMA).

[105] Gude, V. G. 2016. Wastewater treatment in microbial fuel cells–an overview. *Journal of Cleaner Production*, *122*, pp. 287-307.

[106] Ge, Z., Wu, L., Zhang, F. and He, Z. 2015. Energy extraction from a large-scale microbial fuel cell system treating municipal wastewater. *Journal of Power Sources*, *297*, pp. 260-264.

[107] Hashisho, J. and El-Fadel, M. 2016. Membrane bioreactor technology for leachate treatment at solid waste landfills. *Reviews in Environmental Science and Bio/Technology*, *15*(3), pp. 441-463.

[108] Abuabdou, S. M., Bashir, M. J., Aun, N. C. and Sethupathi, S. 2018, April. Applicability of anaerobic membrane bioreactors for landfill leachate treatment: review and opportunity. In *IOP Conference Series: Earth and Environmental Science* (Vol. 140, No. 1, p. 012033). IOP Publishing.

[109] Brito, G. C., Lange, L. C., Santos, V. L., Amaral, M. and Moravia, W. G. 2019. Long-term evaluation of membrane bioreactor inoculated with commercial baker's yeast treating landfill leachate: pollutant removal, microorganism dynamic and membrane fouling. *Water Science and Technology*, *79*(2), pp. 398-410.

[110] Zhang, J., Lü, F., Shao, L. and He, P. 2014. The use of biochar-amended composting to improve the humification and degradation of sewage sludge. *Bioresource technology*, *168*, pp. 252-258.

Chapter 12

Biological Wastewater Treatment Technology

P. A. K. C. Wijerathna[1,2]
M. S. Ekanayake[1,3]
F. S. Idroos[1]
and P. M. Manage[1,2,*]

[1]Centre for Water Quality and Algae Research, Department of Zoology, University of Sri Jayewardenepura, Gangodawila, Nugegoda, Sri Lanka
[2]Faculty of Graduate Studies, University of Sri Jayewardenepura, Gangodawila, Nugegoda, Sri Lanka
[3]Department of Aquaculture and Fisheries, Faculty of Livestock, Fisheries and Nutrition, Wayamba University of Sri Lanka

Abstract

Anthropological activities associated with urbanization result in significant wastewater generation. Xenobiotic pollutants in wastewater treatment have resulted in severe environmental issues. Improper discharge of wastewater into the natural environment has several adverse effects on aquatic and terrestrial ecosystems. Significantly, wastewater contributes to the spread of infectious and non-infectious diseases, which ultimately negatively impact human health. Thus, practical and cost-effective wastewater treatment methods are required for long-term wastewater disposal. Advanced physicochemical methods are being used to treat wastewater in the modern world. However, biological wastewater treatment, as opposed to traditional physicochemical treatment methods, is mainly due to its cost-effective and eco-friendly approach. Biological treatment is an essential step in nearly all wastewater treatment processes. Bioprocessing of waste is taken by syntrophic microbial

[*] Corresponding Author's Email: pathmalal@sjp.ac.lk.

In: Microbial Technologies and Their Applications
Editors: Santosh Kumar Karn and Anne Bhambri
ISBN: 979-8-88697-913-8

communities followed by metabolism and co-metabolism of specific waste substrates. In biological wastewater treatment, the partial conversion of complex waste substrates into non-toxic simpler compounds by interconnected fermentative, acetogenic, and methanogenic microbial groups is critical. Aerobic and anaerobic microbial application techniques are used for wastewater treatment as traditional methods.

In contrast, membrane bioreactors, microbial fuel cells, activated slug systems, sequencing batch reactors, and aerobic and anaerobic lagoon systems are considered novel treatment technologies. The use of extracellular cell-free hydrolytic enzymes to degrade organic waste materials is a novel new technology that has some advantages over the use of microbial biomass. Hence, biological treatment has been in the limelight in recent wastewater treatment processes.

Keywords: wastewater, bioremediation, microbial community, extracellular enzymes, organic contaminants

Introduction

Water is a precious natural resource required for the survival of living beings. Its quality has a critical impact on both human and animal health. Inadequate freshwater supply and poor sanitation are crucial issues in the urbanized world, and they appear to be persisting; consequently, good technologies should be implemented [1]. According to WHO reports, three out of ten people require access to safe drinking water, and six out of ten people lack safe sanitation [2]. Groundwater utility, widely available in many parts of the world, is currently managing the increase in water demand. However, in all developing countries, it has been technologically overexploited [3].

The unregulated release of contaminants from point and nonpoint sources pollutes the water supplies more every day. Increased human population and anthropogenic activities have resulted in excessive waste discharge into water bodies, resulting in nutrient enrichment and subsequent degradation of aquatic habitats and freshwater resources [4]. Anthropogenic activities and global climate change have an enormous strain on many aquatic ecosystems' biological conditions and long-term viability. Soil lead pollution is caused by anthropogenic sources such as mining, smelting, refining, battery manufacturing, fertilizers, pesticides, sewage, waste effluent discharge from industry to water ways, and coal-based industries and power plants [4, 5].

Anthropogical activities caused the increasing quantities of pollutants and the nature of the contaminants leading to adverse impact to the quality of water in aquatic ecosystems [5]. According to urban water consumption, only 15 – 25% is utilized, and the rest is turned into the wastewater for the urban hydrologic system [5-6]. Considerable amounts of fresh water from various industries have become a significant source of water pollution, and water scarcity since these industries discharge their wastewater into the aquatic and terrestrial environments [6].

Several industries release diverse range of toxic pollutants [7, 8]. The paper and pulp industry, photography, oil refineries, artificial nitrogenous fertilizers, herbicides, printing industry, and home wastewater are new sources of pollution and water overuse. However, many laundry detergents contribute to water pollution, including domestic detergent runoff and agricultural runoff from herbicides, pesticides, and some businesses [8]. Because numerous pathogens are commonly associated with faeces and other contaminants, industrial and faecal pollutants' pollution of aquatic systems poses considerable threats to health.

Wastewater Types, Composition and Xenobiotic Nature

Wastewater generation is a complicated process which is contributed by several points and non-point sources. The wastewater sources are categorized into three major forms as domestic, agricultural and industrial sources. Figure 1 describes important wastewater types and different activities involved with wastewater generation.

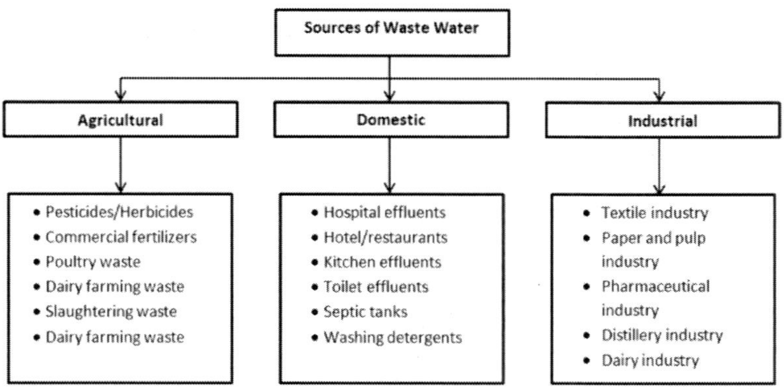

Figure 1. Major sources of waste water generation.

Table 1. Major industries and different pollutants caused for water pollution

Industry	Major pollutants	Reference
Paper and pulp Industry	β-sitosterol trimethylsilyl ether, 1-tetradecane, 2-methoxy phenol, trichlorocatechol, tetrachlorocatechol, chlorophenols, chloroguaiacols, chlorosyringols, chlorocatechols, terpenes, methanol, phenol, alkylated phenols, decalone, benzoic acid, abietic acid, and dehydroabietic acid, hydrogen sulfides, sodium sulfide, methyl mercaptan, sulfur, Fe, Cu, Zn, Ni, Mg Alkylphenols, nonylphenol ethoxylates (NPEs), nonylphenols (NPs)	[12]
Textile industry	Chlorobenzenes, Heavy metals (Sb, Cd, Pb, Hg, As, Ni and Cr), Organotin compounds (DBT: dibutyltin, TBT: tributyltin, and TPhT: triphenyltin) Formaldehyde, Chlorophenols, Enzymes, Starch, Waxes, CMC	[13]
Battery industry	Heavy metals (Cr, Cd, Ni and Pb)	[12, 13]
Dairy industry	Fats, Whey proteins, carbohydrate, Detergents, Cu, Zn, Pb, Fe, Ni	[13]
Pharmaceutical industry	Methanol, toluene, acetone hexanes, Aromatic compounds, chloroform, methylene chloride, toluene, Alcohols (CH_4O), ethylene glycol, different acids (HCl), chloroform, phenol	[12, 13]
Pesticide and fertilizers industry	Urea, carbamates, pyrethroids, Oganocholorine, polyaromatic hydrocarbons, Cd, Hg, Pb, As	[12]
Distillery industry	Melanoidins, amino-carbonyl, polymers, di-n-octyl phthalate, di-butylphthalate, benzenepropanoic acid, 2 -33 hydroxysocaproicacid, toxic metals	[14]

Wastewater has high Biological Oxygen Demand (BOD), very high Chemical Oxygen Demand (COD), Total Solids (TS), different inorganic pollutants such as sulphate, nitrate, nitrite, phosphate, and some xenobiotic pollutants [9, 10]. The heavy metals (Cr, Hg, Cd and Pb) a vast range of organic substances (hydrocarbons, aromatic substances, PAHs, phthalate acid esters, artificial dyes), inorganic compounds (ammonical substances, chlorides, sulphates, and solids) and different pathogenic and nonpathogenic microorganisms (bacteria, viruses, protozoans) are some of the significant components of the wastewater [11]. Such poisonous compounds from industrial contaminants present in the wastewater make it toxic to the ecosystems [11]. Table 1 represent a summary of significant industries causing water pollution and different types of contaminants released by various industries.

Environmental Impacts of Untreated Wastewater

The continuous discharge of untreated wastewater into ecosystems can cause severe environmental threats. The mortality of aquatic wildlife, eutrophication, damage habitat of aquatic organisms and bioaccumulation are some of the issues originating from these wastewaters [15]. The enrichment of water resources with nutrients, primarily nitrogen and phosphorus, by the continuous discharge of effluents leads to the eutrophication, which causes algal blooms, aquatic plant spread, oxygen depletion, extinction of essential species, and finally, the destruction of freshwater ecosystems [16].

The continuous discharge of insufficiently treated wastewater into aquatic bodies has resulted in a significant environmental concern in groundwater pollution. In some nations, wastewater is used for agricultural irrigation, resulting in groundwater pollution [17]. Fluorides, nitrites, and various harmful metalloids are found in many countries' water supplies [14]. Day-to-day increases in organochlorine pesticide concentrations result in the accumulation of hazardous heavy metals such as arsenic, even in the exposed community's body parts (hair, blood and urine) [18]. Water can become a source of life-threatening infections like typhoid, cholera, and other diseases due to contamination [17-20]. Therefore, wastewater discharges could cause toxic, carcinogenic, and mutagenic disorders in terrestrial animals and humans [20].

Wastewater on Human Health

The direct discharge of untreated wastewater into ecosystems has multiple detrimental health effects. Exposure to wastewater poses many significant risks, including the accumulation of heavy metals in fish and edible plants, the spread of waterborne infectious diseases, carcinogenic effects, and the emergence of antibiotic resistance [20].

The accumulation of heavy metals in animal tissue beyond permissible limits has negative health consequences for vital organs such as the kidneys, liver, brain, and blood. Heavy metals cause neurological disorders, gastrointestinal problems, renal impairment, cardiovascular problems, and paralysis, among other things. The accumulation of heavy metals (Cd and Hg in body organs such as the kidneys causes severe damage and contributes to diseases like osteoporosis. Some research has found a link between osteoporosis and toxic metals in the body [20]. Heavy metals have been linked to various tumors, pulmonary disorders, and severe lung problems [21].

Drinking wastewater-contaminated water can expose people to waterborne and foodborne microorganisms, which can cause various illnesses. Among the diseases linked to harmful bacteria in drinking water are Amoebiasis, typhoid and paratyphoid fevers, salmonellosis, shigellosis, and viral hepatitis [22].

Different organic pollutants in wastewater can spread and persist in the environment, causing biomagnification by food chains. PCBs are classified as probably carcinogenic to humans by the International Agency for Research on Cancer (IARC) [23]. Immune system disorders, liver toxicity, immunodeficiency, organ damage, neurological effects, mutagenicity, and endocrine disruptors are all reported side effects of organic pollutant exposure. PCBs accumulate in human fatty tissues because of their nonpolar, lipophilic physical properties and resistance to biochemical degradation [24].

Intake of PAHs may interfere with the activities of cellular membranes and enzyme systems connected with them. The epoxides and dihydrodiols in PAHs can cause for DNA mutations, biochemical disturbances and tissue damages. The hazardous equivalency factor relative to benzopyrene was used to determine the carcinogenicity of the selected PAHs [25].

Antibiotic resistance is another critical health hazard caused due to the frequent exposure to the hazardous waste water [26]. Antibiotics contamination of drinking water may cause the immune system to develop resistance to antibiotics. Moreover, exposure to antibiotics in miner quantities causes the development of antibiotic resistance genes, particularly in the younger generation [27]. Antibiotic resistance genes are induced by the remaining antibiotics residual available in drinking water. When these resistance genes are passed on to harmful bacteria, there is a high risk of developing super bacteria that can resist drugs, triggering infections in humans and potentially infectious disease outbreaks. Even though antibiotic wastewater may not cause illness right away, it is unquestionably a danger to public health. Streptomycin, Ampicillin, and Tetracycline are some of the antibiotics that can transmit antibiotic genes to environmental bacteria [26-28].

Improper handling of oily wastewater from natural and manufactured sources could pose a health risk. Carcinogens generated hydrocarbons related to the petroleum industry, particularly those polluted with polyaromatic hydrocarbons, may damage public health when consumed through food chains [29]. Through inhalation, oil pollutants can influence the regular operation of organs, resulting in a range of disorders. Oil-exposed children have a four-fold increased chance of getting leukemia and a seven-fold increased risk of acute non-lymphocytic leukemia than normal children [30].

Wastewater as a Source for Agriculture

Due to a scarcity of water, the use of waste for agricultural purposes has become a global concern [31]. Reusing wastewater for agricultural purposes is a complicated problem involving many factors, including agricultural production, water quality and purification, hydrological health concerns, cultural concerns of those involved particularly users, and environmental hazards [31].

The United Nations' Food and Agriculture Organization (FAO) has published many agricultural wastewater standards. In 1987, guidelines for wastewater quality for agricultural use were published [32]. These standards linked the salinity and toxicological characteristics of specific ions to the degree of water scarcity. The FAO recommended guidelines for "agricultural reuse of treated waters and treatment requirements" in 1999. Agricultural reuse of water has been classified according to these standards based on the type of crop irrigated. According to the FAO, untreated or partially treated wastewater irrigates 20 million hectares of land in 50 countries, accounting for nearly 10% of washed cultivation surfaces [32].

Around the world, over 800 million farmers work in urban agriculture. About 200 million people in this category engage in open-land market-oriented agriculture, with waste irrigation water being used when high-quality water is unavailable [33]. Water from the toilet irrigates 26% of Pakistan's vegetable crop 80% of Hanoi's vegetable crop is grown in urban and peri-urban areas [34]. In Ghana, informal irrigation uses diluted wastewater from rivers and streams to irrigate an estimated 11,500 hectares, far exceeding the country's claimed formal irrigation area. In Mexico, wastewater irrigates around 260,000 hectares of land [34, 35].

The use of wastewater for irrigation benefits public health, the environment, and the economy. This method is used in various places where clean water is scarce for urban agriculture [36]. Wastewater is high in macronutrients (N, P, and K) and micronutrients (Ca, Mg, B, Fe, Mn or Zn). Furthermore, the various nutrients naturally present in wastewater reduce chemical fertilizers, forming a closed and environmentally friendly nutrient circulation system and overcoming eutrophication issues [37]. Although these nutrients in treated municipal wastewater provide fertilizer value to crop or landscape production, they may be more than plant requirements, resulting in excessive vegetative growth, delayed or uneven maturity, or poor quality. Nutrients such as nitrogen, phosphorus, potassium, zinc, boron, and sulfur are essential [38].

Wastewater is used for aquaculture and crop production in Africa and Central, South, and Southeast Asia (Bangladesh, Cambodia, China, India, Indonesia, and Vietnam) [39]. Purified wastewater is used for various purposes, such as growing fodder for agricultural farms, recharging groundwater resources, and achieving ecological goals such as managing wetlands water resources, wildlife conservation, and creating urban water bodies in many areas. Some places use significant wetland areas as natural treatment facilities, such as Vientiane [40].

Failure to use appropriately treated wastewater has negative health consequences [41]. Some of the key concerns are the spreading of water-borne pathogenic infections and toxicity from a specific ion to the exposed community, which leads to bioaccumulation. Sodium, chloride, and boron are the most dangerous ions in treated wastewater. Boron is typically found in home detergents and industrial plant emissions [42]. Another potential issue linked with wastewater reuse is the presence of high residual chlorine levels in the purified water. It may cause plant damage when sprinklers are utilized to spread this water for irrigational purposes [43]. Plant leaves should not be affected by chlorine below one mg/l; however, significant plant damage can ensue when the chlorine residual exceeds five mg/l [43]. Chloride and sodium levels significantly increase in domestic wastewater, mainly if water softeners are utilized. When treated municipal wastewater is used for crop irrigations, the clogging difficulties with sprinklers and the toxic nature of wastewater to sensitive plants are some significant concerns [44].

Wastewater to Generate Alternative Energy

Energy demand has risen day by day, causing a rapid utilization of fossil fuels. Utilizing fossil fuels at this rate is not sustainable for long periods, primarily because of increasing greenhouse gas emissions. Wastewater contains much organic stuff and acts as a carbon carrier

Bio-electrochemical systems are biological set-up that alters the chemical energy of wastewater into electrical energy. Bio-electrochemical systems that use indigenous Exo- electrogenic bacteria collect clean energy from organic waste [45]. This energy is retrieved as bioelectricity in a Microbial Fuel cell (MFC). In contrast, microbial electrolysis cells extract it as valuable biofuels like ethanol and hydrogen peroxide. Direct waste conversion into pure electrical energy and valuable by-products has been acknowledged as a preferable choice to overcome difficulties in conventional wastewater treatment methods [45]. The application of such MFC technologies lowers the cost of wastewater treatment and improves the long-term viability of renewable energy generation. It recovers electrical energy from wastewater organic contaminants. Electrogenic microorganisms remediate several wastewaters by their very nature while also producing electricity [46].

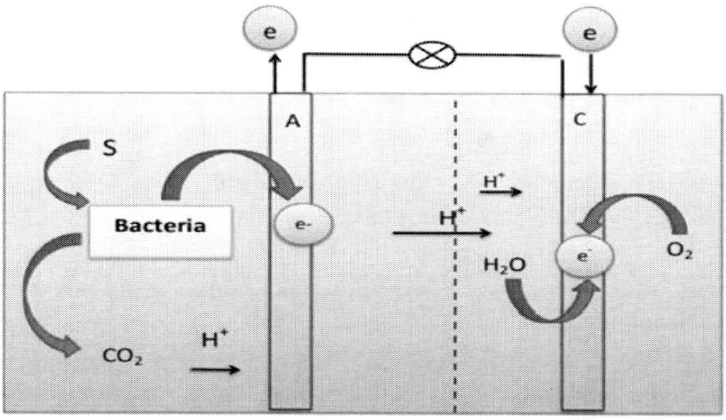

Figure 2. Microbial fuel cell technologies.

Biofuels derived from algae which can cultivate using wastewater, provide another alternative fuel that has no detrimental impact on agriculture. Researchers are currently focusing on increasing biodiesel yield from suitable algae species. Different studies on biodiesel production using less expensive, effective technologies are being conducted [47]. The purification and handling of sludge is a high-cost procedure that can cost up to 50% of overall sewage treatment expenses [48]. As a result, numerous thermochemical technologies transform wastewater into value-added energy products. Some researchers have recorded the potential applicability of hydrothermal carbonization to increase the yield of bio methane from sewage sludge [47, 48].

Further, sludge digestion is already regular practice in some developed countries, such as Netherland, with a potential value of 2,215 TJ). This biogas is transformed into electricity (143 MWh) and heat in a combined heat and power system. The use of adequately managed anaerobic sludge treatment resulted in a reduction in greenhouse gas emissions by minimizing methane leakage and maximizing the usage of biogas [49].

Several notable researchers are studying the thermochemical treatment of sludge, focusing on a different aspect of the conversion process. The fate of specific substances initially present in the sludge must also be investigated during the process, mainly when working with a feed that may contain varying quantities of hazardous chemicals, heavy metals, pathogens, and antibiotics.

Biological Wastewater Treatment Methods

Conventional physicochemical wastewater treatment methods such as oxidation, flocculation, settling, adsorption, ultrafiltration, reverse osmosis, and ozonation have some drawbacks for the high operation cost and requirement of skilled labours [50]. Therefore, biological treatments have become widely accepted between treatment processes to gain additional benefits such as relative cost-effectiveness and environmental friendliness.

Biological treatment practices are divided into two categories based on the amount of oxygen required by the processing medium and anaerobic methods. Organic pollutants in leachates are converted into CO_2 and solid biological products (sludge) through atmospheric oxygen transferred to the wastewater during aerobic processing [51]. Organic matter is converted into biogases, mainly CH_4 and CO_2, and biological sludge during anaerobic treatment. Most microbial treatment techniques are composed of interconnected, mixed microbial communities adapted to efficiently remove individual pollutant substances from the media [52].

Aerobic treatment technologies for suspended or adherent biomass growth include aerated lagoon systems, activated sludge systems, sequence batch reactors, moving bed biofilm reactors, and trickling filter systems. Aerobic biological treatment processes like the Moving-Bed Biofilm Reactor (MBBR), Sequencing Batch Reactors (SBR), and Bio Filters rely on suspended biomass development in which microorganisms are free to move around in the medium [53]. The Moving-Bed Biofilm Reactor (MBBR), Sequencing Batch Reactors, and Bio Filters are examples of these processes. A microbial community in an oxygen-rich environment is required for aerobic treatment operations, capable of carrying out the natural oxidation of organic materials into carbon dioxide, water, cellular material, and other byproducts.

Aerobic Lagoons Systems

Aerobic Lagoons Systems (ALS) are used to meet the requirement of safe leachate disposal Land Fill Leachate (LFL)in a relatively short period and decontaminate the space occupied by leachate storage tanks. ALS has long been recognized as an effective and cost-effective method of removing pathogenic microorganisms and inorganic and organic contaminants from the environment [53]. LFL is treated in a tank under vigorous mechanical aeration in an anaerobic lagoon system.

This is made possible by adding oxygen and then transferring it to a settling tank to separate the sludge resulting from the treatment. To maintain the required microbial cell concentration, a portion of the sludge is separated and recycled in subsequent batches.

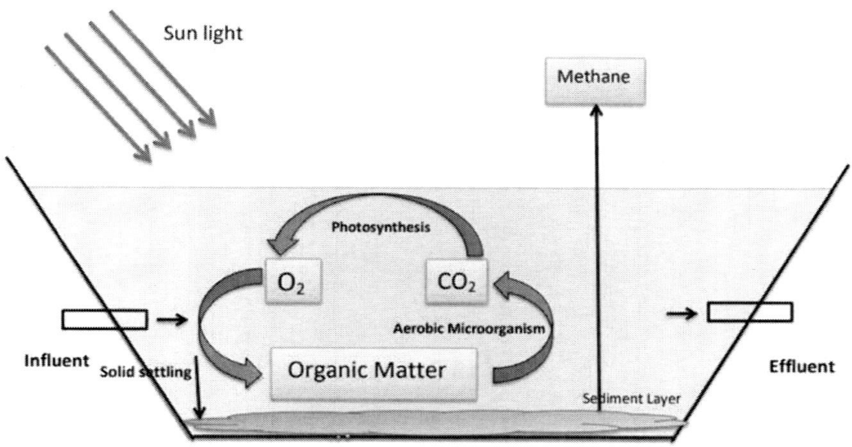

Figure 3. Aerobic lagoon system.

ALS is effective in removing settle- able solids, BOD, pathogens, and ammonia. Further, these lagoon systems are a good nutrient source with a high amount of N and P for agricultural purposes [54]. Another benefit of ALS is its relative economic effectiveness in constructing sewage treatment plants with higher flexibility to quickly change treatment mode. Although ALS have some drawbacks, including the low efficiency of removing heavy metals, the requirement of additional cost for further disinfection of the effluent and the requirement of large land area.

Activated Sludge Systems (ASS)

The ASS is an oxygenated reactor that continuously supplies organic pollutants to be degraded by a microbial community that has developed a solid adaptation to the environment. Organic contaminants from sewage systems are cleaned up by a microbial community primarily composed of different microorganisms, including bacteria, fungi, and protozoa groups, suspended in biomass and act as a bio filter [55]. Currently, present organic substances are utilized by microorganisms in their aerobic biochemical processes, where they are partially converted into CO_2, H_2O, and soluble minerals, and new microbial biomass is formed as well. It has been discovered that the ASS process effectively treats inorganic pollutants, organic carbon substances, and NH_3 in the environment [56].

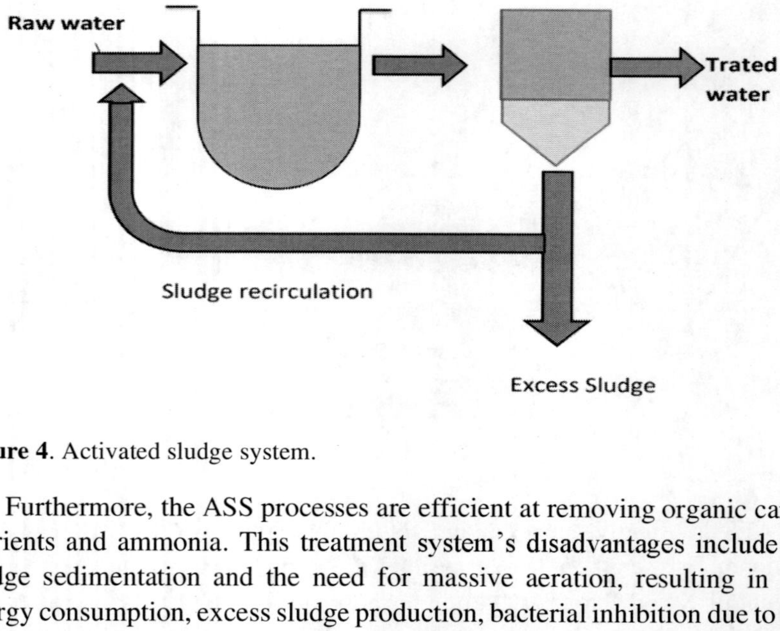

Figure 4. Activated sludge system.

Furthermore, the ASS processes are efficient at removing organic carbon nutrients and ammonia. This treatment system's disadvantages include low sludge sedimentation and the need for massive aeration, resulting in high energy consumption, excess sludge production, bacterial inhibition due to high ammonia concentrations, and microbial inhibition due to high ammonia concentrations and ammonium-nitrogen strength. This treatment system has several advantages, including high sludge sedimentation, low energy consumption, and excess sludge production [56].

Membrane Bio-Reactor (MBR) Technology

A Membrane Bio-Reactor (MBR) is an activated slug process that can be used in aerobic and anaerobic conditions under an efficient filtration process that can hold tiny particles. It is a type of Activated Slug Process. Despite the wide variation in LFL quality, it is used as a very effective treatment method to treat leachate by membrane separation of high biomass concentration despite the high biomass concentration [57]. Conventional MBR processes are composed primarily of two essential components: the bioreactor and the membrane module, which are responsible for the biodegradation and filtration processes, respectively, in the process [57, 58]. In most cases, anaerobic reactors are used to treat young LFL containing a high organic load that has been discharged from landfill sites.

The ability to treat challenging wastewaters and handle high organic loadings, high resistance to shock load, less sludge production, high effluent quality and decomposition of the micro-pollutants. Disadvantages higher technological requirement is needed for infrastructure investment of MBR; membrane module must be cleaned or replaced, increasing membrane cost and short lifespan; a certain membrane fouling during system operation results in flux decline and reduces water production rate of the system [58].

Anaerobic Reactor

Anaerobic reactors are a common technique for landfill leachate treatment that reduces solid waste and recycles methane, which helps to promote global carbon circulation. Anaerobic reactors reduce solid waste while also recycling methane to promote global carbon circulation [58]. Anaerobic digesters produce methane gas as a by-product of the decomposition of organic matter. Hydrolysis, acidogenesis (acetogenesis), and methanogenesis are the three major steps in anaerobic digestion. The first of these steps involves hydrolysis. Some microorganisms that live and thrive in this environment are hydrolytic bacteria, acidogens, hydrogen-producing acetogenins, homo-acetogens, and methanogens.

Microbes to Wastewater Treatment

Microorganisms play a vital role in bioremediation, an economically beneficial and ecologically beneficial substitute for conventional treatment methods. Numerous microbial groups, particularly fungus and bacteria, have been studied for their potential to degrade and decolorize hazardous pollutants present in a variety of industrial wastewaters, especially distillery wastewater [59].

For bioremediation of distillery effluent, free or immobilized cells have been extensively researched. Compared to free-cell culture, microorganisms are better immobilized on non-reactive materials such as polymer substances and polyurethane. A few advantages of carrier pellets are their compact physical structure, good biomass retention, reusability of culture, and ease of separation method [60].

Compared to traditional bacteria, different varieties of bacteria in sewage treatment have become a rising and essential technology in recent decades. This method is owing to the economic effectiveness, efficiency of such operation, and decreases in garbage volume [61]. One of the innovative treatment strategies is bacteria to remove heavy organic pollutants from sewage treatment systems [62, 63]. The anaerobic degradation process comprises significant three reactions where the complex substrates are converted to simpler forms by hydrolysis, followed by their conversion [64, 65].

Integrated Microbial and Phytoremediation Approach for Wastewater Treatment

The application of living organisms to remove environmental pollutants from a contaminated site is known as bioremediation, which is considered a promising alternative for wastewater treatments. Some researchers have studied the potential applicability of *Bacillus cereus*, *Enterobacter ludwigii*, and *Enterobacter* sp. on the biodegradation of two major antibiotics: Amoxicillin (AMX) and Sulfonamide (SDI). Among these three strains, the B. cereus strain was the most potential candidate as it removed 100% of AMX and 80% of SDI within 14 days of incubation [26]. Further, it has recorded the complete degradation of microcystin toxins: MC-LR, by two native bacteria; *Bacillus cereus* and *Rahnellaa quatilis*, at 28°C [30]. Furthermore, it has recorded the decolorization potential of textile dye by indigenous bacteria and discovered that isolated native bacterial strains would be successful use as a key tool for bioremediation artificial textile as complete decolourization has achieved within a short period, along with complete detoxification [8, 9]. The use of plants to remove toxic pollutants from waste effluents is known as phytoremediation. Phytoremediation use several mechanisms; degradation of pollutants through phytoextraction, phytodegradation and phytovolatilization processes or immobilizing pollutants via rhizofiltration followed by phytostabilization. These plant mechanisms reduce the amount of hazardous pollutants through their enzymatic biological processes. Depending on the purpose, they can be applied in both *in-situ* and *ex-situ* remediation processes [66, 67]. Plants that are effective for phytoremediation should have some unique characteristics, such as fast-growing ability, creating a lot of biomass, and absorbing metals. The selected plant species for the phytoremediation should be native species reflective of the ecology conveniently harvestable. Phytoremediation of toxic metals is considered a better alternative to treating wastewater with heavy metals based on the absorption and accumulation of waste substances [68]. Plants contain different mechanisms for survival against toxic metal components present in the surroundings; some rely on effective metal removal strategies that limit heavy metal transport to aerial parts, while others concentrate within aerial parts of the plant as harmless substances [69].

Microbial Consortium for Wastewater Treatment

A single microbial strain cannot completely degrade most complex environmental contaminants into nontoxic and simpler forms. Therefore, the application of several bacterial species together, known as the bacterial consortium, has become a popular novel concept. Microorganisms have different biochemical pathways and syntrophic interactions between individual bacteria in a bacterial consortium resulting in the complete mineralization of complex pollutants [8-10]. Several studies have been conducted on removing a textile dye using a developed bacterial consortium and have obtained promising results for removing a textile dye in textile effluents [70, 71]. Further, some researchers have studied the application of bacterial consortia treatment of wastewater with Bisphenol, a synthetic compound widely used in the plastic industry, and have achieved successful results [72]. Moreover, researchers have studied the biotreatment of COD

(92.8%), total nitrogen (78.1%) and phosphate (99%) in distillery effluent using a Leptolyngbya-based microbial formula. Similar studies have been done for the applicability of a microbial consortium isolated from paper mill effluent [73]. It has recorded promising results for converting cellulose to different by-products such as biogas and organic acids using a consortium of *Clostridium* sp., *Klebsiella* sp., *Rotella* sp. and *Desulfovibrio* sp. and treatment of wastewater. Further studies done by Singh et al. (2022) have developed a potential bacterial combination for treating effluent water containing oil, grease, and different toxic metals. It has recorded the removal of heavy metals of Cr and Ni by 84.93% and 69.25%, respectively, while Oil and grease removal by 71.8%.

Biochemical Degradation Mechanisms

Biological wastewater treatment using diverse microorganisms has long been regarded as a cost-effective and environmentally friendly option. A broad range of microbial communities that can withstand the effluent medium's demanding conditions can degrade organic waste into soluble simpler compounds. Although microorganisms may absorb organic substances and transform them into CO_2, this approach has some disadvantages, including high aeration costs, maintenance costs, and more significant sludge generation [63].

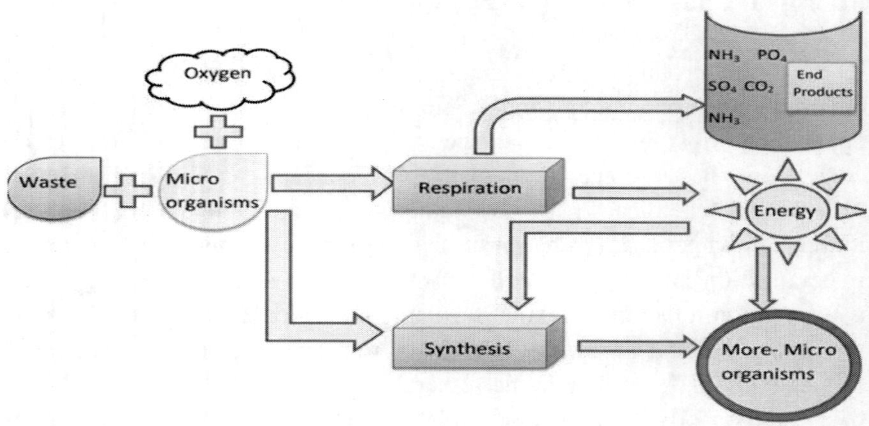

Figure 5. Aerobic waste water treatment mechanism.

Anaerobic Digestion

Under an oxygen-free environment, anaerobic bacteria undertake anaerobic digestion. Hydrolysis, fermentation, acetogenesis, and methanogenesis are the four main steps of anaerobic digestion [74]. Organic macromolecules in the solid waste matter are degraded into small molecules through a series of reactions, resulting in different biogases of acetate. However, with the greater volume of H_2 generated by acidogenic bacteria during the acidogenesis phase, the partial pressure of H_2 increased, making the acidogenesis process thermodynamically infeasible [75, 76]. The syntrophic interaction between two organisms known as methanogenic archaea and fermentative bacteria can eliminate the internal energy gap to produce methane [27].

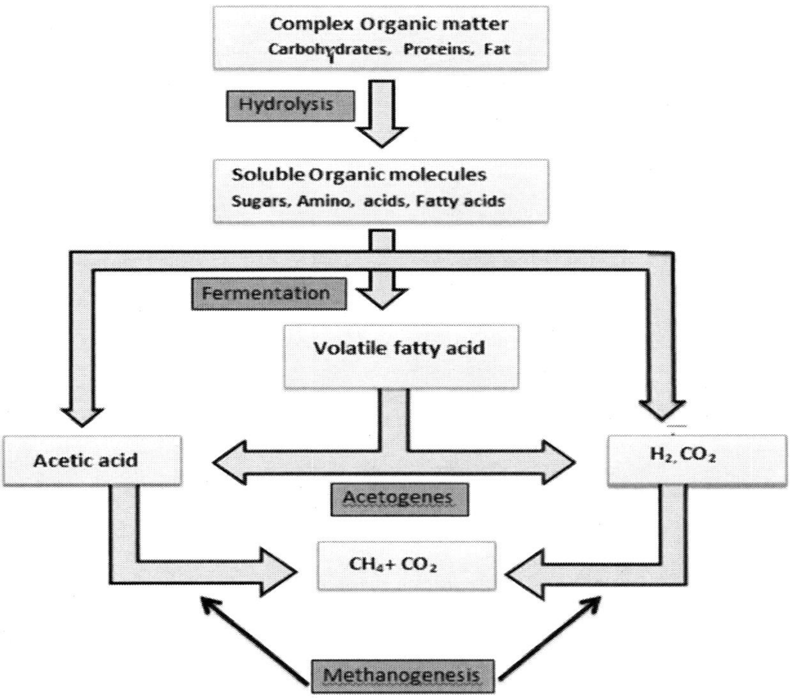

Figure 6. Anaerobic respiration mechanism.

Bio Augmentation and Bio Stimulation in Wastewater Treatment

Bioaugmentation and biostimulation are promising technologies for improving contaminant removal efficiency in the environment. Microorganisms possess a vast range of metabolic pathways to withstand inhospitable environments. Therefore, microorganisms play an important role in site remediation. Even though relative efficiency is based on the pollutant concentrations and environmental physicochemical characteristics, the ability of native microbial communities to break down environmental contaminants can be stimulated by introducing different nutrients or adding specialized microbial species into the native community [77].

Incorporating supplemental microorganisms and their related biodegradation capabilities into effluent treatment processes is defined as bio augmentation [78].

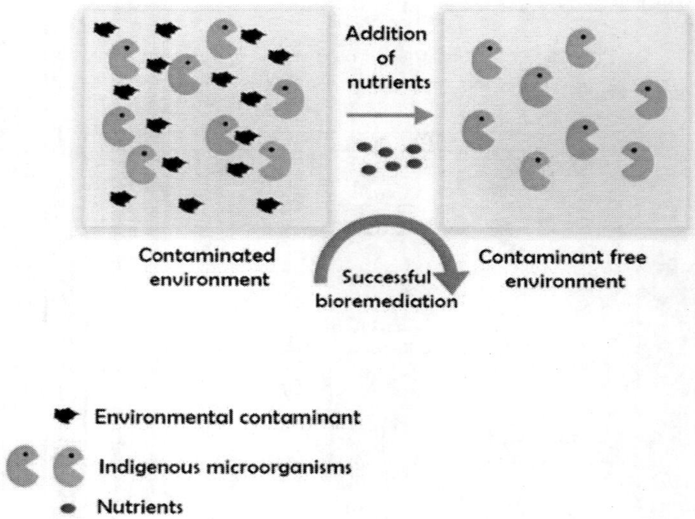

Figure 7. Mechanism of biostimulation.

This intervention aims to increase the biochemical capacity of the native microbial population already existing in the environment/ through exogenously increased genetic diversity. The application of appropriate microorganisms is critical to bio augmentation success. Temperature, pH, dissolved oxygen, nutrition availability, toxicity, and microbial pressures are just a few of the environmental factors borne by the selected strain(s) during the treatment process [79]. Immobilization of microorganisms has been proposed to improve bio augmentation effectiveness [79].

This intervention tries to boost the biochemical capacity of the native microbial community in the environment by increasing genetic variation externally. The use of good microorganisms is crucial for bio augmentation success. Temperature, pH, dissolved oxygen, nutrient availability, toxicity, and microbial pressures are some environmental factors that the selected strain(s) must contend with during the treatment process. Bioaugmentation efficiency could be improved by immobilizing organisms [78, 79].

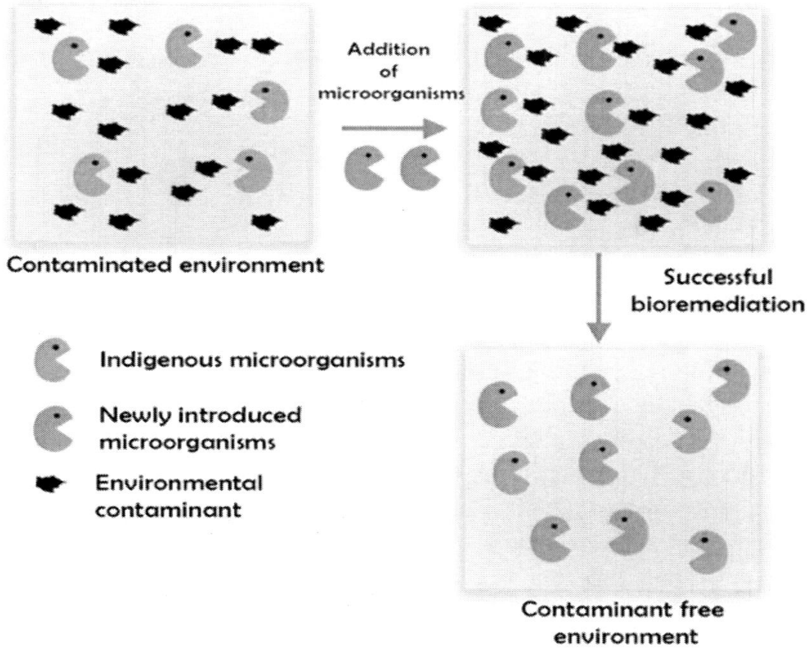

Figure 8. Bioaugmentation.

Some researchers have also attempted to biodegrade hydrocarbon and pesticide compounds using genetically engineered microorganisms. The genetic diversity and metabolic versatility of microorganisms have also been investigated for their ability to improve derivative properties. The results showed that the bioaugmentation process of integrated side-stream nitrifiers improved nitrification efficiency even at low pH.

Biostimulation entails increasing the nutritional content of soils to promote microbial growth and pollutant decomposition, and both methods have proven to be more effective [80]. Despite microorganisms capable of pollutant biodegradation in contaminated soils and aquifers, local environmental conditions in these areas may be unfavorable. In some cases, the naturally occurring microbial community at a polluted site may lack the metabolic potential to degrade and completely mineralize the target pollutant into small, stable molecules like CO_2 or CH_4 and H_2O [26]. This could be due to a single microorganism with all of the enzymes necessary to 'productively' transform the pollutant.

Novel Approaches of Wastewater Treatment Technology

Application of Bio Char

Cell growth, cell separation, and sensitivity to environmental factors such as influent quality fluctuations are all factors that limit biodegradation. To tackle these questions, cell immobilization can be used in microbial treatment, which restricts microorganisms to a specific spatial region, allowing them to maintain biological activity and multiply faster. Because of its high stability, rapid reaction rate, and high activity, microbial immobilization technology (MIT) has rapidly developed and removed pollutants from water/wastewater.

With its porous structure, large surface area, and variable surface compositions, bio char, a carbon-rich material produced by thermal conversion of biomass under anaerobic or oxygen-limited conditions, has been widely used in soil remediation [81]. Because of its high porosity and good adsorption effect, bio char is considered an effective carrier for microbial immobilization, as it can provide a habitat for microorganisms. Several studies have used bio char as a carrier material to immobilize bacteria in waste treatment technology. For the immobilized bacterial species, the biochar surface serves as a complete carbon source and an effective sorbent material. Some researchers explored the use of bio char in the removal of organic micropollutants from wastewater using three major fungal divisions: Basidiomycota, Ascomycota, and Mucoromycotina [80-82]. It also includes fungal bioreactors, fungal biofilms, and immobilization systems. To decolorize Congo red (CR) dye, researchers have investigated using a hybrid treatment system. The biological treatment happened in a packed bed bioreactor with *Terminalia Arjuna* seeds biochar immobilized with *Providencia stuartii*, while the ozonation was done in an ozone reactor. Congo red (CR) dye is almost completely de-coloured using the hybrid approach [83].

Nanotechnology

Nanotechnology is a cutting-edge technology that is used in a variety of industries, including wastewater treatment. Nanotechnology has the potential to remove pollutants and germs from water more effectively. Because nanoscale materials have a higher surface-to-volume ratio, their average properties are altered. Nanotechnology becomes more sustainable and environmentally friendly when nanomaterials are bio-fabricated, and microbes are used simultaneously. Carbon nanotubes (CNTs), activated carbon, graphene, and, to a lesser extent, fullerene are examples of carbon-based nanoparticles [84].

Nanomaterials improve membrane permeability, stink resistance, mechanical and thermal strength, and offer new pollutant degradation functions. Creating green nanomaterials from microorganisms and other organism extracts have paved the way for environmentally friendly pollution remediation. Iron nanoparticles are used in remediation because of their redox potential when reacting with water, magnetic susceptibility, and non-toxic nature [85].

Several studies on the potential use of nanomaterials for wastewater treatment have been conducted. Due to their unique properties (i.e., high specific surface area, desired morphology, catalytic reactivity) for the biodegradation and biosorption of T and EPs, biogenic nanoparticles fabricated using microbes/microorganisms have recently attracted much research interest. The breakdown of pollutants by bacteria has been reported as a viable and safe sewage treatment method. This agent removed 83.1% of chemical oxygen demand, 87.5% NH_4^+, and 92.1% of PO_4^{3-} from wastewater in the study [82].

Additionally, in the Sundarbans, India, *Aspergillus tubingensis* bio fabricated iron oxide nanoparticles from the rhizosphere of *Avicennia officinalis*. The nanoparticles were found to be capable of removing more than 90% of heavy metals from wastewater and could regenerate up to five times. Metal ions were chemically adsorbed on the surface of nanoparticles in endothermic processes [86]. Another study used exopolysaccharides from *Chlorella vulgaris* to co-precipitate with iron oxide nanoparticles. The successful modification of nanoparticles by functional groups of EPS was discovered using Fourier-transform infrared spectroscopy (FT-IR). Furthermore, the nanocomposite removed 91% of PO_4^{3-} and 85% of NH_4^+ [86].

Conclusion and Future Perspectives of Biological Wastewater Treatment

Biological wastewater treatment is a rapidly evolving technology covering a wide range of topics, including recombinant DNA, enzymes, biotechnology, and advanced nanotechnology [87]. As a result, the immobilization of enzymes in various Nanoscaled materials is a rapidly expanding field. Nanostructured materials, such as nano adsorbents, nanometals, nano-membranes, and photocatalysts, are novel nanotechnology techniques for wastewater treatment processes [87].

The feasibility, scalability, safety, and cost of operating microbial incorporated Nano based systems will focus on future biological wastewater treatment research. Many of the current limitations will likely be overcome by innovations addressing these concerns, allowing for more opportunities to improve water reclamation treatment. The restriction of nanotechnology for water applications, such as rules and regulations and potential health risks, are discussed alongside the promised scientific advances.

References

[1] Andersson, M., Beale, S. B., Espinoza, M., Wu, Z., Lehnert, W., 2016. A review of cell-scale multiphase flow modeling, including water management, in polymer electrolyte fuel cells. *Applied Energy*, 180, pp. 757-778.

[2] World Health Organization, 2017. Annual report WHO/UNICEF joint monitoring programme for water supply, sanitation and hygiene. In *2017 annual report WHO/UNICEF joint monitoring programme for water supply, sanitation and hygiene*, pp. 20-20.

[3] Kumar, L. and Bharadvaja, N., 2020. *Microbial remediation of heavy metals. Microbial bioremediation and biodegradation*, Springer, pp. 49-72.

[4] Yohannes, H. and Elias, E., 2017. Contamination of rivers and water reservoirs in and around Addis Ababa City and actions to combat it. *Environ Pollution and Climate Change*, 1(116), pp. 8.

[5] Hoang, H. T. T., Duong, T. T., Nguyen, K. T., Le, Q. T. P., Luu, M. T. N., Trinh, D. A., Le, A. H., Ho, C. T., Dang, K. D., Némery, J. and Orange, D., 2018. Impact of anthropogenic activities on water quality and plankton communities in the Day River (Red River Delta, Vietnam). *Environmental monitoring and assessment*, 190(2), pp. 1-18.

[6] Valek, A. M., Susnik, J. and Grafakos, S., 2017. Quantification of the urban water-energy nexus in México City, México, with an assessment of water-system related carbon emissions. *Science of the Total Environment*, 590, pp. 258-268.

[7] Chowdhary P, Yadav A, Kaithwas G andBharagava, R. N., 2017. Distillery wastewater: a major source of environmental pollution and its biological treatment for environmental safety. *Green Technologiesand Environmental Sustainability*: pp. 409–435.

[8] Ekanayake, E. M. M. S. and Pathmalal M. Manage (2020a). Decolorization and detoxification of CI Direct Blue 201 textile dye by two fungal strains of genus Aspergillus. *Journal of National Science Foundation*, Sri Lanka, Vol. 48(1), pp. 69-80.

[9] Ekanayake, E. M. M. S. and Manage, P. M., 2020c. Green approach for decolorization and detoxification of textile dye-CI direct blue 201 using native bacterial strains. *Environment and Natural Resources Journal*, 18(1), pp. 1-8.

[10] Mahagamage M. G. Y. L, Chinthaka S. D. M, Manage P. M., 2015. Multivariate analysis of physico-chemical and microbial parameters of surface water in Kelani river basin. *International Journal of Multidisciplinary Studies* 1(1):55-61.

[11] Sharma, R., Kumar, R., Sharma, D. K., Sarkar, M., Mishra, B. K., Puri, V., Priyadarshini, I., Thong, P. H., Ngo, P. T. T. and Nhu, V. H., 2021. Water pollution examination through quality analysis of different rivers: a case study in India. *Environment Development and Sustainability*, pp. 1-22.

[12] Singh, P., Kadam, V. and Patil, Y., 2022. Isolation and development of a microbial consortium for the treatment of automobile service station wastewater. *Journal of Applied Microbiology*, 132(2), pp. 1048-1061.

[13] Ekanyaka, E. M. M. S and Pathmalal M. Manage (2020b). Effect of mixed bacterial consortia for decolourization and detoxification of azo dye:CI direct blue 201 textile dye, *Pollution Research journal*, Vol 39(2), pp. 187-195.

[14] Zhu, Y., Chen, Y., Gong, X., Peng, Y., Wang, Z. and Ji, B., 2018. Plastic film mulching improved rhizosphere microbes and yield of rain fed spring wheat. *Agricultural and Forest Meteorology*, 263, pp. 130-136.

[15] Konig, M., Escher, B. I., Neale, P. A., Krauss, M., Hilscherová, K., Novák, J., Teodorović, I., Schulze, T., Seidensticker, S., Hashmi, M. A. K. and Ahlheim, J., 2017. Impact of untreated wastewater on a major European river evaluated with a combination of in vitro bioassays and chemical analysis. *Environmental Pollution*, 220, pp. 1220-1230.

[16] Okereke, J. N., Ogidi, O. I. and Obasi, K. O., 2016. Environmental and health impact of industrial wastewater effluents in Nigeria-A Review. *International Journal of Advance Research in Biological Science*, 3(6), pp. 55-67.

[17] Zohdi, E. and Abbaspour, M., 2019. Harmful algal blooms (red tide): a review of causes, impacts and approaches to monitoring and prediction. *International Journal of Environmental Science and Technology*, 16(3), pp. 1789-1806.

[18] Lakind, J. S., Lehmann, G. M., Davis, M. H., Hines, E. P., Marchitti, S. A., Alcala, C. and Lorber, M., 2018. Infant dietary exposures to environmental chemicals and infant/child health: a critical assessment of the literature. *Environmental health perspectives*, 126(9), pp. 096002.

[19] Gupta, A., Rai, D. K., Pandey, R. S. and Sharma, B., 2009. Analysis of some heavy metals in the riverine water, sediments and fish from river Ganges at Allahabad. *Environmental monitoring and assessment*, 157(1), pp. 449-458.

[20] Rehman, K., Fatima, F., Waheed, I. and Akash, M. S. H., 2018. Prevalence of exposure of heavy metals and their impact on health consequences. *Journal of cellular biochemistry*, 119(1), pp. 157-184.

[21] Adedokun, A. H., Njoku, K. L., Akinola, M. O., Adesuyi, A. A. and Jolaoso, A. O., 2016. Potential human health risk assessment of heavy metals intake via consumption of some leafy vegetables obtained from four market in Lagos Metropolis, Nigeria. *Journal of Applied Sciences and Environmental Management*, 20(3), pp. 530-539.

[22] Pal, M., Ayele, Y., Hadush, M., Panigrahi, S. and Jadhav, V. J., 2018. Public health hazards due to unsafe drinking water. *Air Water Borne Dis*, 7(1000138), pp. 2.

[23] Saha, J. K., Selladurai, R., Coumar, M. V., Dotaniya, M. L., Kundu, S. and Patra, A. K., 2017. Organic pollutants. In *Soil Pollution-An Emerging Threat to Agriculture*, pp. 105-135). Springer, Singapore.

[24] Sinha, D. and Prasad, P., 2020. Health effects inflicted by chronic low-level arsenic contamination in groundwater: A global public health challenge. *Journal of Applied Toxicology*, 40(1), pp. 87-131.

[25] Alegbeleye, O. O., Opeolu, B. O. and Jackson, V. A., 2017. Polycyclic aromatic hydrocarbons: a critical review of environmental occurrence and bioremediation. *Environmental management*, 60(4), pp. 758-783.

[26] Liyanage, G. Y. and Manage, P., 2016. Evaluation of Amoxicillin and Sulfonamide removal by Bacillus cereus, *Enterobacter ludwigii* and *Enterobacter* sp. *Environment and Natural Resources Journal*, 14(1), pp. 39-43.

[27] Wu, B., 2020. Human health hazards of wastewater. In *High-Risk Pollutants in Wastewater*, pp. 125-139. Elsevier.

[28] Ben, Y., Fu, C., Hu, M., Liu, L., Wong, M. H. and Zheng, C., 2019. Human health risk assessment of antibiotic resistance associated with antibiotic residues in the environment: A review. *Environmental research*, 169, pp. 483-493.

[29] Oyibo, J. N., Wegwu, M. O., Uwakwe, A. A. and Osuoha, J. O., 2018. Analysis of total petroleum hydrocarbons, polycyclic aromatic hydrocarbons and risk assessment of heavy metals in some selected finfishes at Forcados Terminal, Delta State, Nigeria. *Environmental Nanotechnology, Monitoring and Management*, 9, pp. 128-135.

[30] Idroos, E. S. and Manage, P. M., 2018. Bioremediation of Microcystins by Two Native Bacteria; *Bacillus Cereus* and *Rahnella Aquatilis*. *Asian Journal of microbial Bio Technology and Environmental Science*, 20 (3), pp. 24-32.

[31] Wheater, H.S. and Gober, P., 2015. Water security and the science agenda. *Water Resources Research*, 51(7), pp. 5406-5424.

[32] Demiroglu, D., 2018. The Effects of Integrated Urban Water Management on the Quality of Urban Life. *Science, Ecology and Engineering Research in the Globalizing World*, pp. 291.

[33] Pescod, M. B., 1992. Wastewater treatment and use in agriculture-FAO irrigation and drainage paper 47. *Food and Agriculture Organization of the United Nations*, Rome.

[34] Ingle, K. N., Fenta, M. C., Harada, K., Ingle, A. S. and Ueda, A., 2022. Wastewater Treatment Plants Advantage to Combat Climate Change and Help Sustainable Water Management. *Water Resources Management and Sustainability*, pp. 73-90.

[35] Khalid, S., Shahid, M., Bibi, I., Sarwar, T., Shah, A. H. and Niazi, N. K., 2018. A review of environmental contamination and health risk assessment of wastewater use for crop irrigation with a focus on low and high-income countries. *International journal of environmental research and public health*, 15(5), pp. 895.

[36] Shahid, M., Khalid, S., Niazi, N. K., Murtaza, B., Ahmad, N., Farooq, A., Zakir, A., Imran, M. and Abbas, G., 2021. Health risks of arsenic buildup in soil and food crops after wastewater irrigation. *Science of the Total Environment*, 772, pp. 145266.

[37] Jaramillo, M. F. and Restrepo, I., 2017. Wastewater reuse in agriculture: A review about its limitations and benefits. *Sustainability*, 9(10), pp. 1734.

[38] Magwaza, S. T., Magwaza, L. S., Odindo, A. O., Mditshwa, A. and Buckley, C., 2020. Partially treated domestic wastewater as a nutrient source for tomatoes (*Lycopersicum solanum*) grown in a hydroponic system: effect on nutrient absorption and yield. *Heliyon*, 6(12), pp 05745.

[39] Nyomora, A. M., 2015. Effect of treated domestic wastewater as source of irrigation water and nutrients on rice performance in Morogoro, Tanzania. *Journal of Environment and Waste Management*, 2(2), pp. 47-55.

[40] Edwards, P., 2015. Aquaculture environment interactions: past, present and likely future trends. *Aquaculture*, 447, pp. 2-14.

[41] Abdallah, C. K., 2018. *Water Quality Assessment of Irrigation Water Used by Informal Peri Urban Vegetable Irrigation Farmers: The Case of Tamale Metropolis*, Ghana (Master's thesis).

[42] Adegoke, A. A., Amoah, I. D., Stenström, T. A., Verbyla, M. E. and Mihelcic, J. R., 2018. Epidemiological evidence and health risks associated with agricultural reuse of partially treated and untreated wastewater: a review. *Frontiers in public health*, pp. 337.

[43] Vardhan, K. H., Kumar, P. S. and Panda, R. C., 2019. A review on heavy metal pollution, toxicity and remedial measures: Current trends and future perspectives. *Journal of Molecular Liquids*, 290, pp. 111197.

[44] Dolores, M., Barrio, H. and Martin, J. M., 2016. *Urban Water Reuse in Agriculture.*

[45] Millar, G. J., Couperthwaite, S. J. and Moodliar, C. D., 2016. Strategies for the management and treatment of coal seam gas associated water. *Renewable and Sustainable Energy Reviews*, 57, pp. 669-691.

[46] Enamala, M. K., Enamala, S., Chavali, M., Donepudi, J., Yadavalli, R., Kolapalli, B., Aradhyula, T. V., Velpuri, J. and Kuppam, C., 2018. Production of biofuels from

microalgae-A review on cultivation, harvesting, lipid extraction, and numerous applications of microalgae. *Renewable and Sustainable Energy Reviews*, 94, pp. 49-68.

[47] Patel, D., Bapodra, S. L., Madamwar, D. and Desai, C., 2021. Electroactive bacterial community augmentation enhances the performance of a pilot scale constructed wetland microbial fuel cell for treatment of textile dye wastewater. *Bioresource Technology*, 332, pp. 125088.

[48] Gaur, R. Z., Khoury, O., Zohar, M., Poverenov, E., Darzi, R., Laor, Y. and Posmanik, R., 2020. Hydrothermal carbonization of sewage sludge coupled with anaerobic digestion: Integrated approach for sludge management and energy recycling. *Energy Conversion and Management*, 224, pp. 113353.

[49] Rulkens W. Sewage sludge as a biomass resource for the production of energy: overview and assessment of the various options. *Energy Fuel*. 2008; 22(1), pp. 9–15.

[50] Arefin, R., Meshram, S. G., Santos, C. A. G., da Silva, R. M. and Pushparaj, J., 2020. Hybrid modelling approach for water body change detection at Chalan Beel area in northern Bangladesh. *Environmental Earth Sciences*, 79(19), pp. 1-18.

[51] Capodaglio, A. G., 2017. Integrated, decentralized wastewater management for resource recovery in rural and peri-urban areas. *Resources*, 6(2), p. 22.

[52] Luo, H., Zeng, Y., Cheng, Y., He, D. and Pan, X., 2020 Recent advances in municipal landfill leachate: A review focusing on its characteristics, treatment, and toxicity assessment. *Science of The Total Environment*, 703, pp135468.

[53] Kleerebezem, R., Joosse, B., Rozendal, R. and Van Loosdrecht, M., 2015. *Anaerobic digestion without biogasiews in Environmental Science and Bio/Technology*, 14(4), pp. 787-801.

[54] Bove, J., Arbat, G., Duran-Ros, M., Pujol, T., Velayos, J., de Cartagena, F. R. and Puig-Bargués, J., 2015. Pressure drop across sand and recycled glass media used in micro irrigation filters. *Biosystems Engineering*, 137, pp. 55-63.

[55] Ramachandra, T. V., Mahapatra, D. M., Bhat, S. P. and Joshi, N. V., 2015. Biofuel production along with remediation of sewage water through algae. In *Algae and environmental sustainability* (pp. 33-51). Springer, New Delhi.

[56] Xu, J., Cao, Z., Zhang, Y., Yuan, Z., Lou, Z., Xu, X. and Wang, X., 2018. A review of functionalized carbon nanotubes and graphene for heavy metal adsorption from water: Preparation, application, and mechanism. *Chemosphere*, 195, pp. 351-364.

[57] Mujtaba, G. and Lee, K., 2017. Treatment of real wastewater using co-culture of immobilized *Chlorella vulgaris* and suspended activated sludge. *Water research*, 120, pp. 174-184.

[58] Abuabdou, S. M., Bashir, M. J., Aun, N. C. and Sethupathi, S., 2018, April. Applicability of anaerobic membrane bioreactors for landfill leachate treatment: review and opportunity. In *IOP Conference Series: Earth and Environmental Science* (Vol. 140, No. 1, p. 012033). IOP Publishing.

[59] Feng, P., Wang, B., Li Liu, D., Waters, C. and Yu, Q., 2019. Incorporating machine learning with biophysical model can improve the evaluation of climate extremes impacts on wheat yield in south-eastern Australia. *Agricultural and Forest Meteorology*, 275, pp. 100-113.

[60] Fito, J., Tefera, N. and Van Hulle, S. W., 2019. Sugarcane biorefineries wastewater: bioremediation technologies for environmental sustainability. *Chemical and Biological Technologies in Agriculture*, 6(1), pp. 1-13.

[61] Bezuneh, T. T., 2016. The role of microorganisms in distillery wastewater treatment: a review. J. *Bioremediat. Biodegrad*, 7(6).

[62] Ibrahim, S., El-Liethy, M. A., Elwakeel, K. Z., Hasan, M. A. E. G., Zanaty, A., Mahmoud, A. and Kamel, M. M., 2020. Role of identified bacterial consortium in treatment of Quhafa Wastewater Treatment Plant influent in Fayuom, Egypt. *Environmental monitoring and assessment*, 192(3), pp. 1-10.

[63] Ayilara, M. S., Olanrewaju, O. S., Babalola, O. O. and Odeyemi, O., 2020. Waste management through composting: Challenges and potentials. *Sustainability*, 12(11), pp. 4456.

[64] Saxena, G., Kishor, R. and Bharagava, R. N., 2020. *Bioremediation of industrial waste for environmental safety*. Springer Singapore.

[65] Rawoof, S. A. A., Kumar, P. S., Vo, D. V. N. and Subramanian, S., 2021. Sequential production of hydrogen and methane by anaerobic digestion of organic wastes: a review. *Environmental Chemistry Letters*, 19(2), pp. 1043-1063.

[66] Idroos, E. S. and Manage, P. M., 2018. *Bioremediation of Microcystins by Two Native Bacteria*; *Bacillus Cereus and Rahnella Aquatilis*.

[67] Awa, S. H. and Hadibarata, T., 2020. Removal of heavy metals in contaminated soil by phytoremediation mechanism: a review. *Water, Air, and Soil Pollution*, 231(2), pp. 1-15.

[68] Bordoloi, N., Tiwari, J., Kumar, S., Korstad, J. and Bauddh, K., 2020. Efficiency of algae for heavy metal removal, bioenergy production, and carbon sequestration. In *Emerging Eco-friendly Green Technologies for Wastewater Treatment*, pp. 77-101.

[69] Yaashikaa, P. R., Kumar, P. S., Jeevanantham, S. and Saravanan, R., 2022. A review on bioremediation approach for heavy metal detoxification and accumulation in plants. Environmental Pollution, p. 119035.

[70] Das, A. and Mishra, S., 2017. Removal of textile dye reactive green-19 using bacterial consortium: process optimization using response surface methodology and kinetics study. *Journal of environmental chemical engineering*, 5(1), pp. 612-627

[71] Thiruppathi, K., Rangasamy, K., Ramasamy, M. and Muthu, D., 2021. Evaluation of textile dye degrading potential of ligninolytic bacterial consortia. *Environmental Challenges*, 4, pp100078.

[72] Eio, E. J., Kawai, M., Tsuchiya, K., Yamamoto, S. and Toda, T., 2014. Biodegradation of bisphenol A by bacterial consortia. *International Biodeterioration and Biodegradation*, 96, pp. 166-173.

[73] Rabelo, C. A. B. S., Soares, L. A., Sakamoto, I. K. and Varesche, M. B. A., 2018. Bioconversion of cellulose into hydrogen, biogas and organic acids using microbial consortium from a pulp and paper mill wastewater treatment plant. *Quimica Nova*, 41, pp. 169-175.

[74] Wang, D., Liu, Y., Ngo, H. H., Zhang, C., Yang, Q., Peng, L., He, D., Zeng, G., Li, X. and Ni, B. J., 2017. Approach of describing dynamic production of volatile fatty acids from sludge alkaline fermentation. *Bio resource technology*, 238, pp. 343-351.

[75] Kaur, G., Johnravindar, D. and Wong, J. W., 2020. Enhanced volatile fatty acid degradation and methane production efficiency by biochar addition in food waste-sludge co-digestion: A step towards increased organic loading efficiency in co-digestion. *Bio resource technology*, 308, pp. 123250.

[76] Wu, B., 2020. Human health hazards of wastewater. In *High-Risk Pollutants in Wastewater*, pp. 125-139. Elsevier.

[77] Adams, G. O., Fufeyin, P. T., Okoro, S. E. and Ehinomen, I., 2015. Bioremediation, biostimulation and bioaugmention: a review. *International Journal of Environmental Bioremediation and Biodegradation*, 3(1), pp. 28-39.

[78] Raper, E., Stephenson, T., Anderson, D. R., Fisher, R. and Soares, A., 2018. Industrial wastewater treatment through bioaugmentation. *Process Safety and Environmental Protection*, 118, pp. 178-187.

[79] Lebeau, T., 2011. Bioaugmentation for in situ soil remediation: how to ensure the success of such a process. In *Bioaugmentation, biostimulation and biocontrol*, Springer, Berlin, Heidelberg, pp. 129-186.

[80] Kumar, V., Shahi, S. K. and Singh, S., 2018. Bioremediation: an eco-sustainable approach for restoration of contaminated sites. In *Microbial bioprospecting for sustainable development*, pp. 115-136.

[81] Godlewska, P., Schmidt, H. P., Ok, Y. S. and Oleszczuk, P., 2017. Biochar for composting improvement and contaminants reduction. A review. *Bioresource Technology*, 246, pp. 193-202.

[82] Lu, L., Yu, W., Wang, Y., Zhang, K., Zhu, X., Zhang, Y., Wu, Y., Ullah, H., Xiao, X. and Chen, B., 2020. Application of biochar-based materials in environmental remediation: from multi-level structures to specific devices. *Biochar*, 2(1), pp. 1-31.

[83] He, S., Zhong, L., Duan, J., Feng, Y., Yang, B. and Yang, L., 2017. Bioremediation of wastewater by iron oxide-biochar nanocomposites loaded with photosynthetic bacteria. *Frontiers in microbi*ology, 8, p. 823.

[84] Ali, I., Peng, C., Khan, Z. M., Naz, I., Sultan, M., Ali, M., Abbasi, I. A., Islam, T. and Ye, T., 2019. Overview of microbes based fabricated biogenic nanoparticles for water and wastewater treatment. *Journal of environmental management*, 230, pp. 128-150.

[85] Manage, P. M., Kannangara, S. D. and Liyanage, G. Y., 2015. Role of native bacteria as a bioremediating agent for naphthalene and phenanthrene. *Proceedings of Academics World 12th International Conference*, Singapore. 20th December 2015.

[86] Govarthanan, M., Jeon, C. H., Jeon, Y. H., Kwon, J. H., Bae, H. and Kim, W., 2020. Non-toxic nano approach for wastewater treatment using *Chlorella vulgaris* exopolysaccharides immobilized in iron-magnetic nanoparticles. *International Journal of Biological Macromolecules*, 162, pp. 1241-1249.

[87] Reshmy, R., Philip, E., Sirohi, R., Tarafdar, A., Arun, K. B., Madhavan, A., Binod, P., Awasthi, M. K., Varjani, S., Szakacs, G. and Sindhu, R., 2021. Nanobiocatalysts: advancements and applications in enzyme technology. *Bioresource Technology*, 337, pp. 125491.

Chapter 13

Exploration and Exploitation of Bacterial Cellulose for Industrial Applications

Pardeep Kumar[1]
Babita Khosla[2]
and Machiavelli Singh[1,*]
[1]Amity Institute of Biotechnology, Amity University Haryana, Manesar, Gurugram, India
[2]Department of Environmental Sciences, Maharshi Dayanand University, Rohtak, India

Abstract

Bacterial cellulose is an aggregated event to form a ribbon like structure from β-1,4-glucan chain and form hydrogel film at air surface of culture media. Cellulose is a food source for herbivores, and it requires microbial degradation for the digestion in the alimentary tract. Bacterial cellulose has very high economic value due to its various usages in industries like paper, plastics, synthetic polymers, adhesive, explosive, moisture proof coating and thickening agents for foods. Cellulose is most abundant in plants due to its presence in plant cell walls but its separation and isolation from plants is very cost effective. During isolation of cellulose some other organic monomers and polymers are also filtered out and it is very difficult to separate them.

The isolation of cellulose does not meet the requirements of industries. The dependency on plats for cellulose is also harmful for the environment. To meet with the industrial requirement of cellulose we need another sustainable source of production of cellulose. The exploration and exploitation of bacterial cellulose can reduce the dependency on plants. The bacterial species that belong to the *Gluconacetobacter* genus produce cellulose. The bacterial cellulose is a

* Corresponding Author's Email: machiavellisingh@gmail.com.

In: Microbial Technologies and Their Applications
Editors: Santosh Kumar Karn and Anne Bhambri
ISBN: 979-8-88697-913-8

nanomaterial that possesses exclusive properties like high polymerization degree, crystallinity, surface area, tensile strength and flexibilities, water holding capacity. Bacterial cellulose is pure form and free from lignin and hemicelluloses compared to plants.

The bacterial cellulose has wide application due to their non-cytotoxic behavior and attracting interest in therapeutics industries. Nowadays the bacterial cellulose captures more strength in industries and to sustain the industrial demand it also requires bacterial strains that produce highly pure and more amount of cellulose. The *Komagataeibacter xylinus*is an acetic acid bacterium able to produce bacterial cellulose from various carbon sources. The K2G30 strain has the ability to produce much higher amounts of bacterial cellulose in mannitol and glucose-based media. The genomic analysis of K2G30 confers presence of one *bcsAB* gene and three *bcs* operons that are part of bacterial cellulose production.

The bacterial cellulose can be produced from various carbon sources and various types of organic waste. The optimization of the upstream process on the basis of genomic analysis can produce a higher concentration of bacterial cellulose. The downstream process also determines the end quality and quantity of the product. By the help of genome editing and recombinant DNA technology can modify the genome of various bacterial cellulose production bacteria and develop new strains that produce higher concentration of bacterial cellulose.

Keywords: bacterial cellulose, K2G30 strain, *Gluconacetobacter*, *Komagataeibacter xylinus*, genomic analysis, gene editing technology

Introduction

The large-scale production of biomaterial from microorganisms compared to synthetic one attracted the industrial community in recent years. The genetic modified organisms are getting more attention compared to synthetic biology. The major point is carbon sources for bacterial cellulose production and standard conditions that ensure good growth of microbes and produce large amounts of bacterial cellulose. The industrial waste can be a good carbon source for novel strain for production of bacterial cellulose. The conversion of glucose into cellulose by *Gluconacetobacter xylinus* is best fit to screening of various aspects of production of bacterial cellulose. The development of the best screening model for analyzing all parameters of cellulose production by bacteria play a crucial role in understanding the biological process of cellulose production. The understanding of utilization of carbon source is also the most influential factor because only growth of bacteria cannot provide well insights of production of desired product.

Table 1. Bacterial strain for production of bacterial cellulose

Strain	Carbone source	Nitrogen source	Reference
Gluconacetobacter hansenii UAC 09	Sucrose	-	[1]
G. xylinus ATCC 10245	Molasses	Corn syrup	[2]
Gluconacetobacter xylinus ATCC 53524	Sucrose	Peptone yeast extract	[3]
G. xylinus ATCC 23770	wheat straw hydrolysate	Peptone yeast extract	[4]
*Gluconacetobacter*sp.4B-2	Sucrose	Peptone yeast extract	[5]
G. hansenii PJK KCTC 10505BP	Glucose	Beer fermentation broth	[6]
G. xylinus BCRC 12334	Glucose	Peptone yeast extract	[7]
Acetobacter Xylinum ATCC 23769	Glucose	Peptone yeast extract	[8]
G. xylinus ATCC 23770	cotton cloth hydrolysate	Peptone yeast extract	[9]
Gluconacetobacter xylinus ATCC 23770	fiber hydrolysate	Peptone yeast extract	[10]
Gluconacetobacter xylinus ATCC 13693	Glucose	Peptone yeast extract	[11]

The plant-based cellulose is mainly contaminated with hemicelluloses and lignin that reduces the pure quality of cellulose. The bacterial cellulose is the purest form of cellulose with no contaminates and no other unusable byproducts.

Bacterial cellulose is known for its crystallized and high tensile strength that makes it a good alternative to natural and synthetic cellulose. The chemically highly designed, structured and well mixed with other microfibers of bacterial cellulose provides various industrial applications. This kind of approach increased the application of bacterial cellulose in biomedical and electronic developments. The understanding of patterns of cellulose pellicle formation increases utilization in other fields that are still not well explored.

The narrow knowledge is a defining feature of large scale production of bacterial cellulose from bacteria. Development of novel approaches or genetically modified organisms tackle the issue of large-scale production of bacterial cellulose. The deep insights of major biological processes and standard conditions of bacterial growth can ensure the higher production of material compared to normal selected bacteria. The development of observable standard approaches is more helpful to study production of desired products on a small scale and further for large scale production. The basic statistics that ensure the quality and quantity of desired products must be analyzed and compared on small scale production to get better strains that produce good amounts of desired products.

Nanoscale Details

Biomaterials require advanced understanding of the nanoscale environment. In recent years nanotechnology became a more advanced field that required the most sophisticated knowledge of any material that was used for various applications in industries. To reach this compatibility required understanding of every parameter of desired nanofibers. Bacterial cellulose is cell driven biomaterial so, microbes grown in the culture media required basic environment or optimal condition to grow and at cellular level with nanoscale environment aspects provides better understanding of biophysical environment. Cell communication is the most influential factor in production of bacterial cellulose. The impacts of cell physiology and communication are driven by surrounding biochemistry. The nano-environment can impact the biochemistry of the surrounding because cells can modulate the nano-environment to establish good communication and attachment to other cells. This biochemistry plays a huge role in quality and quantity in production of bacterial cellulose. The biochemical nature of culture media can also affect the production due to the type of carbon, nitrogen source and availability of other essential nutrients for growth and proliferation of bacterial cells.

Essential Properties of a Biomaterial

The biocompatibility of any biomaterial has some basic physical, biochemical, and mechanical characteristics. To be a most suitable biomaterial it must have these tunable characteristics. The bacterial cellulose possesses these all-tunable characteristics. Because its production is not only focused on one species of bacteria it can be produced by many species of bacteria that depend on utilization of nutrient media. It is also important for selecting a strain that allomorphs of cellulose we want to produce from bacteria. The crystallization of cellulose allomorphs is also one another factor that determines the stability of bacterial cellulose [12]. The understanding of degradation and suitability is a quality driven factor in biomaterial designing [13]. The macroscopic characteristics defined by surface chemistry are also crucial for bacterial cellulose.

Nanostructure of Plants and Bacterial Cellulose

The plant based cellulose and bacterial cellulose have very minute differences in arrangement of micro fibril and structure that makes them to possess different characteristics of biomaterials. Bacterial cellulose is more pure than plant based cellulose. The bacterial cellulose has better properties than plant-based cellulose that attract the vision of modern industries to focus on bacterial production of cellulose. The bacterial cellulose has more water retention capacity due the smaller size of micro fibril and porosity. The approaches of production and growth medium can change the arrangement of microfibrils at nanoscale level [14]. The type of strain of bacteria also causes differences in various properties of biomaterials. The arrangement of micro fibril and crystallization can be altered by changing in growth condition [15].

Mechanical Properties

There are many parameters that can be used to measure the strength and stability of biomaterials. Biomaterial must stand with these parameters considered as good quality material. The Young's modulus in parameters used to measure tensile strength or stiffness of biomaterial sheets. Bacterial cellulose has tensile strength of 250MPa and Young's modulus more than 15GPa [16]. The extensive biochemistry of hydrogen bonding provides better Young's modulus, high thermal stability and tensile strength [17]. On the basis of these mechanical parameters genetic engineering provides better strains of bacteria that leads to higher production of bacterial cellulose. The mechanical properties with the understanding of molecular biology and bioprocessing of growth of bacteria can play a vital role in producing model strain that is helpful in understanding the cellulose producing capacity of other bacterial species and microbes like fungi and yeast.

Understanding of 3D Structure of Biomaterials

The 3D modeling of biomaterials can provide better insights of a biomaterial strength and internal biochemistry. Hydrogen bonding is the most crucial factor of stability of any biomaterials. The cellulose is one of biomaterials in which mechanical parameters will be tunable according to the needs of the type of materials [18]. The tuning of mechanical parameters requires a good understanding of the 3D model or structure of any biomaterial. The 3D structure of bacterial cellulose can be helpful in inserting macro, micro scale with nanoscale level features or tuning internal biochemistry. The based on intended application of material properties can be tuned at micro and macro level.

The fabrication and modeling of bacterial cellulose required advanced technology such as Guided assembly based biolithography (GAB) used in tuning the nanoscale topography at the surface level of cellulose [19]. The mold is added on the surface structure of bacterial cellulose and these molds guide the modeling of bacterial cellulose structure as per the requirement of fabrication. These molds control the overall topography of bacterial cellulose that is helpful in controlling the cellular activity in tissue regeneration [19]. The tuning of topography at surface level is another most advanced field that needs to be more explored for multipurpose application in various industries.

Engineering Control of Bacterial Cellulose Production in *Komagataeibacter rhaeticus*

The strain *K. rhaeticus* GEM produces high cellulose in low nitrogen source and tuning of this strain according to need provides a genetic based control over production of cellulose [20]. We already know that reprogramming DNA can perform new tasks, and this reprogramming of DNA can sift the purpose of bacteria in another field. Genetic engineering in biotechnology has become the most advanced field that solves many problems of industries within a small fraction of change in DNA. The biotechnological approaches for bacterial cellulose production require control over at molecular level that provides the knowledge about how a bacterial strain deals with surrounding or at which level consume nutrients for growth and at which level produce desired product.

The development of new strains that is highly capable of producing desired products required a well-advanced understanding of genetic mechanisms. Those genes that are specific for bacterial cellulose production must be listed out and compared and incorporated to model bacteria to get deep insights of bacterial cellulose production. The *Acetobacteraceae* family of bacterial genus *Komagataeibacter* produces cellulose [20]. The AcsA and AcsB operon protein help in synthesizing cellulose from UDP-glucose [21] and AcsC, AcsD helps in secretion of cellulose and formation of pellicle of cellulose beside bacterial cells [22]. This type of mechanism can be used to understand the basic genetic circuit of bacterial strain for producing cellulose and for further engineering of strain.

In situ Fabrication of Bacterial Cellulose

The development of any functional material required involvement of biotechnology, nanotechnology and chemistry, physics. The types of functional material directly based on their purposes. The bacterial cellulose based functional material attracted attention due to their extensive properties with the excellent biodegradability and biocompatibility. The fabrication of bacterial cellulose functional material is done by physical coating and chemical modification. Another method of functionalization are metal oxide modification, ionic liquid replacement and nanoparticle coating.

The microbes-based functionalization is another field that needs to be explored in vitro to develop cost effective in situ fabrication of bacterial cellulose. The fertilized ovules of *Gossypium hirsutum* were selected for the *in-situ* functionalization of cellulose under in vitro fabrication condition [23]. This is functionalized fabrication done with the help of carrier molecule glucose and introduced functional molecules on plant cellulose fiber by the cotton ovules. This type of strategy can be developed for the bacterial cellulose at the fermentation level [24]. The genetic engineering can provide the genetic circuit for in vitro fabrication at in situ functionalization. The glucose is the most effective molecule that can play a vital role in producing functional bacterial cellulose by in situ fermentation [24].

Production of Bacterial Cellulose

Bacterial cellulose is generally produced by agitated and static fermentation. Both the method used depends upon the application scenario because both methods produce different morphology. During the agitated fermentation irregular small pellets are produced and fully suspended in culture media. In static fermentation gel like pellicle is formed on the surface of culture media and different methods described in Table 2. The static fermentation yields high cellulose due to genetic stability of strain. The advantage of agitated fermentation is it is easily transferred to large scale industrial production. The most important factor in both fermentation methods is culture media that provide all essential nutrient and growth support. The quality and availability of nutrients in media will determine the bacterial cellulose mechanical property and biocompatibility. The raw material for developing Carbon and Nitrogen source (described in Table 1) can be collected from the food and agricultural industrial waste The waste fiber sludge is the product of lignocellulosic biorefineries and pulp industries. This is rich in cellulose and less lignin contamination that makes it a good source for bioconversion for production of bacterial cellulose and other hydrolytic enzymes [10]. The waste fiber sludge is easily enzymatically hydrolyzed and does not require any kind of pretreatment that leads to production of high concentration of glucose and inhibitory compound in less concentration. The inhibitory compounds may be helpful for bacterial strain in production of bacterial cellulose [10]. After the production of bacterial cellulose another important aspect is purification of bacterial cellulose fibers. The various purification methods are already determined, so the most preferred one is used and pure bacterial cellulose fibers are isolated.

Table 2. Methodology of production of bacterial cellulose

Methods	Properties of methods and cellulose
Shaking culture [25, 26]	Increased oxygen delivery and cellulose produced in varying particle size and shape
Static culture [27]	Takes two week in production and produce cellulose in form of hydrogel sheet
Airlift bioreactor [26, 28]	Efficient oxygen supply and produce cellulose in pellet form
Trickling bed reactor [29]	High oxygen concentration and low shear force and cellulose produced in irregular sheets
Rotating disk reactor [30]	Homogenous cellulose produced

Production of Bacterial Cellulose on Immobilized Support

The immobilization is process of providing support for stability of material. Immobilization can be used during fermentation process for enhanced production of biomaterials. One study reported production of bacterial cellulose immobilized support on *Luffa aegyptiaca* support by *Gluconacetobacter kombucha* bacteria [31]. This type of technique not reported before so this field can be more explored. The biomaterial production directly based on the production methodology and culturing conditions. During the fermentation by microbial immobilization technique microbes attached to surface of support and form a biofilm. The *Luffa aegyptiaca* is rich in carbohydrates and fibers and can be used as carrier for animal and plant cells immobilization due to low biodegradability and high stability on various chemical parameters like pH, temperature [31]. The immobilized cell culture for production of cellulose must be further investigated to develop advanced functionalized biomaterial and for enhanced applicative use in various fields of life science and in engineering industries.

Applications

Bacterial cellulose has a vast range of application in many fields like biomaterials, biomedicines, cosmetics, and food items. In recent times bacterial cellulose uses have increased in more advanced fields like engineering, electronics, material science etc. the various applications explored in following sections.

Textile Industries

The synthetic fibers are non biodegradable so they cause pollution in the environment and plant-based fibers also consume a heavy amount of energy. An Australian company Nanollose ltd., develops eco-friendly textile from bacterial cellulose [32]. They developed viscose-rayon fibers from bacterial cellulose as an alternative of plant-based fibers [32]. They also developed a commercial level production setup to produce textile grade bacterial cellulose from coconut water. The Nullarbor^{TM} fibers can be used in all types of clothing, sports, and home furnishing [32]. They mention that viscose rayon fibers have many advantages to plant-based fibers. This is a sustainable alternative of a plant based fiber. Bacterial cellulose can be used in coating fibers and yarns [33]. The use of bacteria can be explored by using fabric during the culturing process.

Biomedicine Industry

Bacterial cellulose is already used in dental implants, wound dressing, artificial skin, vascular grafts, hemostatic materials, drug delivery, biosensor and diagnosis and scaffolds for tissue engineering [34-36]. The well extensive characteristics like large surface area, water holding capacity, permeability for gas and liquid, mechanical properties make bacterial cellulose as advanced material used in biomedical applications [37]. The contact lenses based on the bacterial cellulose are under development and still require study on their stability and toxicity [38]. Skin regeneration or wound healing is the most significant application of bacterial cellulose. There are some standard properties for an ideal wound dressing are moisture maintained on lesion, allow gas exchange, reduce protein and electrolyte loss, prevent infection and must accelerate wound healing [39]. Bacterial cellulose was first used for wound dressing due to their extensive characteristics that make it a most satisfied biomaterial [40]. After studying these extensive characteristics many wound dressing devices developed and commercialized under the trademarks of Bionext®, Membracell®, Gengiflex®, Xcell®, Nanoderm™ etc. [41]. Bacterial cellulose also has the ability to deliver drugs and can improve controlled release drug delivery mechanisms [42]. Bacterial cellulose can be used as a carrier in vaccine development because it can be a safe and effective drug carrier in transportation of drugs into the stomach.

Bacterial cellulose can be used in vascular grafts fabrication due to their extensive mechanical strength and biocompatibility. The composites of bacterial cellulose have capacity to enhance proliferation of retinal pigment epithelium and keratinocytes and grafts developed from bacterial cellulose may reduce the rejection rate of transplanted cornea [43]. The extensive functionalizations suggested to develop 3D structure more sophisticated to cell growth to be used as artificial glaucoma or treatment of glaucoma [44, 45]. The bacterial composites support the growth of corneal stromal cells while restoring full vision of the patient [43]. Bacterial cellulose-based contact lenses can be used as drug carriers as effective treatment of eye infections.

They may be used as wound dressing after eye surgery for various purposes like use of antibiotics.

Bacterial cellulose can be used in the replacement as artificial vessels, due to their extensive molding ability and similar characteristics of small diameter blood vessels. Bacterial cellulose can be molded into different composites during the synthesis and develop an optimal substrate for cell proliferation and attachment [43]. The extensive 3D structure of bacterial cellulose provides the idea of guided bone regeneration. The development of biocompatible grafts from bacterial cellulose can fill the defected place and induce tissue regeneration. These types of biocompatible grafts reduce the problem of rejection during transplantation. The surface phosphorylated bacterial cellulose with higher porosity helps in forming complexes with calcium that enhances nutrient rates during bone regeneration.

Bacterial cellulose-based diagnostics is the recent approach that needs to be explored with the help of advanced technology and biological processes. The extensive properties of bacterial cellulose fibers can replace synthetic materials used in diagnostics. The synthetic cellulosic composites used in antibody conjugation diagnostic assays do not have the much mechanical strength compared to bacterial cellulose. Bacterial cellulose can be better candidates for development of biosensors. The fabrication of piezoelectric sensor with nanoscale accuracy at functionalized bacterial cellulose surface [46]. Such approaches may be used for the preparation of ultra thin films and fabricated over many substrates [47, 48]. This approach makes bacterial cellulose surface more sensitive for recognition and analyzing techniques, such as atomic force microscopy (AFM), X-ray photoelectron spectroscopy (XPS), quartz crystal microbalance (QCM) and surface Plasmon resonance (SPR) [46]. Bacterial cellulose immobilization is another approach that can be used as direct deposition of bacterial cellulose nanocrystals by spin coating for preparing 3D structure with greater surface area to anchor a higher amount of antibodies or desired molecule [49]. Bacterial cellulose nanocrystal shows the ability to anchor higher amounts of IgG antibodies in diagnosis of dengue [50]. The utilization of bacterial cellulose ultra thin films can provide sophisticated devices of biomedical diagnosis.

Food Industries

Bacterial cellulose is the purest form of biomaterials that is generated through the fermentation process. Bacterial cellulose has extensive use in food industries and its application directly based on the methodology used during the production. Bacterial cellulose is generally recognized as safe food additives by FDA [51]. It has extensive use in various applications like in low cholesterol diet, food and beverage additives, food packaging etc. [52]. The jelly-like pellicle of bacterial cellulose produced by static fermentation used as raw materials for food ingredients and food dessert [53]. The thickeners and suspension agent produced from the agitated fermentation with nature of hydrocolloid and it well fit as function to suspend particulates [54]. Bacterial cellulose is also used as a new flavor and texture in food [51]. It also has the ability to suspend low viscosity insoluble particulates [55]. Its extensive degree of crystallinity ensures it is enzyme resistant and has at higher temperature undiminished suspension stability [54, 56]. These characteristics make bacterial cellulose as cost-effective irreplaceable suspension agents. All these extensive properties of bacterial cellulose show a great future ahead.

Cosmetics Industries

Bacterial cellulose is the most sophisticated compound that can be used in cosmetics due to its natural production and biodegradation in the environment. The recent use of bacterial cellulose in developing moisturizers, anti-aging compounds and whitening effect masks are explored. The high level of porosity in bacterial cellulose can be used as loading of nutrients, and ingredients for therapeutic actions to cure skin disease [57].

Conclusion and Future Perspective

Bacterial cellulose is an ultrapure biomaterial and has extensive properties that make it irreplaceable cost-effective biomaterial. The synthesis rate of bacterial is much high than plant cellulose via photosynthesis. It can be synthesized from the industrial and agricultural waste that also reduce the pollution and reduce the cost of production. There is a great potential in utilizing waste fibre sludge for production of bacterial cellulose. It possesses extensive mechanical strength with high water holding capacity, high permeability, low viscosity, good tolerance to acid, salts and ethanol, high suspension stability. It is biocompatible, biodegradable, and renewable biomaterial. By the fabrication and functionalization extensive composites produced. The commercial use of bacterial cellulose is including biomedical area, household chemical, cosmetics, foods, textile and composites materials.

The developing cost-effective bacterial cellulose requires extensive industrial infrastructure. The industrial production of bacterial cellulose and application need improvement. The production quality and quantity can be enhanced by the help of genetic engineering to develop high yield bacterial cellulose producing strains and also development of fermentation equipment and culture media. The industrial waste from food and agriculture sources are cost effective nutrients that can be explored to be used in the fermentation process.

The biomedical application of bacterial cellulose still needs to be explored in application in tissue engineering and diagnostics. The future of bacterial cellulose in biomedicine is vast in the pharmaceutical and cosmetics industries. The application of bacterial cellulose in biomedicine are such as emulsified hydrogel reducing use of surfactants, immobilization for enhanced stability and use, as drug delivery carrier, anchoring of antibodies for developing nano-engineered sensors for diagnostics and artificial skin or wound regeneration therapeutics.

The bacterial cellulose is a sustainable biomaterial that offers many opportunities in the area of biomedical field and industrial biotechnology to develop value added products. Green chemistry is in recent trends in developing bacterial cellulose-based biomaterials. To overcome the reduction of environmental pollution bacterial cellulose must remain in focus in research.

References

[1] Rani, M. U., Udayasankar, K., & Appaiah, K. A. A. (2011). Properties of bacterial cellulose produced in grape medium by native isolate *Gluconacetobacter* sp. *Journal of Applied Polymer Science*, 120(5), 2835–2841.https://doi.org/10.1002/app.33307.

[2] El-Saied, H., El-Diwany, A. I., Basta, A. H., Atwa, N. A., & El-Ghwas, D. E. (2008). Production and characterization of economical bacterial cellulose. *BioResources*, 3(4), 1196–1217. https://doi.org/10.1007/s00226-008-0180-y.

[3] Mikkelsen, D., Flanagan, B. M., Dykes, G. A., & Gidley, M. J. (2009). Influence of different carbon sources on bacterial cellulose production by *Gluconacetobacter xylinus* strain ATCC 53524. *Journal of Applied Microbiology*, 107(2), 576–583. https://doi.org/10.1111/j.1365-2672.2009.04226.x.

[4] Hong, F., Zhu, Y. X., Yang, G., & Yang, X. X. (2011). Wheat straw acid hydrolysate as a potential cost-effective feedstock for production of bacterial cellulose. *Journal of Chemical Technology & Biotechnology*, 86(5), 675–680. https://doi.org/10.1002/jctb.2567

[5] Pour ramezan, G., Roayaei, A., & Qezelbash, Q. (2009). Optimization of culture conditions for bacterial cellulose production by *Acetobacter* sp. 4B-2. *Biotechnology (Faisalabad)*, 8(1), 150–154. https://doi.org/10.3923/biotech.2009.150.154.

[6] Ha, J. H., Shehzad, O., Khan, S., Lee, S. Y., Park, J. W., Khan, T., & Park, J. K. (2008). Production of bacterial cellulose by a static cultivation using the waste from beer culture broth. *Korean Journal of Chemical Engineering*, 25(4), 812–815. https://doi.org/10.1007/s11814-008-0134-y.

[7] Wu, J.-M. & Liu, R.-H. (2012). Thin stillage supplementation greatly enhances bacterial cellulose production by *Gluconacetobacter xylinus*. *Carbohydrate Polymers*, 90(1), 116–121. https://doi.org/10.1016/j.carbpol.2012.05.003.

[8] Ha, J. H. & Park, J. K. (2012). Improvement of bacterial cellulose production in *Acetobacter xylinum* using a byproduct produced by *Gluconacetobacter hansenii*. *Korean Journal of Chemical Engineering*, 29(5), 563–566.

[9] Hong, F., Guo, X., Zhang, S., Han, S-f., Yang, G., & Jonsson, L. J. (2012). Bacterial cellulose production from cotton-based waste textiles: enzymatic saccharification enhanced by ionic liquid pretreatment. *Bioresource Technology*, 104, 503–508. https://doi.org/10. 1016/j.biortech.2011.11.028.

[10] Cavka, A., Guo, X., Tang, S.-J., Winestrand, S., Jonsson, L. J., & Hong, F. (2013). Production of bacterial cellulose and enzyme from waste fiber sludge. *Biotechnology for biofuels*, 6(1), 1–10.

[11] Keshk, S. & Sameshima, K. (2006a). Influence of lignosulfonate on crystal structure and productivity of bacterial cellulose in a static culture. *Enzyme and Microbial Technology*, 40(1), 4–8. https://doi.org/10.1016/j.enzmictec.2006.07.037.

[12] Gibson, L. J. (2012). The hierarchical structure and mechanics of plant materials. *J. R. Soc. Interface* 9, 2749–2766. doi:10.1098/rsif.2012.0341.

[13] Modulevsky, D. J., Cuerrier, C. M., and Pelling, A. E. (2016). Biocompatibility of Subcutaneously Implanted Plant-Derived Cellulose Biomaterials. *PLoS One* 11, 1–19.doi:10.1371/journal.pone.0157894.

[14] Jozala, A. F., Pértile, R. A., dos Santos, C. A., de Carvalho Santos-Ebinuma, V., Seckler, M. M., Gama, F. M., & Pessoa, A., Jr (2015). Bacterial cellulose production by *Gluconacetobacter xylinus* by employing alternative culture media. *Applied microbiology and biotechnology*, *99*(3), 1181–1190. https://doi.org/10.1007/s00253-014-6232-3.

[15] Bi, J. C., Liu, S. X., Li, C. F., Li, J., Liu, L. X., Deng, J., & Yang, Y. C. (2014). Morphology and structure characterization of bacterial celluloses produced by different strains in agitated culture. *Journal of applied microbiology*, *117*(5), 1305–1311. https://doi.org/10.1111/jam.12619.

[16] Yamanaka, S., Watanabe, K., Kitamura, N., Iguchi, M., Mitsuhashi, S., Nishi, Y., & Uryu, M. (1989). The structure and mechanical properties of sheets prepared from bacterial cellulose. *Journal of Materials Science, 24*, 3141-3145.

[17] Feng, X., Ullah, N., Wang, X., Sun, X., Li, C., Bai, Y., Chen, L., & Li, Z. (2015). Characterization of Bacterial Cellulose by *Gluconacetobacter hansenii* CGMCC 3917. *Journal of food science*, *80*(10), E2217–E2227. https://doi.org/10.1111/1750-3841.13010.

[18] Stumpf, T. R., Yang, X., Zhang, J., and Cao, X. (2018). In situ and ex situ modifications of bacterial cellulose for applications in tissue engineering. *Mater. Sci. Eng. C* 82, 372–383. doi:10.1016/j.msec.2016.11.121.

[19] Bottan, S., Robotti, F., Jayathissa, P., Hegglin, A., Bahamonde, N., Heredia-Guerrero, J. A., Bayer, I. S., Scarpellini, A., Merker, H., Lindenblatt, N., Poulikakos, D., & Ferrari, A. (2015). Surface-structured bacterial cellulose with guided assembly-based biolithography (GAB). *ACS nano*, *9*(1), 206–219. https://doi.org/10.1021/nn5036125.

[20] Florea, M., Hagemann, H., Santosa, G., Abbott, J., Micklem, C. N., Spencer-Milnes, X., de Arroyo Garcia, L., Paschou, D., Lazenbatt, C., Kong, D., Chughtai, H., Jensen, K., Freemont, P. S., Kitney, R., Reeve, B., & Ellis, T. (2016). Engineering control of bacterial cellulose production using a genetic toolkit and a new cellulose-producing strain. *Proceedings of the National Academy of Sciences of the United States of America*, *113*(24), E3431–E3440. https://doi.org/10.1073/pnas.1522985113.

[21] Wong, H. C., Fear, A. L., Calhoon, R. D., Eichinger, G. H., Mayer, R., Amikam, D., Benziman, M., Gelfand, D. H., Meade, J. H., & Emerick, A. W. (1990). Genetic organization of the cellulose synthase operon in *Acetobacter xylinum*. *Proceedings of the National Academy of Sciences of the United States of America*, *87*(20), 8130–8134. https://doi.org/10.1073/pnas.87.20.8130.

[22] Lee K-Y, Buldum G, Mantalaris A, Bismarck A (2014). More than meets the eye in bacterial cellulose: Biosynthesis, bioprocessing, and applications in advanced fiber composites. *Macromo lBiosci* 14(1):10–32

[23] Natalio, F., Fuchs, R., Cohen, S. R., Leitus, G., Fritz-Popovski, G., Paris, O., Kappl, M., & Butt, H. J. (2017). Biological fabrication of cellulose fibers with tailored properties. *Science (New York, N.Y.)*, *357*(6356), 1118–1122. https://doi.org/10.1126/science.aan5830.

[24] Gao, M., Li, J., Bao, Z., Hu, M., Nian, R., Feng, D., An, D., Li, X., Xian, M., & Zhang, H. (2019). A natural in situ fabrication method of functional bacterial cellulose using a microorganism. *Nature communications*, *10*(1), 437. https://doi.org/10.1038/s41467-018-07879-3.

[25] Watanabe, K., Tabuchi, M., Morinaga, Y., & Yoshinaga, F. (1998). Structural Features and Properties of Bacterial Cellulose Produced in Agitated Culture. *Cellulose, 5*, 187-200.

[26] Wang, J., Tavakoli, J., & Tang, Y. (2019). Bacterial cellulose production, properties and applications with different culture methods - A review. *Carbohydrate polymers*, *219*, 63–76. https://doi.org/10.1016/j.carbpol.2019.05.008.

[27] Rani, M. U., Rastogi, N. K., & Appaiah, K. A. (2011). Statistical optimization of medium composition for bacterial cellulose production by *Gluconacetobacter hansenii* UAC09 using coffee cherry husk extract--an agro-industry waste. *Journal*

of microbiology and biotechnology, *21*(7), 739–745. https://doi.org/10.4014/jmb.1012.12026.

[28] Chao, Yp., Sugano, Y., Kouda, T, F. Yoshinaga, M. Shoda. (1997) Production of bacterial cellulose by *Acetobacter xylinum* with an air-lift reactor. *Biotechnology Techniques* 11, 829–832 (1997). https://doi.org/10.1023/A:1018433526709.

[29] Lu, H., & Jiang, X. (2014). Structure and properties of bacterial cellulose produced using a trickling bed reactor. *Applied biochemistry and biotechnology*, *172*(8), 3844–3861. https://doi.org/10.1007/s12010-014-0795-4.

[30] G., S., R., M., & H., B. (2002). Inclusion of solid particles in bacterial cellulose. *Applied Microbiology and Biotechnology*, *58*(6), 756-760. https://doi.org/10.1007/s00253-002-0978-8.

[31] Sameeha Syed Abdul Rahman, T. Vaishnavi, G. Sai Vidyasri, K. Sathya, P. Priyanka, Ponnusami Venkatachalam & Sugumaran Karuppiah (2021). Production of bacterial cellulose using *Gluconacetobacter kombucha* immobilized on *Luffa aegyptiaca* support. *Sci Rep* **11,** 2912. https://doi.org/10.1038/s41598-021-82596-4

[32] Nanollose Ltd (2020). Available online at: *https://nanollose.com* (accessed May 17, 2022).

[33] Fernandes, Marta, António P. Souto, Fernando Dourado, & Miguel Gama. (2021). "Application of Bacterial Cellulose in the Textile and Shoe Industry: Development of Biocomposites" *Polysaccharides* 2, no. 3: 566-581. https://doi.org/10.3390/polysaccharides2030034.

[34] Rajwade, J. M., Paknikar, K. M., & Kumbhar, J. V. (2015). Applications of bacterial cellulose and its composites in biomedicine. *Applied microbiology and biotechnology*, *99*(6), 2491–2511. https://doi.org/10.1007/s00253-015-6426-3.

[35] Anton-Sales, I., Beekmann, U., Laromaine, A., Roig, A., & Kralisch, D. (2019). Opportunities of Bacterial Cellulose to Treat Epithelial Tissues. *Current drug targets*, *20*(8), 808–822. https://doi.org/10.2174/1389450120666181129092144.

[36] Carvalho, T., Guedes, G., Sousa, F. L., Freire, C., & Santos, H. A. (2019). Latest Advances on Bacterial Cellulose-Based Materials for Wound Healing, Delivery Systems, and Tissue Engineering. *Biotechnology journal*, *14*(12), e1900059. https://doi.org/10.1002/biot.201900059.

[37] Zhong C. (2020). Industrial-Scale Production and Applications of Bacterial Cellulose. *Frontiers in bioengineering and biotechnology*, *8*, 605374. https://doi.org/10.3389/fbioe.2020.605374.

[38] Coelho, F., do Vale Braido, G. V., Cavicchioli, M., Mendes, L. S., Specian, S. S., Franchi, L. P., Lima Ribeiro, S. J., Messaddeq, Y., Scarel-Caminaga, R. M., & O Capote, T. S. (2019). Toxicity of therapeutic contact lenses based on bacterial cellulose with coatings to provide transparency. *Contact lens & anterior eye: the journal of the British Contact Lens Association*, *42*(5), 512–519. https://doi.org/10.1016/j.clae.2019.03.006.

[39] Portela, R., Leal, C. R., Almeida, P. L., & Sobral, R. G. (2019). Bacterial cellulose: a versatile biopolymer for wound dressing applications. *Microbial biotechnology*, *12*(4), 586–610. https://doi.org/10.1111/1751-7915.13392.

[40] Curvello, R., Raghuwanshi, V. S., & Garnier, G. (2019). Engineering nanocellulose hydrogels for biomedical applications. *Advances in colloid and interface science*, *267*, 47–61. https://doi.org/10.1016/j.cis.2019.03.002.

[41] Abeer, M. M., Mohd Amin, M. C., & Martin, C. (2014). A review of bacterial cellulose-based drug delivery systems: their biochemistry, current approaches and future prospects. *The Journal of pharmacy and pharmacology*, *66*(8), 1047–1061. https://doi.org/10.1111/jphp.12234.

[42] Kusano Sakko Inc (2020). Available online at: *https://www.kusanosk.co.jp/lab/2016* (accessed May 18, 2022).

[43] Picheth, G. F., Pirich, C. L., Sierakowski, M. R., Woehl, M. A., Sakakibara, C. N., de Souza, C. F., Martin, A. A., da Silva, R., & de Freitas, R. A. (2017). Bacterial cellulose in biomedical applications: A review. *International journal of biological macromolecules*, *104*(Pt A), 97–106. https://doi.org/10.1016/j.ijbiomac.2017.05.171.

[44] Chiaoprakobkij, N., Sanchavanakit, N., Subbalekha, K., Pavasant, P., & Phisalaphong, M. (2011). Characterization and biocompatibility of bacterial cellulose/alginate composite sponges with human keratinocytes and gingival fibroblasts. *Carbohydrate Polymers*, *85*(3), 548-553. https://doi.org/10.1016/j.carbpol.2011.03.011.

[45] Zaborowska, M., Bodin, A., Bäckdahl, H., Popp, J., Goldstein, A., & Gatenholm, P. (2010). Microporous bacterial cellulose as a potential scaffold for bone regeneration. *Acta biomaterialia*, *6*(7), 2540–2547. https://doi.org/10.1016/j.actbio.2010.01.004.

[46] Picheth, G. F., Sierakowski, M. R., Woehl, M. A., Pirich, C. L., Schreiner, W. H., Pontarolo, R., & De Freitas, R. A. (2013). Characterisation of ultra-thin films of oxidised bacterial cellulose for enhanced anchoring and build-up of polyelectrolyte multilayers. *Colloid and Polymer Science*, *292*(1), 97-105. https://doi.org/10.1007/s00396-013-3048-0.

[47] Tammelin, T., Saarinen, T., Österberg, M., & Laine, J. (2006). Preparation of Langmuir/Blodgett-cellulose surfaces by using horizontal dipping procedure. Application for Polyelectrolyte adsorption studies performed with QCM-D. *Cellulose*, *13*(5), 519-535. https://doi.org/10.1007/s10570-005-9002-7.

[48] Mohan, T., Spirk, S., Kargl, R., Doliska, A., Vesel, A., Salzmann, I., Resel, R., Ribitsch, V., & Stana-Kleinschek, K. (2012). Exploring the rearrangement of amorphous cellulose model thin films upon heat treatment. *Soft Matter*, *8*, 9807-9815. https://doi.org/10.1039/C2SM25911G.

[49] Pirich, C. L., De Freitas, R. A., Woehl, M. A., Picheth, G. F., Petri, D. F., & Sierakowski, M. R. (2015). Bacterial cellulose nanocrystals: Impact of the sulfate content on the interaction with xyloglucan. *Cellulose*, *22*(3), 1773-1787. https://doi.org/10.1007/s10570-015-0626-y.

[50] Pirich, C. L., de Freitas, R. A., Torresi, R. M., Picheth, G. F., & Sierakowski, M. R. (2017). Piezoelectric immunochip coated with thin films of bacterial cellulose nanocrystals for dengue detection. *Biosensors & bioelectronics*, *92*, 47–53. https://doi.org/10.1016/j.bios.2017.01.068.

[51] Shi, Z., Zhang, Y., Phillips, G. O., & Yang, G. (2014). Utilization of bacterial cellulose in food. *Food Hydrocoll.* 35, 539–545. https://doi.org/10.1016/j.foodhyd.2013.07.012.

[52] Azeredo, H. M. C., Barud, H., Farinas, C. S., Vasconcellos, V. M., & Claro, A. M. (2019). Bacterial cellulose as a raw material for food and food packaging applications. *Front. Sustain. Food Syst.* 3:7. https://doi.org/10.3389/fsufs.2019.00007.

[53] Hainan Yeguo Foods Co., Ltd (2020). Available online at: *http://www.yeguo.com* (accessed May19, 2022).

[54] San-Ei Gen F. F. I., Inc (2020). Available online at: *https://www.saneigenffi.co.jp/closeup/san.html* (accessed May19, 2022).

[55] Swazey, J. M. (2014). Surfactant Thickened Systems Comprising Microfibrous Cellulose and Methods of Making Same. U.S. Patent No 8,772,359 B2. Washington, DC: U.S. Patent and Trademark Office.

[56] Torres, F. G., Arroyo, J. J., & Troncoso, O. P. (2019). Bacterial cellulose nanocomposites: An all-nano type of material. *Materials science & engineering. C, Materials for biological applications*, *98*, 1277–1293. https://doi.org/10.1016/j.msec.2019.01.064.

[57] Chantereau, G., Sharma, M., Abednejad, A., Vilela, C., Costa, E., Veiga, M., Antunes, F., Pintado, M., Sèbe, G., Coma, V., Freire, M., Freire, C., & Silvestre, A. (2020). Bacterial nanocellulose membranes loaded with vitamin B-based Ionic liquids for dermal care applications. *Journal of Molecular Liquids*, *302*, 112547. https://doi.org/10.1016/j.molliq.2020.112547.

Index

A

B

C

W

About the Editors

Dr. Santosh Kumar Karn

Dr. Santosh Kumar Karn is an Associate Professor & Head, at Department of Biochemistry & Biotechnology, Sardar Bhagwan Singh University, Balawala, Dehradun, Uttarakhand, India. Dr. Karn actively involved in Teaching & Research. S.B.S. University recognized and appreciated Dr. Karn for their outstanding contribution in Teaching & Research (2022) He taught several papers in B.Sc./B.Tech Biotech, M.Sc./M.Tech. Biotech, M.Sc. Biochemistry, M.Sc. Microbiology and in Ph.D. Dr. Karn awarded for Prestigious International Young Scientist Award (2011) and Visiting Researcher Award (2013) by the Chinese Academy of Science (CAS) Beijing. In the year (2016) he awarded for Presidential International Fellowship Initiative (PIFI initiative) by CAS, Beijing & National Natural Science Foundation of China. Further, Dr. Karn received Publons peer-review award (2017 and 2018) as one of the top 1% peer reviewers in Microbiology/Biotechnology internationally by Clarivate Analytics, USA. He published about 100 research paper in the journal of international repute & high impact factor next, reviewed for 80 journals and 250 papers of top-notch journals of microbial biotechnology. Prof. Karn is an Associate Editor of the Journal of Environmental Biology, and Frontier in Microbiology, Frontier in Environmental Science, etc. Dr. Karn ventures to inspire students with his enthusiasm for the subjects, cultivate critical-thinking and decision-making for innovative approaches.

Anne Bhambri

Anne Bhambri has an M.Sc. in Biochemistry from S. B. S. University, Dehradun, and received Silver Medal at post-graduate level and is currently a doctoral student. She works on biological nutrient removal using specific microbes guided by S. K. Karn, Department of Biochemistry & Biotechnology, S. B. S. University, Dehradun, India. Till now she published 20 research papers in leading international journals (Scientific Reports (Nature's Publishers), London, UK; Elsevier, Netherland; Nova Science Publishers, USA; Corrosion Review, Germany; Springer, Germany; Chemistry & Ecology (Taylor & Francis Group), London; Wiley-Scrivener, Chichester, UK; Bentham Science, UAE; Geomicrobiology journal (Taylor & Francis Publishers), United Kingdom etc.) and two proceedings, one in Emerging Scientist- ES2022 organized by Asian Council of Science Editors jointly with Academy of Excellence in Bioscience and second in National Conference organized by National Institute of Technology, Warangal (2021). She presented a paper in National Conference and received "Best Oral Presentation Award" from National Institute of Technology, Warangal (2021) and also "Best P.G. Thesis Award" in another National Conference Oral Presentation (2019). Currently working as a "Project Associate" in the research project sponsored by M/s Xcel Life Science, California, USA. Recently in (2022) she received Young Women Scientist Excellence Award by USERC, Dehradun, Government of Uttarakhand for her contribution in Science & Technology. Recently she published work in Frontier in Microbiology and edited two books with Nova Science Publisher, USA. She has broad interest in soil and water pollution and their remediation, medicinal mushroom cultivation and Phytochemical extraction, Environmental Hydrology and Nanotechnology, Bioremediation etc.